T0300128

Introduction to
Surveillance
Studies

Introduction to
Surveillance
Studies

J. K. Petersen

CRC Press
Taylor & Francis Group
Boca Raton   London   New York

CRC Press is an imprint of the
Taylor & Francis Group, an **informa** business

CRC Press
Taylor & Francis Group
6000 Broken Sound Parkway NW, Suite 300
Boca Raton, FL 33487-2742

© 2013 by Taylor & Francis Group, LLC
CRC Press is an imprint of Taylor & Francis Group, an Informa business

No claim to original U.S. Government works

Printed in the United States of America on acid-free paper
Version Date: 20120820

International Standard Book Number: 978-1-4665-5509-9 (Hardback)

Library of Congress Cataloging-in-Publication Data

Petersen, Julie K.
    Introduction to surveillance studies / Author, J. K. Petersen.
    pages cm
    Includes bibliographical references and index.
    ISBN 978-1-4665-5509-9 (hardback)
    1. Electronic surveillance. 2. Privacy, Right of. I. Title.

TK7882.E2.P47 2013
621.389'28--dc23   2012027056

Visit the Taylor & Francis Web site at
http://www.taylorandfrancis.com

and the CRC Press Web site at
http://www.crcpress.com

CONTENTS

Preface

When the author first approached publishers in the mid-1990s to produce a book about surveillance, the general reaction was indifference. The topic was considered little more than a prop for thrillers or science fiction movies. That it might become one of the most prevalent technologies and subjects of controversy in less than 15 years was not apparent to many in the industry. The author queried for two years with no takers. Even CRC Press was reluctant at first.

Nevertheless, the author persisted, feeling that the increasing miniaturization of surveillance tools, the proliferation of computer networks, and the globalization of commerce and communication was going to change surveillance in significant ways that would soon impact everyone.

While there were a number of shorter volumes on specific aspects of surveillance and some that provided practical guidance for building surveillance components, a survey book was overdue—no such volume existed—and there was a need for surveillance technologies to be categorized, as this had not been done. It was also time for surveillance "tools" such as drug-sniffing dogs, radio transmitter-equipped bees, and DNA profiling, to be acknowledged as part of the surveillance landscape.

While reluctant, CRC Press listened and, in time, said yes, we'll take a chance on this and, two years later, in 2000, it reached the shelves. CRC Press is to be credited with publication of the first comprehensive survey and history of surveillance technology, *Understanding Surveillance Technologies,* a book that is now acknowledged as a standard-setting text often quoted online and in professional journals.

Understanding Surveillance Technologies also served to set the record straight on many incorrect historical attributions regarding communications inventions and surveillance devices, information that has since been corrected on the Web, but which was barely mentioned online and not easily accessible to the author in the 1990s. The author's research did not come from sources like Wikipedia (which didn't exist at the time) or second-hand news reports, but from original patent documents, articles and letters written by inventors, pioneer users, vintage AARL journals, scientific white papers, and reams of original government documents available through the Freedom of Information Act (which were frequently and sometimes almost entirely obfuscated with felt-pen censorship strike-outs).

Many government projects that are in the open now were secret at the time and the author even stumbled upon classified information that clearly could not be included in the book and which is still not public knowledge. In time, with the Internet opening so many locked closets, they probably will come to light and then, perhaps, can be discussed and documented in a systematic way.

Which brings us to the current volume. The idea to write an introductory text focusimg on the newest changes and technologies had crossed the author's mind, but I was busy with other projects and tabled the idea. I'm happy to say that CRC Press editor Mark Listewnik not only had the same thoughts, but also enough enthusiasm for the project to set a fire under me to get it done as he ferried it through the acquisitions process. For this, I thank him, and the other supporters at CRC Press who understand that surveillance is no longer a fringe curiosity, but a significant aspect of our lives.

J.K. Petersen

Focus of this Book

Content

Introduction to Surveillance Studies emphasizes recent technologies and emerging trends such as miniaturized devices, wireless communications, analytics, data mining, and social networks. It also includes a brief history to provide context for the evolution of today's high-tech devices.

The text is appropriate for those new to surveillance and those seeking to further their basic understanding. It can be used as an introduction to surveillance and privacy by journalists, urban planners, public servants, trainees in surveillance, law enforcement, military service, forensics, research science, and technology.

It is also an appropriate text for those studying sociology, constitutional freedoms, privacy, and the impact of technology on communications and the structure of human society.

Introduction to Surveillance Studies covers both the pros and cons of surveillance technologies. The middle ground has all but disappeared. Law enforcement agencies use surveillance to do their jobs. Companies and residents use it to protect property. Medical practitioners use it to track viral and bacterial infections, and wildlife conservators to detect and deter poaching and illegal fishing. Surveillance is a potent tool for protecting resources and assets.

Commentary

At the same time, surveillance can be extremely invasive. The smaller and more powerful a technology becomes, the more it inserts itself into the lives of people who are doing no wrong and who may treasure their privacy. It increasingly limits our freedom to go about our business without being watched and recorded and, year by year, casts not only a wider net, but also a tighter one.

Most surveillance systems are in place to gather marketing information or to deter crime. A high proportion of juvenile offenders come from one-parent families where the parent is working and there is no one at home to keep a child company after school. Those endeavoring to reduce and prevent crime have questioned whether the money is better spent on surveillance technologies or youth programs that teach better coping skills through participation in community center programs that reduce reliance on gangs for a child's social needs.

This book presents both sides of the issues because those

who use and promote surveillance technologies are, on the one hand, trying to do their jobs but, on the other, are equally at risk of losing privacy, freedom of movement, and freedom of speech. Even within law enforcement agencies, there is debate over whether technological surveillance is a good allocation of funds. Some feel it is better to have officers working in the field rather than watching cameras all of the time, and others are concerned about liability if something is recorded but missed by the person monitoring the cameras.

Humans are very adaptable—it is one of our most important survival mechanisms—but it is possible to adapt to the point where there is no turning back and no room to go forward. Perhaps, in the end, before surveillance erodes personal freedom entirely, the technology that ensnares us can be retooled to set us free.

SECTION 1
INTRODUCTION AND OVERVIEW

INTRODUCTION

BRIEF HISTORY

INTRODUCTION

- *emergence of surveillance devices*
- *use and abuse of spy devices*
- *public controversy and accountability*

Introduction

There was a time when surveillance meant little more than spying from behind a bush. When humankind learned to fabricate metal and glass, it became possible to observe objects farther away, or smaller than the unaided eye could see, and surveillance entered the technological realm.

Spy tools were originally luxuries, available only to nobility and the wealthiest merchants. They spread to navigation and trade and, by the 1800s, to the middle class. Since the mid-2000s, however, there has been a significant increase in the availability and sale of surveillance devices and the consumer market has boomed. Surveillance has shifted from human eyes and security forces to sensors and software.

The attack on the World Trade Center in 2001 struck a nerve in human vulnerability and created a climate where people willingly gave up privacy and freedom in exchange for greater safety or, at least, the perception of safety. It spawned an entrepreneurial swing toward defense projects, shifted priorities of venture capitalists, and gave governments license to move tax dollars into the purchase of high-end technology such as body scanners, biometrics equipment, increased border surveillance, and computer databases. Money was also allocated to open airspace to uninhabited aerial vehicles, most of which are designed for surveillance. Consumer spending on spy gadgets also increased with advances in microelectronics. Low-cost video cameras and other computerized data-capture devices are available to anyone and many are sold for entrepreneurial or voyeuristic purposes rather than for security.

The microchip is an Orwellian milestone in the evolution of surveillance technologies. It furthered global communication but also enabled unprecedented intrusion into personal lives with gadgets that are inexpensive and more powerful than ever before.

surveillance *noun*

1. observing, sensing, or otherwise determining the presence or influence of persons, activities, or phenomena of interest, especially as regards protection of assets, territory, property, family, personal safety, power, commercial opportunities, or social relationships.

Surveillance has the character of vigilance and observations or related data may be logged or recorded.

etymology: from Latin *vigilare;* and French *sur-* (over/above) and *veiller* (to be on watch, to stay up), thus, to watch over. *Veilleur de nuit* is a 'watcher of the night'—a night watchman.

The term *surveillance* covers many forms of observation and information-gathering. From wired secret agents to drug-sniffing dogs and portable DNA kits, there are many ways to gather and record data. Some forms of surveillance are so common, we don't give them a second thought, like checking a peephole if someone is at the door, or glancing at Caller ID to see who phoned.

Other forms of surveillance are sophisticated and pervasive. In most urban centers, it is no longer possible to walk down a street or buy groceries without being watched and recorded—sometimes as often as 300 times a day. Cameras blanket the streets, gaze down from ceilings, and are mounted on roving vans that peer into yards and sometimes windows, snapping private moments. Planes and satellites are watching from the skies. Many of the images captured by these technologies are stored for extended periods or uploaded to the Internet.

Cradle-to-Grave Surveillance

Surveillance begins before we understand the meaning of the word. Babies are fingerprinted, footprinted, and cheek-swabbed to collect DNA samples. Some have been implanted with rice-sized radio transceiving chips to monitor their whereabouts. By the time a child is three, he or she will be recorded tens of thousands of times, perhaps even displayed on video-sharing sites for the entertainment of the masses. The child has no choice and no control over how this happens or how it might affect his or her future.

Street cams, employer security systems, and satellites chart our activities on a daily basis, and sometimes minute-by-minute. Some employers have tried to make off-duty monitoring a requirement of the job, a trend of concern to privacy advocates.

Surveillance technologies are not limited to devices that monitor our physical attributes and movements. Data repositories linked to online forums, search engines, and social networks house a growing body of information that enables data aggregators (those who collect data from various sources) to construct a sophisticated picture of a person's personal habits, relations, and economic status—one that is routinely sold without the person's knowledge or consent.

In the past, surveillance was mainly aimed at observing business rivals or protecting property. Even national defense serves these aims, if one considers competing countries and national territory as a broad analogy. Now it is increasingly used to gather information to sell to third parties or to track potential consumers, so they can be targeted for individualized marketing. These days, cameras above retail checkouts are less about preventing theft than they are about profiling. Supermarket and department store "rewards" cards serve a similar purpose, by logging consumer buying habits and preferences. Some of these activities come back to the customer in terms of discounts and improved service, but much of the information is sold to third parties. Another big venue for consumer surveillance is online social networking sites—a significant resource for consumer demographics.

Social networking began as a way for people to keep in touch. Since their modest beginnings, online "water coolers" have become hotspots for advertisers to target groups and individuals, and for data miners to glean personal, insurance, and marketing details on millions of unwitting users. People have lost their jobs because a work rival or disgruntled acquaintance reported an innocent holiday photo posted in a personal section of a site that the user thought was inaccessible to the general public. On some sites, censors are paid to snoop through private areas.

Free email services are a boon for contacting loved ones or business associates, but are also widely abused. Not only can administrators see a person's correspondence and contacts, regardless of company policy, but there are also hackers who make a point of breaching computer systems and selling the "cracking" software to data miners. In addition to information on the Internet, *mal*icious soft*ware* (malware) can surreptitiously invade people's home and work computers—unleashing programs that quietly snoop through hard drives and send the information to other locations.

We are in the birthing stage of the surveillance movement. In the future, it is conceivable that every cranny of every home will be mapped in exacting detail, if not by the current owner, then possibly the previous homeowner seeking to sell the property, or a contractor wanting to promote newly built properties. The same is true of our daily activities. No matter how much effort an individual makes to protect personal privacy, he or she has little control over what others innocently or maliciously say or post online relating to him or her.

Data Security

Password and security experts work diligently to create encryption systems to protect sensitive information and yet, despite their efforts, data is rarely safe from intrusion or distribution.

The weakest link is not encryption, or physical access, but the human factor. Frailty, greed, lack of foresight, bigotry, and revenge have all, at one time or another, motivated misuse of surveillance equipment or the information gathered. Further, if information is uploaded to the Internet, it is impossible to take it back. Humans and software robots regularly crawl the Net, hunting for additions or changes, and instantly grab and store what they find.

Whether we like it or not, we are being watched, numbered, and cataloged as commodities, our personal lives logically reconstructed and stored by companies we've never heard of and people we've never met. The information is combined with details from other sources to be cross-referenced, freely distributed, or sold to subscribers.

Large-Scale Users of Surveillance

Governments, corporations, and law enforcement agencies are heavy users of surveillance. Through the ages, they have sought to increase their information storehouses even when the data was not of immediate use. In the past, humans who had committed no crime were kept in dungeons for years, sometimes decades, on the remote chance they could be used for political ransom or prisoner exchange in the unforeseeable future. Some countries still do this.

While our methods have changed, our motives have not. Both the U.K. and U.S. governments have given frequent assurances that personal information, such as driver's license data, DNA profiles, and other sensitive information on law-abiding citizens will not be sold to third parties or, in the case of DNA, be retained in their systems, and yet they have been found selling driver's license information and have incrementally and repeatedly changed policies regarding DNA collection to the point that almost everything is retained, on the off-chance someone might commit a crime or be politically dangerous in the hypothetical future. Manacles that bound people to dungeon walls two centuries ago could potentially re-emerge as data and biometric storehouses.

Information is a tool of warfare. A great deal of government and law-enforcement information-gathering is legitimate. These institutions are trying to maintain a competitive edge for their nations and to keep the domestic peace.

In other instances, the information-gathering is for political gain or repressive purposes. Honesty is not universal, and democracy is not planet-wide—there are regions where governments still dominate the populace. Even in democracies, new leaders are elected every few years and the political climate can change overnight—there's no guarantee democracy will continue without public vigilance.

Balance of Power

One of the biggest challenges facing a surveillance society is that of maintaining two-way observation. If those in power can surveil the public, without the public having the same technology and right of access to surveil those in power, over time, the inequity can create a top-heavy imbalance that threatens constitutional rights.

Governments don't automatically balance government, corporate, and individual rights without public input. Individuals and institutions interpret their needs through a narrow filter—it takes effort to consider the big picture.

In America, lobbying[1] has a strong presence in the halls of Congress. Corporate lobbies have increased their influence on government since the early days of the republic to the point where some are calling it a corporatocracy.

That's not to say every aspect of surveillance is negative or damaging. There are significant benefits to responsible use of surveillance, including

- caretaker cams that monitor infants, handicapped individuals, and pets,

- security systems to protect livestock, property, and ATMs,

- safety systems in industrial zones, hazardous areas, and toxic waste sites,

- forensic tools for locating and analyzing evidence for solving crimes,

- street and highway cams that warn of traffic accidents or bottlenecks, ferry line-ups, and delays at border crossings,

- global positioning systems (GPSs) that facilitate rescue operations, wildlife tracking, and navigation,

- aerial and marine monitors that aid in search and rescue, promote significant scientific discoveries, and help deter poaching and smuggling,

- environmental surveillance that warns of earthquakes, volcanic eruptions, floods, and tornadoes,

- specialized satellite data for monitoring human rights violations and military actions against unarmed civilians,

- military surveillance that prevents ambushes and keeps servicemembers out of harm's way,

- scientific surveillance to locate and monitor archaeological sites, global and local climate changes, ecology, and our place in the universe, and

- health surveillance and medical diagnostics that provide early warnings and statistical information that can save countless lives.

Surveillance is a tool and, like any tool, it can be used for good or evil.

---

[1] Lobbying is advocate pressure on government decision-making and, in the U.S., a constitutionally protected right for persons to peacably petition for a "redress of grievances." Since establishing lobbying rights in the First Amendment, corporate lobbies not only petition the government on grievances, but expend large sums of money to influence or promote those in power (or those seeking power) so they will structure laws and policies in their favor.

BRIEF HISTORY

KEY CONCEPTS

- *human desire for power and a sense of security*
- *key developments leading to current technology*
- *public controversy and accountability*
- *increasingly generalized nature of surveillance*

Introduction

Surveillance has a fascinating history. As a species, we are insatiably curious—always wanting to see farther and better. Prehistoric humans climbed trees to survey territory, shake spears at rivals, or impress potential mates. In doing so, they were expressing essentially the same impulses that motivate surveillance today.

Managers are often conservative about adopting new technologies, reluctant to allocate funds to something that is unproven, but once those technologies show their worth, many are eager to incorporate them into the workflow, especially if they can decrease costs or open new windows to profit. In recent years new technologies have changed the way we work and the ways in which we carry out surveillance.

Information about our surroundings and associates confers a sense of power and security. When a person or nation acquires a new invention with commercial value, or a new means of engaging in defense or warfare, other countries see it as a challenge or threat and "ramp up" their acquisitions, as well.

The space program, which has yielded many of our most significant tracking, communication, and surveillance systems, was a response to the Soviet launch of the Sputnik satellite in the 1950s. Satellite communications, GPS location services, microminiaturization, intelligent vehicle systems, and many more technologies came out of the U.S. space program, as a direct response to Sputnik.

This section briefly summarizes some of the technological milestones that led to where we are today.

Significant Discoveries Leading to Surveillance Technologies

With the exception of metal and glass, most inventions underpinning surveillance devices were developed in the last 300 years, many of them in the last 30 years. In the early days, many came from small entrepreneurial laboratories, but the majority are now developed in commercial or academically funded labs.

A common misconception is that the military is responsible for the bulk of surveillance inventions. It is understandable, considering the importance of surveillance to military effectiveness. However, a study of early patents shows that most devices adapted for surveillance were invented to facilitate daily life, to aid navigators and the handicapped (e.g., those with poor vision), to promote entrepreneurial ventures (especially manufacturing and gathering of market demographics), to aid scientific research, and to protect property. While it is true that the military has taken many existing inventions and significantly developed them, it sometimes takes years or decades before the military utility of a new invention is fully understood or desired.

As one example, the invention of radar, an indispensable surveillance technology used in submarines, aircraft, and ground transportations systems, occurred in the late 1800s, motivated by a desire to prevent collisions of ships at sea. Decades later, the technology was offered to European military organizations and rejected. It was not until WWII that radar's usefulness was fully acknowledged for military purposes.

Thus, the military has made substantial contributions to the development of existing technologies like radar and computer networks, but given that budgets are taxpayer funded, it is sometimes conservative in adapting unproven technologies or inventing new ones that might not come to fruition fast enough to aid in a military crisis.

The model for computer networks is also older than generally accepted. Many have stated that computer networks were invented in the 1970s, but stock ticker machines that transmitted stock listings through telegraph lines are a rudimentary form of data network that evolved into teletype machines by the early 1900s.

Wireless technology is almost a century old, as well. By 1916, at least one patent had been submitted for wireless telegraph and telephone receivers.[2]

Teletypes were similar to early email—they enabled people to share conversations, business tips, and political news over long distances in the mid-20th century. Teletypes were also interfaced with computers and used as terminals for long-distance text-based gaming by the 1960s.

Fire

Fire is a natural phenomenon. The heat from volcanic eruptions, decaying plant matter, or heat concentrated by reflective objects can trigger combustion in grass or dry shrubs. If it becomes intense, it can spread to trees. There is a desert

---

[2] Frank E. Summers, *Receiver for Wireless Telegraph and Telephone Circuits*, filed 28 April 1916, U.S. Patent #1,338,756.

plant in the Mediterranean region (*Dictamus albus*) that exudes a vapor that readily ignites and burns so rapidly, the plant is scarcely harmed. Sufficiently hot sun exposure may trigger spontaneous combustion in plants such as these. Pistachios have a fat/water balance that promotes heating if they are packed closely in containers. While compost heaps don't frequently ignite, steam can be seen emanating from them as waste materials decay.

Humans observed these natural fires and sought to replicate them. Eventually they found they could create combustion through friction. Once they had control of heat, it was possible to dramatically change the character of natural materials like ore and sand. This led to mining and the invention of mechanical devices and glass.

Significant Discoveries and Inventions

Almost any general-purpose process or component can be adapted for use with surveillance devices but certain key inventions/discoveries stand out as important, including

- the harnessing of fire,

- glass and lenses,

- the discovery of the electron,

- the vacuum tube,

- energy-storage components such as capacitors and general-purpose batteries,

- recording mechanisms such as chemical and magnetic surfaces,

- equipment for sending and receiving acoustic waves,

- devices to generate and sense wavelengths invisible to the human eye,

- the transistor,

- air and space travel,

- software to harness components utilizing micro-electronics, and

- hardware and software to interconnect electronic devices.

Glass and Lenses

Glass was observed sometime between 5000 and 3000 BC, at about the time of the early Bronze Age. As humans developed sea trade, mining, and the shaping of metal they discovered sand could melt with other materials to form translucent objects.

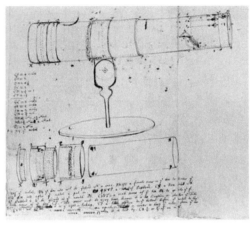
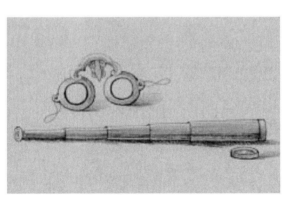

Invention of the lens led to many practical tools to see with greater precision, or at greater distances. In this quest, telescopes became larger and longer, but there were tradeoffs in clarity and the amount of light needed to use them. It was not until the 20th century that new technologies such as array designs and microelectronics significantly improved telescopes in ways other than simply increasing size. Top left: Drawing of Galileo Galilei showing his telescopic invention to three maidens, 1655. Top right: A sketch by Isaac Newton of a reflecting telescope and its components, late 17th century. Bottom left: Spectacles and a spyglass, ca. 1878. Bottom right: Diagram of the Lick Refractor telescope, one of the most ambitious lens-building projects up to that time. It took more than a dozen failed attempts before an acceptable lens was fabricated and installed and, even then, the length of the barrel had to be adjusted to enable it to focus, 1889.

Glass was initially used for trade and adornment but, gradually, as manufacturing processes became more sophisticated, was enlisted for tool-making.

Ancient manuscripts suggest that optics arose in the Middle East and moved westward as trade routes opened between east and west. The production of glass and mirrors and the development of lenses followed, giving rise to spy glasses, spectacles, mirrors, and telescopes.

Lenses not only help us see farther, but with greater clarity and, by the 16th century, opened the door to microscopic worlds.

Since then, scientists have found innovative ways to replace glass lenses for some applications. A water droplet can act as a lens and, unlike glass, can be dynamically shaped by electrical impulses to focus at different distances. Plastic can be shaped more readily than glass can be ground and is suitable for lenses that don't require the same properties as glass. These innovations have led to significant improvements in surveillance cameras in recent years.

Discovery of the Electron

One of the most significant milestones in human history is the discovery of the electron. It is a fundamental component of our universe and the basis for electrical conductivity, chemical reactions, and the functioning of computer circuits. Electrons balance the effects of positively charged neutrons and facilitate bonding between nearby atoms.

Since electrons are exceedingly small, we do not observe them directly. We study them by their influence on their surroundings.

The discovery of electrons and invention of simple vacuum tubes, which eventually were developed into cathode-ray tubes, are closely related. An understanding of electrons is also important to contemporary microelectronics, as some electrons may disrupt sensitive components.

Vacuum Tubes

The vacuum tube started from humble beginnings.

It was originally a simple container from which air was pumped. But our air-filled atmosphere doesn't like a vacuum, it wants to even things out, so porous materials tend to let air back in. Glass is better at holding a vacuum and one can peer inside to observe phenomena different from those in air-filled environments.

Scientists noted that a vacuum did not transmit sound. This led to a better understanding of the difference between sound and electromagnetic waves. A longitudinal sound wave is created by compression and decompression of molecular material (as a rough analogy, picture a Slinky™ stretching and compressing). If there's nothing to compress, as in a vacuum, there is no sound.

The cathode-ray tube (CRT) evolved from a simple vacuum tube to a sophisticated display device. Inventors added wires to the historic tubes to facilitate the flow of electrons. In time, with other components, control was improved so it was possible to generate and intercept electromagnetic waves and, eventually, to display images.

Left: Vacuum tubes were constructed in a great variety of shapes and sizes for different purposes until transistors and microelectronics made them obsolete (except for specialized applications). Right: Inventor Brent Daniel displays a historic portable super-heterodyne radio-wave receiving unit. Seven vacuum tubes can be seen in the cabinet. Inventions like this eventually led to military field radios and consumer portable radios. [Left: Vacuum tubes in the American Radio Museum courtesy of Classic Concepts, © 1997. Right: Harris & Ewing photo courtesy of the Library of Congress, public domain.]

The CRT is familiar to most people as a large, heavy display device for older TVs and computer monitors. It has been largely superseded by plasma, transistor, liquid crystal displays (LCD), and light-emitting diodes (LEDs), but it was the most widely used display device for over 100 years. Most CRTs could display only grayscale images until the 1960s, but colored computer monitors became more widely available in the late 1970s.

 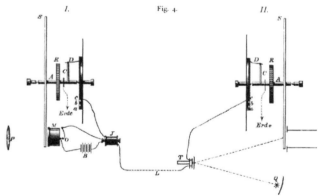

The Nipkow disc, called an *electric telescope* by the inventor, was one of the forerunners of television. The disc would spin to filter light through a system of spiral holes to stimulate photosensitive selenium cells behind the disc. This example is from the American Antique Radio Museum. [Photo courtesy of Classic Concepts ©1998, used with permission; diagram courtesy of Nipkow article in *Elektrotechnische Zeitschrift*, 1885, pp. 433–440, copyright expired.]

Cathode-ray monitors have a "gun" about a foot behind the viewing screen that shoots electrons at the monitor's inside surface to stimulate phosphors to display images. This necessitates a large enclosure so the gun has enough distance to reach all areas of the imaging surface.

Newer flat-screen monitors work on a number of principles, depending on the size and primary use, with LCDs being prevalent, but some embody thin phosphor films to emit colored light when bombarded with electrons from numerous more-closely spaced guns rather than the single gun found in a traditional CRT. These tiny high-resolution display devices show great promise for wearable surveillance systems, tight-space industrial tasks, and medical applications.

Capacitors

A capacitor is a component that stores energy, usually in the form of electrical energy, but a component doesn't have to be sophisticated to be effective. A rock can be considered a capacitor in the sense that it absorbs heat from the sun during the day and radiates it into the soil at night as it cools. Many wine growers leave stones in the soil to help moderate the temperature around the roots of the grape vines.

A battery is a form of electrical capacitor sold for many purposes such as powering a flashlight, cordless phone, or portable computer. Automobiles are equipped with alternators that charge a large storage battery when the engine is running so reserve power is available when needed to turn on lights, run the radio, arm an alarm, or start the engine running.

Solar panel-based electrical systems illustrate capacitance. The panels collect energy from the sun during the day and the energy is stored until it is needed by the associated electrical system for use as heat or light.

Capacitors are widely used in the electronics industry. They are common components on computer circuit boards for storing energy and filtering current. Some components have capacitance as a byproduct of their main function. Consumer warnings have long been attached to CRTs to warn people of the possibility of electrical shock from electrical charges built up from their use.

How much energy a capacitor can store depends upon its size, composition, and structure. Whether a capacitor can be recharged also depends on how it is made, as is the speed with which it can be recharged. Recently developed *ultracapacitors* are energy-storage components that can be recharged faster and more times than regular rechargeable batteries. They are also more temperature-resistant than regular batteries, which makes them suitable for powering devices in the field.

Data-Recording Technologies

The capability to record information is a key aspect of surveillance. Computer hard drives are now taken for granted for their ease of use and large storage capacity, but many of the earliest "recording" surfaces were paper and metal. Holes can be punched in paper to store songs for player pianos, patterns for weaving, or computer software programs. Metal can be used the same way, to record songs for music boxes or historic record players.

The U.S. Census Bureau has surveyed population statistics since the 1700s and, as the immigrant population grew, was increasingly pressed to gather, sort, and record data and generate statistics. Since 1802, it has also recorded information related to certain occupations. Due to the pressing need for automation, it was one of the first organizations to adapt punched paper cards to store results. The cards were then processed with the help of tabulating machines invented by Herman Hollerith. Hollerith recognized the commercial value of his invention and established the Tabulating Machine Company in the late 1800s, a firm that later evolved into IBM.

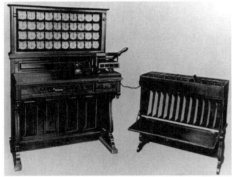

As the population of the American colonies grew, the Census Bureau needed a more efficient way to tabulate census results and, in 1888, held a competition for a better mechanism. As a result, card punching machines (top right) and Hollerith tabulators (top left) were installed. The Hollerith card reader (bottom left) was an electro-mechanical device with spring-loaded conductive pins that passed through holes previously punched in a series of cards to make electrical contact with wells of mercury underneath. This, in turn, used magnetism to advance the counters on the upper dials. The system still required people to complete the process, however. Since there was no way to record the position of the dials for each card, an alarm was sounded so a clerk would know when to write down the results. Half a century later, computers adapted punch cards as a model for storing both data and computer algorithms. [Photos courtesy of the Census Bureau, public domain.]

Paper-based data technologies have limitations, however. A great deal of paper is needed to store large amounts of data, and paper is easily shuffled out of order, lost, or damaged by moisture or fire. Metal is more durable, but also heavier and difficult to punch.

The search for better recording technologies led to the development of vinyl music and speech records, magnetic tapes, and magnetic floppy diskettes. In the 1970s, floppy diskettes were large and flimsy and subject to damage leading to data loss. They were soon superseded by smaller, firmer diskettes with higher storage capacities. These were still common in the mid-1980s but were gradually replaced by hard drives. More recently, the idea of putting pits in a platter, like early metal records, re-emerged as CDs and DVDs, with lasers creating the pits and coatings creating and protecting the surface.

DVDs have considerable advantages over tape media (data on magnetic tapes can be damaged by nearby magnets). The platters are more durable and easier to search. A tape is a "serial" technology—the data is stored end-to-end. If you are searching for something that is recorded near the end, it could take several minutes to find it. A rotating disc can be traversed more quickly from beginning to end, and a location directory sector enables search keys to be saved to find specific parts of the data more quickly. It is also easier to "freeze" a digital image than a signal from an analog tape.

As computer storage devices improved, the Census Bureau was no longer the primary organization compiling information on the populace. Companies and governments began amassing data on people for profiling, marketing, and demographic purposes.

Sending and Receiving Acoustic Signals over Distance

Cupping your hands around your mouth and shouting is one way to send sound over distance, but it is limited to how loud you can shout and it is not very private. It was discovered that electricity and conductive wires could send sounds over longer distances, making it possible to transmit telegraph signals, usually via Morse code (a simple code of long and short pulses). People longed to send voice, however, so inventors sought ways to convert sound into electrical signals that could be turned back into sound at the receiving end. Inventors like Antonio Meucci, Elisha Gray, and Alexander Graham Bell invented and refined the telephone, a device that was quickly adopted by a public eager to talk to distant friends and relatives.

Wiretapping and eavesdropping began almost as soon as telegraph and telephone systems were invented. Telegraph signals were tapped en route and phone operators, who manually connected calls, listened in on local gossip. Campaign workers and businessmen tapped wires to get information on competitors or investment transactions. The invention of automatic telephone switching systems was a direct result of a disgruntled businessman devising a way to thwart a competitor whose wife, a phone operator, had been listening to his business-related calls.

The next obstacle was to get rid of the wires. In the 1800s and early 1900s, it was impractical and difficult to run wires through thick forests and high mountain passes. Some areas were so rocky, there was no way to dig a hole to sink a pole and local tribes were fond of snatching the copper wire to make bracelets. Inventors looked for ways to send sound through air.

One of the most important devices for intercepting early radio transmissions was the *crystal detector*, a small device that resonated in tune with radio waves by means of a crystal with natural rectifying properties (a characteristic in which current flows more readily in one direction than the other). While an important step, crystal detectors had limitations. The crystal had to be carefully selected and oriented, could only pick up a small portion of the radio-wave spectrum at a time, and could only be heard clearly through amplifying headphones. Inventors looked for ways to improve tuning and better amplify the sound.

The breakthrough came with the Audion triode tube, a vacuum tube based on the Fleming two-element tube, but which included a third element—a grid to control the flow of electrons between the cathode and the anode. This led to wireless telegraph receivers and tube-based radios that could be heard by several people in a room at the same time.

Left: A crystal radio set, which used a fine *cat whisker* wire to sense and transfer the oscillations of a radio-wave-sensitive crystal mounted in the round "cup" to a set of headphones. Right: In the 1940s, oscillators for crystal radios were ground by hand and production was limited to about 500 units a day. [Left: Radio magazine illustration, copyright expired. Right: Edward Gruber FSA/OWI photo courtesy of the Library of Congress, public domain.]

At first, the use of radios was unregulated, but radio waves are not unlimited—a strong signal can clobber a weak one if they are broadcasting on similar frequencies. Radio showed great commercial promise, so the U.S. Congress gave regulatory authority to the Department of Commerce, a responsibility that was transferred a few years later to the Federal Communications Commission (FCC). Regulations were instituted to control who could transmit and at what power levels, with specific frequency ranges delegated to different groups of radio users. The laying of communications cables to connect distant outposts throttled into high gear and call IDs were issued to authorized users.

Radio technology significantly changed politics. Before telegraphs and radio broadcasts, the only way a political candidate could talk to people was to get out on the *stump* and meet them personally, speaking to audiences from a podium or staircase. With the advent of radio, tens of thousands could be reached at the same time and a candidate could read from a written transcript, rather than having to talk off-the-cuff or memorize lines. It also enabled voters to hear election results as they happened, rather than waiting for the morning newspaper to find out who had won. In terms of surveillance, radio made it possible for political opponents to "listen in" on what their competitors were doing, not only on public airwaves, but on private transmissions.

Soon law enforcement agencies were using radio communications to dispatch patrol cars and communicate with agents in the field. Journalists and radio hobbyists were eager to listen to interesting events and obtained *scanners*—radio receivers that could scan a wide range of frequencies—heralding the first wide-scale surveillance of police operations.

About a decade later, with the outbreak of World War II (WWII), surveillance of German radio communications through an American radio station on the eastern seaboard resulted in formation of the Foreign Broadcast Monitoring Service (within the FCC) and widescale technological surveillance of foreign powers. Near the end of the war, the Foreign Broadcast Monitoring Service was transferred to the Military Intelligence Division of the War Department and, soon after, to the Central Intelligence Group. After the war, it became the Foreign Broadcast Information Service.

Wartime use of portable radio systems made it apparent that rough terrain was a hindrance to radio signals. This led to the development of radio tower repeater stations to amplify and rebroadcast signals over long distances. Technicians also experimented with bouncing signals off the Moon, which eventually led to the use of satellites as repeater stations.

The Discovery of Invisible Wavelengths

Some birds, insects, and reptiles can sense lightwaves beyond the outer edges of light that is visible to humans, and experimentation with prisms and thermometers in the 17th century revealed their presence to humans.

Infrared is the longer wavelength portion of the optical spectrum between radio waves and visible light. Ultraviolet is what we describe as heat and too much of it can be harmful (e.g., sunburn). It took a while before scientists understood how light was a spectrum of wavelengths and how those wavelengths, even if unseen, could be harnessed. For a long time, they thought a prism was creating colors rather than separating out wavelengths so we could see the colors associated with specific frequencies.

Many surveillance sensors now depend on optical wavelengths outside the human-visible spectrum. Infrared is frequently used for motion sensors and remote control devices and ultraviolet is a valuable tool for revealing security features on financial documents or otherwise-hidden clues such as latent blood stains in forensics investigations.

Recently, scientists have been experimenting with "bending" both visible and nonvisible light, and with the electrical stimulation of nanotubes, to create "cloaking" devices that may someday be used for stealth applications.

THE TRANSISTOR

The Atlas of computing, that carries the electronic world on its shoulders, is the transistor. It is hard to think of another invention more significant to technological advancement.

During WWII, vacuum tubes were an essential component of radio systems, radar, and early computer systems, but they tended to burn out, much like light bulbs, and were bulky, power-hungry, and expensive.

Even before the war, in the early 1930s, inventors such as Julius Lilienfeld were looking for alternatives to the vacuum tube. Lilienfeld created a new means of controlling electron flow in the form of electronic transistors. By 1947, scientists at Bell Labs were working along the same lines, bringing the transistor to the fore, and the field of modern-day electronics was born. Soon after, solid-state electronic components were developed and the door to compact circuit boards, portable radios, and portable computers had been opened. Within 30 years, desktop computers would become household items and vacuum tubes largely forgotten except for specialized applications.

COMMUNICATIONS

Communications with Radio Waves

The transistor made it easier to develop compact, powerful radio communications components, and experiments with bouncing radio signals off the Moon in the 1940s demonstrated the feasibility of long-distance communications—a discovery that would later result in the launch of communications satellites capable of wireless radio, telephone, and television transmissions. Radio operators around the world were encouraged to listen for signals when Sputnik was launched in 1957 and American and Canadian communications satellites soon followed.

Over the years, different parts of the radio spectrum were found to be better for certain kinds of communications, and these have been allocated in a number of ways. In some cases, broad guidelines are issued; in others, very specific parts of the spectrum are designated (as for military use) and commercial ventures are permitted to petition for access to certain frequencies (e.g., commercial satellites).

Since radio was first invented, there has been an ongoing competition for airwaves between amateur radio users and commercial interests. Many amateur enthusiasts are scientists and engineers (astronauts are often radio hobbyists) and have created or discovered new uses for parts of the spectrum that were thought to be intractable or worthless. As soon as this happens, companies want access to the frequencies so they can capitalize on their commercial potential. If they are successful in convincing the FCC to give them access, spectrum allocations are shuffled and sometimes new "junk" bandwidths are passed on to amateur operators.

Amateur radio enthusiasts (HAMs) also served as the first emergency broadcasters, letting other operators know if there had been a natural disaster, a collision near the coast, or a downed plane in the neighborhood. They also alerted each other about impending storms. The birth of the Amateur Relay Radio League (ARRL), one of the oldest and most respected radio organizations, began in the 1930s to facilitate emergency communications. In 1952, the Radio Amateur Civil Emergency Service (RACES), comprised of registered radio operators, was established as a result of discussions between the ARRL and the Office of Civil Defense so there would be a standby ready in case war powers were invoked and civilian services taken off the air.

Air and Space Communications

Kites, lighter-than-air balloons, and historic airplanes made it possible to launch cameras into the air and see the world from new heights. Rockets made it possible to escape Earth's gravitational field to view not only events on the ground, but also the Earth itself. Since then, probes have been sent to other planets, and electronic craft have walked on Mars. All these achievements have contributed to surveillance technology in one way or another.

Left: Aerial surveillance has been in use longer than most people realize. This photo illustrates testing of a lens for airplane surveillance and map-making by the Army and Navy in the 1930s. Lights at the far end are directed toward the photographic plate seen in the center. Right: A satellite image of Syria and Lebanon captured by the multinational Terra satellite managed by NASA's Goddard Space Flight Center. [Photo courtesy of Library of Congress, 1937, public domain. Satellite image courtesy of NASA GSFC LANCE/EOSDIS MODIS Rapid Response Team, 2012, public domain.]

The "space race" began when the Soviet Union launched the Sputnik satellite. It wasn't a secret—the imminent launch was announced worldwide so that radio operators could listen for communications from the satellite—but Americans were concerned about losing technical superiority and responded in kind.

One of the most significant communications-related achievements of the American space program was the launch of *SCORE* (Signal Communications by Orbiting Relay Equipment) in 1958, as a response to Sputnik. The orbiting satellite made it possible to relay voice and coded communications over great distances without cables or series of relay stations. The project was undertaken in considerable secrecy until it was orbital, at which point it broadcast a message from President Dwight Eisenhower through one of two onboard tape recorders. The tape recorder also had the capability of receiving and transmitting a new message. Space communication developed quickly. Within two years, it was possible to transmit television images through the TIROS-1 satellite.

Canada also initiated a satellite program in 1958, aimed at studying both space travel and radio communications. The launch of the Alouette-I satellite in September 1962 provided valuable information on the characteristics of radio-wave propagation in the ionosphere.

In the 1960s, the U.S. *TRANSIT* system, the forerunner to GPS, provided six satellites available, in part, for navigation by naval submarines. Over the next decade, TRANSIT evolved into the *NAVSTAR* GPS. In the 1980s, restrictions were lifted and GPS became broadly available for civilian use.

The U.S.S.R. launched the first GLONASS satellites in 1982, with ground control handled in Moscow, and telemetry and tracking in St. Petersburg. GLONASS was probably inspired by *NAVSTAR* GPS and is similar in concept and operation to GPS. It took time to become fully operational, but was widely deployed by the mid-1990s.

As GLONASS evolved, a newer constellation of satellites was being developed and launched to create the first modern GPS network, operated by the U.S. Air Force. Atomic clocks were an important component in newer systems but timing errors were deliberately inserted into the radio signals to limit accuracy for nonmilitary uses. When this security measure was lifted in the early 2000s, commercial firms significantly improved the accuracy of their services and consumer GPS markets grew.

Europe was not yet a unified nation, but as plans for the EC (and later the EU) solidified, the European Space Agency came into its own and began planning its own GPS-style system with plans to launch in the 2010s. When the system is fully operational, it is expected to include 30 satellites in medium Earth orbit.

Since the 1960s, there has been a steady increase in the number of manufactured bodies orbiting the Earth. Some are passive satellites, used to bounce signals from one part of the globe to another, others are active, using relays, filters, sensing devices, and transmitters to capture, process, or resend information such as aerial photographs, phone and television transmissions, GPS coordinates, and coded covert communications.

In October 2011, the European Commission and European Space Agency (ESA) launched two Galileo satellites to initiate the EU's own high-resolution dual-frequency global navigation system, based on more precise atomic clocks than current U.S. systems. It is expected to be interoperable with existing GPS and GLONASS, and broadly deployed by about 2015. One popular application for geolocation (right) is vehicle mapping systems. [News illustrations courtesy of the European Space Agency (ESA), released (http://www.esa.int)]

Global space programs have had a significant impact on communications and surveillance technologies. It is difficult to escape Earth's gravitational pull to get out into space and to operate without nearby power sources, so space vessels need to be compact, reliable, radiation-resistant, and capable of operating with minimal power. Many have thrusters to adjust attitude and altitude and most use a combination of solar power and batteries to collect and store energy. How much time is spent on the dark side depends upon the distance from Earth and the kind of orbit. Some orbit around the Earth and others orbit in relationship to a particular location on Earth (geosynchronous). The specialized needs of space flight have spawned many inventions that have terrestrial applications for solar energy, protective surfaces, and micro-electronics that are discussed in more detail later.

Software to Harness Micro-Electronics and Interconnect Computers

By themselves, transistors and integrated circuits are of limited use. When you combine them with software to control the circuits and data buses and other means to interconnect separate components, their applications seem limitless.

Telegraph, stock ticker, and teletype machines served as the first data networks, transmitting business information and personal messages over distances. In the early days of the stock exchange, runners would physically courier messages between people working and trading at the exchange. This was slow compared to communication over wires, so inventors sought to automate the process.

In the early 1870s, Theodore Foote and Charles Randall invented one of the earliest printing telegraphs and Henry van Hoevenbergh developed a means to synchronize a network of stock tickers through a mechanical central transmitter. These early efforts to get machine parts to talk to each other set the stage for data networks and communications switching systems.

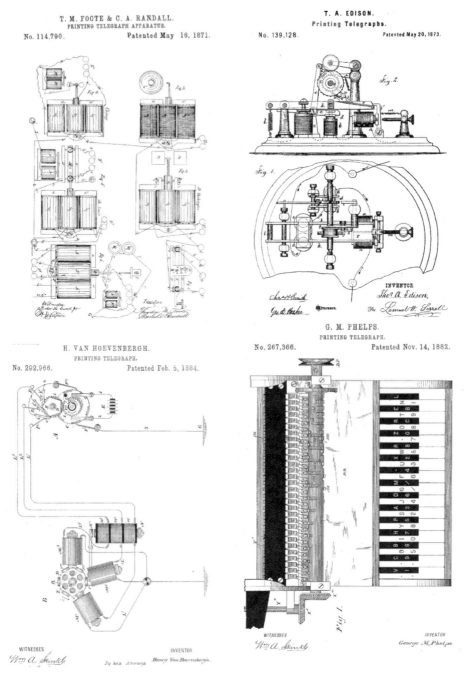

Printing telegraphs were important forerunners to teletype machines, computer terminals, and centralized data networks. Inventors Theodore Foote and Charles Randall patented one of the first printing telegraphs in America and many incremental improvements were added by various inventors over the next four years. A decade later, when more specialized business applications for printing telegraphs caught on, especially at the stock exchange, many inventors jumped in, including Thomas Edison, Henry van Hoevenbergh, and George Phelps. Phelps' designs were considered some of the best. [U.S. Patents 114,790 (May 1871), 139,128 (May 1873), 292,966 (Feb. 1884), and 267,366 (Nov. 1884), public domain.]

In the early 1920s, the concept of telegraph switching systems was described in publications such as *Telegraph and Telephone Age* and inventors such as Clarence Lomax took the concept and began designing practical implementations of telegraph switching systems. It was a primitive form of "telegraph Internet."

Historic Police Communications Networks

By 1929, Gilbert S. Vernam had envisioned a centralized data network for law enforcement (U.S. Patent #1,798,235) and developed a printing-telegraph switching system that could be used by police departments to intercommunicate hub-and-wheel style via printing telegraphs and switchboards connected by telegraph lines. Vernam's system could send individual messages to connected branches, or broadcast general messages, alarms, or orders to all stations simultaneously. These could be sent from a central office all the way down the chain of administration, or from a local precinct to branch offices within its jurisdiction.

Vernam pointed out that while the central office would need two-way communication, not all endpoints required two-way capabilities, other than a signal acknowledging receipt of a message—a concession that had practical advantages in terms of cost. A similar concept was applied to time-share computer systems thirty years later and even now, ISPs stipulate different bandwidth allocations and speeds depending on whether users are uploading or downloading to the Internet.

Building on network concepts from the world of telegraphy, the development of programmable computer networks made it possible to adapt general-purpose machines to a variety of tasks by changing the underlying software programs.

From Calculators to Computers

Music boxes, looms, historic metal and vinyl records, and computer software all have something in common—a need to record the song, pattern, or software program so it can be used again without reprogramming it.

The use of wood with protruding pins or punched metal to record music was already invented by the 1400s when carillons became popular. By using something that revolved, either a rolling drum or a spinning platter, the song or other recorded information could be played over and over without restarting it. The only limitation to continual play was the length of a spring, or the stamina of the person turning a crank to power it. Music boxes were all the rage by the 17th and 18th centuries and recorded songs were played on mantle clocks.

When computers were developed, engineers didn't immediately think of using punched metal or paper to program them. Most were programmed by hardwiring the electrical circuits or by using wires with plug ends that could be repositioned.

In the days of room-sized vacuum-tube computers, programming began with simple arithmetic calculations Wires were connected between circuits to enable them to "talk" to one other in much the same way that telephone switchboard circuits were connected by human operators to facilitate voice communications.

Reprogramming was a tedious process, since every new application required long bulky wires to be moved each time the program changed, and useful programming constructs such as procedures, subroutines, and pointers were impractical. A more efficient method was needed for creating and storing computer algorithms. For inspiration, inventors studied historic ways to record patterns, such as music box cylinders, player piano rolls, Jacquard-loom weaving cards, and punch cards adapted by the Census Bureau to record and tabulate information thirty years earlier. This led to computer punch cards, a means of storing computer programs that was used for several decades.

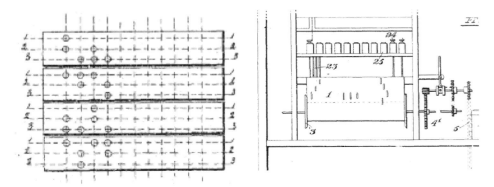

Left: Holes were punched in a series of cards to store patterns for Jacquard looms, a system that was developed in the early 1820s. This provided a way to re-use patterns without having to manually retool the loom. Right: This 1913 patent illustrates a recording machine for creating player piano rolls. As the piano was being played, holes were punched in the long rotating paper roll so that the piano could replay the song later without anyone touching the keys. This is the same concept used in music boxes and Hollerith tabulating cards. [U.S. Patent #152,654, J. Laird & W. Rutherford, June 1874; U.S. Patent #1,076,338, Edwin S. Votey, Oct. 1913, public domain.]

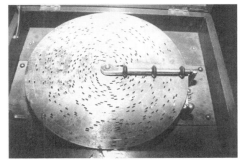

Coding systems based on bumps or holes, which are analogous to digital ones and zeros, enable sounds to be stored and replayed and can be used to store computer data or algorithms, as well. Left: Carmen-Marsch music cylinders from a table-top music box were like cartridges—interchangeable. The pegs would hit the sound mechanism as the cylinder rotated, reproducing a tune. Right: A flat metal cylinder encoded with punched holes rather than raised pegs is similar to player piano rolls or punch cards used on computers until the 1970s. [Historic artifacts from the American Radio Museum. Photos copyright ©1998, Classic Concepts, used with permission.]

Long strands of wires with bare ends, or with electrical jacks similar to those used in historic telephone switchboards, would interconnect circuits on pioneer calculating machines such as this ENIAC computer from the 1940s. Computer programs were altered by rearranging the wires. These wired behemoths, costing the equivalent of millions of dollars, were primitive compared to modern handheld calculators, but were a thousand or more times faster than mechanical or electromechanical calculators of the time. [Photo courtesy of the U.S. Army, public domain.]

Computer programmers needed ideas for more efficient ways to store and retrieve data without physically rearranging wires, so they looked back to the early 1900s, when Census Bureau operators (shown above) used punch cards and Hollerith Machines (the forerunner to IBM computers) to store and tabulate census results. This data-storage method was seen as a way to save and retrieve not only computer data, but also computer instructions, known as algorithms, that tell the computer what to compute and how to do it. [Hollerith Tabulating Machine photos by Waldon Fawcett courtesy of the Library of Congress, ca. 1908, public domain.].

Punch cards were more flexible than rearranging wires, but they had disadvantages as well. It was hard to punch them accurately and keep them in order, and bends and tears would confuse the electronics, causing cards to be misread. Cards were also subject to damage from moisture and fire and were easily lost.

To overcome these limitations, inventors developed magnetic storage, so that a program could be read back through a tape machine similar to an audio tape recorder, except that the data was bits and bytes rather than sound.

Tape had limitations, too. It was a serial technology. If you wanted to reload a program near the end of a tape, it could take a long time to advance to that part of the tape to find it. Floppy diskettes overcame these limitations by going back to a spinning platter, as was used in record players, to enable individual sections of code to be found more quickly using file pointers rather than serial access.

Both tape and floppy diskettes were more durable than paper, and took far less space, but were still vulnerable to damage from magnetic sources or "bit decay."

CDs, DVDs, and hard drives gradually made tape and floppy diskettes obsolete and greatly increased the amount of space available for data storage.

It needs to be emphasized that the evolution of computer storage is a highly significant aspect of surveillance because the ability to store vast amounts of information changes the way people do surveillance.

In the past, when photos were taken with film cameras, photographers carefully planned and framed their shots because a roll was only long enough to accommodate 24 or 36 pictures. If the roll were used up and something important happened, there was no way to record the event.

The same was true of surveillance before mass storage devices. Those doing surveillance had to carefully time their audio, film, or video recordings to catch the most significant moments, or the recording medium would run out.

All that has changed. Digital cameras enable many thousands of pictures to be taken, and additional memory cards are easy to carry and swap into a camera. Many cameras, such as digital video, have hard drives that can store hours of video, compared to three minutes of Super-8mm film.

The quality of the images is significantly improved as well. Film used to be higher resolution than digital photos, but that is changing, and new cameras can instantly adjust for light, contrast, and focus, something that took time to measure and set accurately in the past. New cameras are also capable of recording depth, something unheard of in the early days of photography. Light-field cameras, discussed more later, can be focused *after* the picture is taken.

Now that it is technically easy and significantly more affordable to build large image libraries, the government and many commercial firms maintain vast data repositories such that the idea of recording every word and every action is theoretically possible. Now surveillance is less about "timing" to conserve resources as it is about storing everything and developing sophisticated programs to locate information that may become relevant at a later date.

MINIATURIZATION

Miniaturization isn't just convenient, it can dramatically speed up electrical transmissions, making it possible to develop more powerful circuits.

As mentioned, the transistor was a milestone invention that made it possible to do away with large vacuum tubes and miles of wires. With microcircuitry, developers closely associated computer components so that they could quickly transmit electrical signals, and computer programs were stored on cards so they could be reused or readily altered without overtly changing the hardware. All these innovations led to smaller and smaller computers with more powerful capabilities, as well as teletype machines with more advanced circuitry.

During the 1960s and 1970s, computers were improved to include internal memory to store information, dip-switches to program them without having to physically burn a new chip each time, and tape drives to store and retrieve programs magnetically. From that point on, more efficient, higher-capacity storage devices and interconnection became obsessions and, by the mid-1960s, gamers were interfacing time-share teletype machines to computers through phone lines to serve up text-based games.

Before the 1970s, keyboards and monitors as we know them were uncommon. Now they were added to computers to make it easier to see what was being programmed. By the mid-1970s, hobbyists were eagerly snapping up desktop computers with almost no CPU power or memory, compared to present-day computers, for the equivalent of three month's wages. Despite their limitations, they were put to work as word processors, accounting machines, and gaming devices. Meanwhile, research leading to the development of Ethernet communications protocol and the Internet were underway.

SECURING DATA OVER DISTANCES

The Birth and Spread of Desktop Computers

Desktop computers were invented prior to the 1970s, but they were mostly "one-of" projects or prototypes with limited production. The market for personal computers hadn't yet bloomed due to their prohibitive cost. Some exciting work on ergonomics and graphical user interfaces was being done at Xerox Parc, but it wasn't yet making an impact on the general consumer market.

By the mid-1970s, however, young entrepreneurs excited about the potential of the technology, such as Steve Jobs and Bill Gates, began to capitalize on the opportunity to bring small computers to every desktop. Jobs was charismatic, ambitious, and driven by a desire to take exciting ideas, like those developed at Parc, and put them in the hands of consumers. He paired up with hardware guru Steve Wozniak to create Apple computers. Gates was equally driven and ambitious and saw opportunities to take technologies like Dartmouth BASIC and Gary Kildall's CP/M and sell them to the business market. He partnered with Steve Allen to create Microsoft.

These success stories represented a new breed of entrepreneur. Until that time, a large portion of the consumer market was run by older business executives. The TRS-80, sold by Tandy, Radio Shack, had 80% of the desktop market share, but the young entrepreneurs gradually gained ground and overcame the dominance of Tandy computers. The success of a younger generation who, in turn, hired many young programmers and hardware technicians, represented a shift not only in the computing world, but in the business world. Technology had opened a window of opportunity for youth to be more than messengers or apprentices.

The Altair computer was the first to excite electronic hobbyists, despite its humble appearance and limited capabilities, and the electronic arcade game Pong was a big hit in tourist and entertainment venues. Suddenly there was an electronic alternative to mechanical pinball machines. Then the TRS-80, Apple, and Atari computers came out at prices that made them affordable for university, small businesses, and as Christmas presents. They weren't inexpensive, but they were accessible, thus launching a new industry. Businesses were slow to adopt desktop computers—they were reluctant to give up legacy mainframe equipment—but hobbyists loved them.

By the 1980s, governments and businesses were beginning to depend on desktop computers for day-to-day operations and the need for data redundancy (to protect data in case of electrical failure, file corruption, or operator error), was growing. The transition to desktop computing also threatened data security.

Older mainframes and miniframes were often housed in separate rooms or facilities, with entry/exit security, and only system administrators or authorized personnel had access to them. The new generation of desktop computers was wide open—vulnerability to snooping, hacking, and theft greatly increased.

Dependence on desktop computers by government agencies and businesses, and the desire to interconnect them for communications and the sharing of data, also created concerns about the "cyberspace highway" being vulnerable to intrusion from remote locations.

Encryption schemes were developed, with new, stronger ones appearing every few months, partly because software programmers enjoy the challenge of creating stronger encryption and partly because hackers and crackers keep breaking them.

Despite their weaknesses, encryption made it possible to protect communications. Businesses were happy, but law enforcement and intelligence agencies became concerned that criminals could engage in secured communications with no possibility of interpretation. The days of traditional wiretaps were numbered.

Clipper Chip Controversy

Subsequent events led to one of the most controversial debates in voice communications and desktop-computer history.

A quiet plan was underway by the National Security Agency (NSA) to develop a computer component called the Clipper Chip to encode voice communications, with an analogous Capstone Chip for data communications. The idea behind the chips was to enable people to encrypt their communications but for government agents to "unlock" the data if the need arose. It's important to remember that there was far more secrecy surrounding the existence and operations of Intelligence Community organizations in the days before the World Wide Web.

When the White House announced the existence and adoption of the Clipper Chip in 1993 and 1994, alarm bells went off in the computing populace and resulted in an electronic petition to oppose Clipper. Citizens were concerned that access to private communications might promote surveillance not only of criminals, but of the general public,. They also pointed out that the classified nature of the encryption algorithm, called Skipjack, made it impossible for security experts to evaluate its strength and possible vulnerabilities. One of the strongest arguments against the chip wasn't based on privacy, however, but on profitability.

Commercial interests argued that it might hobble American competitiveness in the global market if foreign companies developed stronger encryption. There was also concern that it would limit American opportunities to export commercial encryption products or products relying on internal proprietary encryption.

By 1995, the Clipper was dying partly because of opposition, partly because independent developers were creating and releasing their own encryption schemes. By 1998, Skipjack was declassified and Clipper mostly forgotten.

Packet Switching

Another strategy to facilitate the transmission and protection of computer data was the development of *packet switching.*

A *packet* is a small unit of data. Breaking information into smaller units has advantages. As a rough analogy, think of data as goods in a freight train, being carried to different destinations. If a train was one mile-long car, it would be difficult to break it apart and send specific things to different destinations. It is also less secure. The train is more likely to be extensively damaged if the entire train is derailed and falls in a canyon rather than the few cars that were hit by a bomb from brigands planning to rob the train. By breaking the train up into smaller units, the goods can be organized in a variety of ways, easily rerouted, and reassembled at other locations. Even if the data is corrupted en route, getting part of the message is better than none.

Packets make it easier to duplicate information, in part or in whole. Thus, packet switching is more flexible and can be made more secure than a long unbroken stream of data.

Packet switching also caught the attention of radio hobbyists seeking ways to get better quality radio transmissions over longer distances by using digital rather than analog broadcasting. When you broadcast over airwaves, you are at the mercy of weather and interference from more powerful transmitters. Distance also degrades the signal. With packet switching, it became possible to send radio messages around the world clearly and accurately through the Internet.

By 1969, the ARPANET, an important development leading to today's Internet, employed packet switching for data transmission.

Computer bulletin board systems (BBSs) started springing up in the late 1970s, mostly because of the growing popularity of the TRS-80, Atari, and Apple computers. Acoustic modems, which looked like a pair of suction cups, made it possible to cradle a telephone handset into the modem to transmit computer data over telephone lines as sound.

By the mid-1980s, FidoNet enabled coast-to-coast digital communication by using a hub-and-wheel analogy for spreading information from a pair of central communications nodes to local BBSs.

By 1985, Fido had grown and routers were in place. Many computer hobbyists intercommunicated via Fido before they had access to the Internet and "eavesdropping" on text-based chats became a popular hobby among system administrators. Hacking into personal and government computer systems also made the news and spawned movies like *War Games*. Institutions grew concerned about data security.

THE WORLD WIDE WEB

In the 1980s the Internet was almost entirely text-based. With the exception of a few ASCII drawings (images constructed from alphanumeric characters), the Internet remained a "command-line" environment, inhabited mostly by scientists, computer science students, computer hobbyists, and intelligence agencies. It was not until the early 1990s, when Tim Berners-Lee embued the Internet with an accessible, easy-to-learn-and-apply graphical user interface that the rest of the world jumped in and the *World Wide Web* (WWW) was born.

Since that time, the Web has grown in leaps and bounds, especially in the areas of social networking and commerce, as have the elements of society that surveil the Net and its users. It is now one of the primary sources of communications, personal information, and ways to peer into other people's business.

Many sites collect demographics on their users (or buy them from others) and not all of it is out in the open. It was alleged through a contract employee leak in February 2012 that social-networking site Facebook was paying third-world workers to delete certain objectionable content from areas of the site that users thought was private. A corporation has a right to control content, especially if it is making changes to comply with laws, but the action implies that workers were given access to areas of the site containing personal information without the company first alerting users. Facebook is probably not the only site hiring people to manage personal data behind the scenes.

Certain companies are aggressively gathering data on everyone and everything in order to provide "screening" information for businesses seeking to hire new employees. Since they are charging fees for providing the information, they are motivated to make the background information as detailed and comprehensive as possible.

One would expect companies to make their own judgments about who they hire, but apparently two of the large research firms for information on job applicants have become common resources for employers to do background checks. In some instances, companies seeking to hire have even asked job applicants for passwords to their private online accounts. These issues are discussed further in the *Computers and Internet Surveillance* chapter.

Facilitation of Crime

The Internet has provided unprecedented opportunities for communication and scientific collaboration, but also enables criminals to intercommunicate, seek out information on how to commit crimes or build bombs, and how to evade arrest. It further provides opportunities for stalkers to seek victims, and con artists to find marks.

Some online con schemes are very sophisticated and many Trojan horse programs are difficult to detect, even for seasoned computer users. Others are obvious cons but even so, people can be incredibly gullible when offered large sweepstakes winnings (to a sweepstakes they never entered), large "commissions" on "royal" treasure troves (Nigeria scams), or offers for fake prizes, coupons, or goods that are never shipped.

The Net also makes it easier for white collar criminals to launder money, offshore their assets, and locate potential victims for schemes like mortgage or securities fraud. Sometimes popular gaming sites are front-ends for money laundering, as well. The users don't know what the site managers are doing behind the scenes and serve as a form of unwitting screen to distract investigators.

A software-based universe provides unprecedented opportunities to insert malware into people's systems, such as spyware that can snoop on address books, personal files, financial documents, and passwords that might provide access to online banking or other sensitive account information.

The computer-using community can also be very quick to steal, if stealing is easy. Pirate sites that offer passwords, copyrighted software programs, and media have extremely high traffic counts and sometimes even earn large sums from paid subscribers. The downside of widescale theft is a reduction in jobs. When the Internet went mainstream many small, formerly viable software companies closed their doors due to reduced revenues from piracy, and the entertainment industry can't keep making entertainment unless actors, artists, photographers, directors, publicity departments, and scriptwriters get paid.

This battle over copyrights has created a constant state of adjustment on the Web. When video-sharing sites like YouTube™ were first initiated, substantial uploads of copyrighted content occurred. Now a certain amount of policing and legal licensing is being implemented on this and other sites. While everyone loves free content, rent and food are not free. If people work to create good products, they are entitled to be paid. Not every business model can be revamped to depend solely on advertising or on a minority of paid subscribers. It will work for some businesses, but not for all.

As Technologies Change, Methods Change

In the 1980s and 1990s, the growing potential for criminal opportunities over national and international networks alarmed law enforcement and intelligence agents who quickly lobbied Congress for broader surveillance powers. The granting of these powers also introduced broader observation of law-abiding citizens than ever before because the capabilities of technology have changed the way people do surveillance.

In the past, law enforcement had a more focused nature. Police with limited time, finances, and observational devices carefully set priorities about who to watch and how to deploy staff to watch them.

Now, with widespread, broad-based communication, highly capable cameras that cost less each day, and high-capacity, low-cost mass-storage devices, security-related surveillance has changed from targeting individual people or small groups under suspicion of committing crimes, to gathering vast amounts of audio/visual and textual data on everyone and everything that happens.

Much of this information is uploaded from local law enforcement agencies to national databases, in case it might lead to something more specific later. Not all national databases are government-managed—some are maintained by commercial interests. Thus, focused policing is gradually shifting to a "net everything and see what you get" approach. This trend is illustrated by the way DNA and license plate numbers are collected, which is discussed in more detail later.

The Generalizing Nature of Surveillance

Changes in the way we do surveillance can be illustrated by changes in the way we fish. In the past, fishermen would throw out a line and hope to catch a cod or snapper.

Now there are electronic fish-finders that give a detailed image and map of the surrounding marine landscape, including Global Positioning System (GPS) location statistics so the hot fishing spots can be found again. Small devices chart water currents, temperature, and anything else that might influence the feeding habits and migration patterns of fish. It's even possible to design tiny radio transmitters that could be spread inside fish food so they are swallowed and show the location of specific fish.

The information is digital, so maps and other data can be stored for later reference. The "fish reconnaissance" information can be uploaded and shared on fishing forums to create a shared library.

Just as we have Google Streetview™, there may eventually be Internet fish-views—free or commercial services that map every underwater valley and cave and all the creatures that inhabit them.

Much of this marine surveillance has already occurred. The U.S. Geological Service has done a remarkable job of electronically charting marine topography to create publicly accessible maps of the ocean landscape. These have been invaluable resources for scientists, ecologists, and those patrolling national borders. Deep-sea submersibles help increase our understanding of the deepest regions of the planet, as well.

If you combine these extensive underwater terrain maps with the fish-finding data gathered by amateur and professional fishermen, a data-rich picture emerges.

Then take the idea a step farther. Add electronic tracking tags for long-term surveillance of the movements of individual fish and it becomes possible to watch them grow and know where they are at any moment of the day. Tags with a satellite transmitter are already available for scientists to study larger fish like marlin or tuna. They are attached by shooting a dart, and additionally transmit information about location, temperature, light intensity, and depth.

Eventually there will be no place for marine life to hide and fishermen with access to the information can pick and choose their targets. The fish in the ocean become "fish in a barrel."

The techniques used to monitor fish are not substantially different from techniques used to monitor people. Some people are now tagged with RFID chips for entry/exit access control, covert tracking, or to find them if they are wandering. Combine this with information from databases and video recordings and humans become fish in a barrel.

Many people quote authors like George Orwell and Edgar Rice Burroughs when describing the current trends.

Edgar Rice Burroughs, *Swords of Mars*

Such surveillance probably meant little in itself, but taken in connection with the gloomy and forbidding appearance of the building and the great stealth and secrecy with which we had been admitted, it crystallized a most unpleasant impression of the place and its master that had already started to form in my mind.

The one mechanism we have for preventing the populace from becoming fish in a barrel is the justice system. How, when, and why we use technology is ultimately our own decision.

Laws to enable law enforcement to do its job and for commercial interests to exploit new technologies are being counterbalanced with laws to protect privacy. In the first few months of 2012, substantial revamping of laws occurred. Some gave back privacy that was gradually being eroded. Others provided broader powers to agencies carrying out surveillance. The back-and-forth nature of this difficult balancing act continues today and is discussed in other chapters.

That's not to say we should be disheartened. Perhaps the Pandora's box of technology that unleashed the surveillance society can be used to reclaim lost privacy and freedom.

SECTION 2
TECHNOLOGIES AND TECHNIQUES

OPTICAL

AERIAL

AUDIO

RADIO-WAVE

GPS

SENSORS

COMPUTERS AND INTERNET

DATA CARDS

BIOCHEMICAL

ANIMAL

BIOMETRICS

GENETICS

OPTICAL SURVEILLANCE

KEY CONCEPTS

- *biological senses as a model for visual technologies*
- *variety of optical technologies*
- *video analytics*

Introduction

Humans sense their environment through sight, sound, touch, smell, and taste. Of these, sight and sound are the most common models for surveillance technologies and sight is particularly important since many eavesdropping laws prevent recording of sounds.

Animals have inspired the development of many new technologies. Animals have senses humans don't have. Bats can "see" objects through echo location and dolphins can see through certain materials (like sand) using biological sonar. They can "ping" an object with sound and interpret the shape of the returning echoes. Some insects and birds can see optical wavelengths beyond human vision. Many birds have exceptional visual acuity and can focus on small objects from great heights. Many nocturnal animals see in the dark.

Recent advances make it possible to sense and create visual records of phenomena that humans can't sense without specialized devices. These technologies enable us to see in the dark, in environments that might otherwise kill, and in places where our presence isn't welcome.

Eyes or optics-related nerves in the brain can be modified with devices to enhance sight—many surveillance technologies started out as ways to improve eyesight in people who are visually impaired.

Sometimes other senses can be used to send visual information, like an innovative device developed by scientists in Wisconsin that electrically stimulates the tongue to recognize light and dark areas within a room so blind people can find objects and building structures like doorways and elevators. Such a device, if attuned to nonvisible frequencies, might provide a new kind of night or heat vision, as well.

VISUAL SURVEILLANCE

Human Vision

Not all surveillance is high-tech. Cameras are common, but a person watching from the shadows is still a very prevalent surveillance scenario. Private detectives and law enforcement officers do much of their work with their eyes. Those planning or committing crimes use their eyes, as well. Even those who carry cameras use their eyes to determine where to aim the camera.

One example of careful visual surveillance by organized crime members is illustrated by the methods of a jewelry crime ring that was arrested in March 2012. The thieves carefully watched jewelry stores to identify traveling jewelry sales reps and their habits, so they could stalk them when they were carrying items of value. They would then intimidate them or violently ambush them in their cars to steal the jewels.

Camera Vision

Sometimes cameras are carried to record events or to photograph evidence that will change, such as evidence that is bagged and moved from the original crime scene. Photographic records are an important aspect of forensic surveillance, detective work, and archaeological studies. Building and vehicle-mounted security cameras are also an increasingly common aspect of visual surveillance.

Images from security cameras are frequently used to identify or locate suspects. Left: This image was released by the Bureau of Alcohol, Tobacco, Firearms and Explosives (ATF) working in cooperation with local police departments and other national agencies to solve an arson crime in a Kansas City restaurant. Right: Theft from a licensed gun dealer in Riverside, CA, was released by the ATF in efforts to find the individual who broke a glass door and pried open security bars in order to steal several handguns and rifles. Stolen guns are often used in subsequent crimes. [News photos courtesy of the ATF, 2009 and 2010, released.]

Surveillance cameras provide many kinds of information. They can reveal patterns of behavior or features of individuals. Even if it is not possible to identify a suspect, due to blurred images or concealments, such as masks or sunglasses, general body types and *modus operandi* (methods of operating) can sometimes be distinguished. Sophisticated high-resolution cameras can be integrated with databases to recognize faces, but many facilities have old cameras, or simple ones. Even these, however, sometimes provide enough information to solve crimes or locate missing persons, such as an abducted child.

Securing Explosives Storage Facilities

"Consider installing fences, floodlights, alarms, security cameras, locked gates, or other security devices at the site to better monitor the location..."

Recommendations from the Bureau of Alcohol, Tobacco, Firearms and Explosives (ATF) on deterring crime

The Bureau of Alcohol, Tobacco, Firearms and Explosives (ATF) recommends the installation of security cameras and prominently posted signs that a facility is monitored by electronic surveillance. Police departments are also increasing the numbers of video cameras that are installed in urban centers, particularly in parks and high crime areas. [News photo courtesy of the ATF, released.]

Many police departments are installing neighborhood security cameras and tiny spy cameras with taxpayer funding provided in part through grant programs from the Department of Homeland Security, administered via individual states.

It has been reported that approximately 80% of the New York Police Department's intelligence work is federally funded.

As a result of the *USA Patriot Act of 2001*, the Homeland Security Grant Program (HSGP) was established in 2003 to provide funding to law enforcement personnel for weapons, training, and surveillance equipment. In 2010, the program was set at $1.8 billion and the "highest priority" for use of the money was to protect the U.S. from terrorist threats. Securing borders, naturalizing new Americans, and improving disaster-preparedness and response were also listed.

Each year the grant program goes through adjustments in its terms and how funds are allocated.

In 2012, close to $1 billion was allocated to four programs:

- the State security program,

- the Urban Areas initiative,

- the Tribal Security initiative, and

- Operation Stonegarden (a program to support border security).

As can be seen from examples given later, many departments have used grants to install cameras to help police deal with local crime "hot spots." As part of the Urban Areas initiative, cameras have also been installed in schools.

Part of the reason grant money is used for cameras is reflected in the terms of the grants themselves. Much of the wording favors physical improvements and training, rather than operational costs. In practical terms, it means the money cannot be used for additional security guards or even to continue to fund existing guards, so communities opt for cameras.

Another reason for increased sales of cameras is commercial security firms using the enticement of federal grants as a selling point when promoting their products.

In surveying the recent installation of cameras by law enforcement in many cities, the following motivations stand out as common:

- The Department of Homeland Security made federal grants available to police departments specifically for the purchase of security devices. Local institutions were often reluctant to say no in case funding was not offered again in the future.

- Local departments with inadequate funding for more patrol officers sometimes install the cameras to try to give the department more "eyes and ears" on the beat to fill in the gaps that can't be filled by staff.

- Local citizens concerned about crime in their neighborhoods, or an increase in crime, sometimes lobby the local police to install cameras.

- City mayors sometimes encourage the installation of cameras. At times, this appears to be a genuine concern for public safety and is supported; at other times, the decision is opposed by law enforcement or the local public.

A few examples of local grant expenditures illustrate different approaches to using money for surveillance cameras.

- The New Haven, CT, police department instituted plans to install surveillance cameras in areas with higher crime rates. The cameras provide 360-degree views of streets and side-walks. The police noted that they would not be used to look in windows.[1] The plan was to use the surveillance images after a crime occurred and, later, to add streamed video to substations and patrol cars equipped with laptops.

- In Fort Worth, TX, grant money was allocated to a variety of cameras, including security cameras for downtown, as well as covert "sprinkler-head" cameras, and spy cameras hidden in common implements like neckties and dry erasers.

- Hagerstown, MD, police installed three security cameras in July 2011, hoping to reduce crime. Residents in one of the problem neighborhoods complained of theft and drug dealing within view of their doorsteps and more cameras were requested. Images are streamed to monitors at the police department and laptops and cellphones carried by officers in the field.

- In Raymondville, TX, there was concern about increased drug dealing, gang activity, and graffiti, so cameras with zoom capabilities and a range of about 500 yards were installed in 2011 to help police watch problem areas. The police chief pointed out that part of the increase in crime may be due to the economy and lack of programs for youth.

School Security

Many recent surveillance camera installations are in public areas and hot spots for crime in urban areas, but incidents like the shootings at Columbine High School and Virginia Tech, which left dozens of people dead, have provoked an increase of security cameras in schools, as well.

- The 2009 Safe Schools/Healthy Students Initiative awarded almost $33 million in grants to help schools create safer, healthier learning environments.

- In 2010, more than 160 law-enforcement agencies were awarded U.S. Department of Justice (DoJ) grants to improve

---

[1] Incidental viewing of windows is difficult to prevent when cameras have a 360-degree view, but police conferred with residents about privacy issues before installing the cameras.

security on school campuses. Some of the money was used for emergency preparedness training. In other areas, access security was added to school entrances. Surveillance cameras have been added to schools and buses, as well.

- The DoJ Office of Community-Oriented Policing Services (COPS) offers *Secure Our Schools* funds that can be used for security cameras. Metal detectors are also installed in some schools. The COPS program awarded approximately $16 million in school security grants in 2010.

- Another *Secure Our Schools* grant was awarded to the Milton, MA, police department to equip Milton public schools with security cameras to provide "a comprehensive approach to preventing school violence..."

- A school in Hollywood, FL, is equipped with 60 cameras connected to Wired Equivalent Privacy (WEP) access points that can be observed in realtime from a central surveillance control room, or monitored by laptops from police patrol cars within the school's perimeter.[2]

- The New Jersey Schools Development Authority provided a grant to Hopewell Valley Regional School District to cover about a third of the cost of new video cameras for Timberland Middle School. Close to $200 million in grants has been provided to schools for health and safety-related issues and to help alleviate overcrowding.

Local Funding Sources

As illustrated by school security funding, not all funding is federal. Individual states provide grants for security cameras, and sometimes local communities fund them, as well.

- In Ashland, PA, applications were made to a community foundation for grants to equip four public parks with video cameras.

- In Hartsville, SC, the Byerly Foundation provided a grant to install a computerized security system downtown, consisting of six cameras and laptops for monitoring the feeds. City officials expressed a desire to expand the system in the future and to integrate the system with security cameras being purchased in Oakdale through a half-million-dollar Community Development Block Grant.

---

[2] Captain Thomas Sanchez, "School watch: digital surveillance technology for school security," *The Police Chief,* March 2012.

- In Houston, TX, approximately $250,000 in grants was provided by the Department of Housing and Urban Development for cameras with recording capabilities to protect the Kelly Village housing community. Live monitoring of video feeds was not within the budget, a situation that is common to many communities with installed cameras.

Effectiveness of Cameras

There is an ongoing debate as to whether cameras deter or displace crime. In 2011, Syracuse, NY, police agents reported reduced crime rates in areas where nine video surveillance cameras had been mounted on poles. Crime rates across the city as a whole, however, increased.

There are many studies that question whether cameras deter crime. It appears they might—in some areas. In a study of two high-crime areas of Chicago, cameras didn't appear to reduce crime in one area, but the second area showed decreases. However, decreases can sometimes be temporary. Short-term studies have to be followed up with longer-term studies to distinguish fluctuations from long-term cause and effect.

One can also ask whether the money devoted to installing and monitoring cameras might be better spent on youth programs, which have a good potential to prevent crime, or other forms of aid for law enforcement.

Some contend that steadily increasing the number of cameras will incrementally reduce crime. Others, like Lancy La Vigne, Director of the Justice Policy Center in Washington, D.C., have pointed out that there may be a point of diminishing returns where each additional camera has less impact on crime reduction. La Vigne also pointed out that installation of cameras may have negligible impact in low crime areas.

These issues are discussed further in the *Effectiveness of Surveillance* chapter.

CHARACTERISTICS OF OPTICAL IMAGING DEVICES

Suitability

Optical devices have a wide range of properties that make them more suitable for some applications than others. Cost, rather than suitability, is sometimes the basis for choosing a security system. This section discusses improvements in recent technology and some of the newer specialized devices.

Resolution

Resolution is a term loosely used to describe the amount of information that is conveyed by a visual image in a given space. In fact, "resolution" depends partly on the way in which the information is displayed and on *perceptual resolution*—how the image is perceived and interpreted by the human brain.

A common way to describe resolution is in pixels, a *pixel* being a discrete spot of light on a display surface. How clear the image looks depends not only on how many pixels are displayed over a given surface, but on the shape of the pixels, how closely they are spaced, their intensity, and the wavelengths (and hence colors) that are represented. The size at which the image is displayed is important as well. A 100 × 100-pixel image might be recognizable if printed or displayed as a small thumbnail, but may be pixelated or staircased (aliased) at a larger size. Sometimes image enhancement with fractal software algorithms can fill in missing information, but contrary to what TV spy shows suggest, enlarging a low-resolution image rarely makes it more recognizable.

Resolution may also be described in scanlines. Not all display devices use discrete dots or light up every part of the screen at the same time. Sometimes the image is built from a series of lines or other patterns that are refreshed faster than human brains can follow, making it appear as though it is a unified image. Older television sets work this way—the image is built from one diagonal corner to the other, in a continuing pattern, until the screen is filled, and then starts again. The image decays as it is built and overwritten to give the perception of objects and motion. This is why old TV screens are difficult to photograph with a still camera. The camera freezes a moment in time so that you see the scanlines and part of the image will be faded or missing.

Perceptual Resolution

For surveillance purposes, how much resolution is needed to make an object or event recognizable?

This depends on many factors, including the kind of display, the speed of motion, the angle of the camera, and whether the image on the screen is familiar to the viewer. In the following series of portraits, each is shown at three different resolutions. A familiar face may be recognizable at lower resolutions than the one that is less well known.

Which face did you recognize fastest? The amount of detail needed to recognize an image depends on both eye-brain perception and the technical specifications of an image. A familiar image can usually be recognized in lower resolution than one that is unfamiliar and a simple image is often more recognizable than one that is complex. Here three different presidents are shown at three different resolutions. Which one is easier to recognize at smaller sizes? [Photos of Presidents Tyler, Buchanan, and Lincoln courtesy of the Library of Congress, public domain.]

The complexity of an object makes a difference as well. A picture of an apple is easier to recognize than the head of a hooded fleeing thief. For moving pictures, frame rates may also affect whether the image can be recognized or events readily understood.

IMAGE PROCESSING

Image processing involves converting pictures into other forms or formats to make certain aspects more clear or useful. Brightening, sharpening, or combining images are common image-processing tasks. Image processing can be used to visualize nonvisible wavelengths. For example, we cannot see infrared or ultraviolet light directly, but we can record them and convert those images to something useful for analysis.

In the past, image processing was a separate step from taking a picture. Now many cameras offer image processing capabilities for artistic and scientific purposes, and some have been adapted for surveillance tasks. For example, there are "smart cameras" that process images in near-realtime. See *Video Analytics*.

OPTICAL SURVEILLANCE BEYOND THE VISIBLE SPECTRUM

Cameras are not restricted to wavelengths visible to humans. They may also image in infrared or ultraviolet on the outer boundaries of the visible spectrum.

Infrared

Infrared is part of the electromagnetic spectrum adjacent to visible light that we perceive as red. Specialized cameras can be used as heat-sensing devices within the "thermal infrared" regions. While humans tend to visualize in shapes and colors, objects can also be assessed according to how they emanate heat, which means daylight isn't necessary to sense the environment. Some species of snakes have biological thermal sensors to help them recognize predators or prey.

In the past, infrared cameras tended to be low resolution and lacking in detail. Now, with long-wavelength focal-plane-array cameras, there has been significant improvement in the amount of detail that can be captured. Infrared imaging is especially helpful for detecting or inspecting heat-radiating objects such as humans trespassing restricted areas or jumping borders in the dark. It can also be used in industrial applications to inspect structural parts that may have defects or points of wear. It is helpful for medical imaging, as well, since areas that are hot may indicate injury or infection.

Infrared sensing can be used at crime scenes to detect blood stains. Luminol™ has been used as a law enforcement tool for years to reveal stains that might not otherwise be visible, but spraying a stain may spread or dilute it. Infrared sensing is a less invasive way to investigate clues. By selectively filtering wavelengths, a textural pattern of stain shapes or levels may emerge. Researchers at the University of South Carolina have developed a camera that can detect blood diluted to as little as 1 part per 100 parts water and is effective in distinguishing blood from other common household liquids.

Ultraviolet

Ultraviolet is part of the electromagnetic spectrum adjacent to visible light that we perceive as violet-blue. If you've ever been to a party where they turn out the lights except for one *black light* that gives everyone's T-shirts and teeth a ghostly glow, then you have seen how ultraviolet light emphasizes objects differently from sunlight—a property that can reveal otherwise invisible details.

Some objects emit ultraviolet radiation naturally and some emit ultraviolet and other frequencies when they are submerged, hot, or moving quickly. One way to detect missiles is by looking for tell-tale emissions.

There are powders and gels that enable people to mark items so that the marks are invisible unless illuminated through a fluorescence process. Many police departments now routinely check stolen items for UV marks.

A Customs and Border Protection (CBP) lab worker holds up a *black light*—a light that emits frequencies that stimulate ultraviolet-sensitive materials like special inks and powders to glow (fluoresce). Ultraviolet lights are used for many forensic purposes by CBP and police departments to check for marked articles, signs of tampering, or incriminating traces of powder or gel that a thief might have on fingers or clothes. [CBP photo by James Tourtellote, 2006, released.]

Articles can be marked for security and ownership identification with ultraviolet powder, paste, or ink (left) and the marks revealed later with a UV flashlight (right). This is helpful for marking valuables and can also be used for gate re-entry at amusement parks and temporary events, in the form of a hand- or card-stamp. People selling goods online often use UV to mark products to verify that a returned item is the original and not a substitute. In low-light conditions, a UV light reveals the otherwise invisible mark. [Photos copyright © 2012 Classic Concepts, used with permission.]

Fluorescing powders can be used to mark a thief. If something coated with a fine powder or paste is stolen, traces of the powder will remain for a time on the thief's hands or clothing and will glow if stimulated to fluoresce.

Most ultraviolet detection systems work best under low light conditions so the fluorescent glow can be seen, but there are systems designed for military use called *solar-blind ultraviolet detectors* that can detect UV emissions in daylight.

Video

Video is a series of related images taken over time and displayed sequentially. Video technology and the illusion of motion is based on human perception.

If similar images are presented in sequence, at about 12 frames per second (fps) or faster, it looks like motion. Thus, a series of similar comic drawings, presented 20 times per second, appears as a moving cartoon. Beyond 30 fps, it becomes difficult for humans to process additional information, but faster frame rates are still useful and may be analyzed for information that we can't perceive with our senses alone. Imaging systems in fast-moving aircraft use high frame rates.

Moving images used to be recorded on film by exposing it briefly to light and then processing the film with chemicals to reveal and stabilize the image, with a series of *frames* creating a motion picture.

Super-8mm and 8mm motion-picture film cameras were widely available in the 1960s and 1970s, but were limited in a number of ways. The cameras were bulky, and most consumer film canisters only supported about three minutes of recording. The film had to be sent to a lab for processing, a service that could take a week or two, and which was not inexpensive. Once processed, they needed to be edited and hand-spliced to provide longer projection times.

In the 1980s, tape-based video cameras overcame some of these limitations. Recording times were increased and some camcorders could replay the recorded tape right away. But VCR tapes had limitations, too. Video tapes are analog, which makes it a challenge to edit the video, or to "freeze" individual frames, and recording times are limited to about 1/2 an hour. As mass storage devices and fast computers gradually became available in the 2000s, and microelectronics shrank the size of camcorders, digital video began to be recorded to analog tapes and, soon after, to be recorded directly as digital data to built-in hard drives.

The distinction between still and video cameras has almost disappeared. Most "still" cameras now have a video mode and video cameras are often equipped with an extra button for shooting still shots.

An important aspect of video that pertains to surveillance is the amount of space it takes to store visual records. Not only must the information fit on available storage media, but it must be physically stored somewhere in case it becomes relevant later. A few years ago, an hour of high quality audio/video could easily take up 21 gigabytes of hard drive space, at a time when terabyte drives were not yet consumer items. Now, an hour of high-resolution video, shot at 30 fps, can fit on a CompactFlash™ memory card.

THE CAMERA'S ROLE IN SURVEILLANCE

Cameras are devices that "sense" visual phenomena and, in most cases, display them in a form that is recognizable to humans. Many store images on recording media such as film, tape, CDs, memory sticks, or hard drives. As mentioned, the trend is away from film and tape toward digital storage devices small enough to fit inside a camera housing.

Cameras extend our vision, or capture images in places we can't access, or which cannot be monitored by humans full-time.

The primary uses for cameras in surveillance include

- detecting and signaling when something changes or moves,

- imaging objects or events of interest, or which look or behave differently from what one might expect, and

- recording visual phenomena, sometimes in wavelengths invisible to humans. This is particularly important if the event is difficult to assess at the time it happens (as in many scientific or security applications).

Most consumer cameras use a single lens to create a two-dimensional image but humans are used to seeing things in three dimensions through a pair of eyes spaced a couple of inches apart. This gives two viewpoints from slightly different angles that the brain resolves into a single image with a sense of depth.

Some creatures, like fish, have round protruding eyes; others, like flies, have honeycomb-like visual sensors spread over a curved surface facing in several directions to produce a picture of what is happening in a broader field of view. These have served as models for fish-eye lenses and, more recently, multi-lens arrays.

Fish-eye lenses (about 8 to 10 mm) and wide-angle lenses give a broader field of view than typical lenses in the 50 to 200 mm range, but they also introduce distortion that increases toward the periphery. This may be an acceptable compromise for surveillance cameras but may create problems for other applications such as scientific imaging or scanning. Image processing algorithms and special lens arrays have been developed to provide larger imaging surface with less distortion. [Image courtesy of NASA, released.]

Most cameras capture wavelengths similar to those humans perceive, but some can image wavelengths on either side of the optical spectrum known as infrared and ultraviolet. Infrared is used in night-vision cameras, but is typically blocked by glass. New developments in broadband infrared imaging enable it to see through glass as well. Ultraviolet facilitates imaging in murky environments such as fog.

The U.S. Space and Naval Warfare Systems Command (SPAWAR) is tasked with maintaining U.S. Navy "Information Dominance"—a phrase that is catching on in many security sectors. To do this, they evaluate, procure, develop, and provide logistical support to a wide variety of surveillance technologies including optical, infrared, and ultraviolet cameras.

Left: Naval vessels and their associated aircraft have to cope with rough weather and poor visibility and use a combination of infrared, thermal, and optical imaging to see through fog and haze. This multisensor array incudes both visible-light and infrared cameras. Right: A sample illustrates how image processing can be applied to optical data to help evaluate objects or terrain obscured by fog, smoke, or sand. [News photos courtesy of U.S. Navy SPAWAR, released.]

 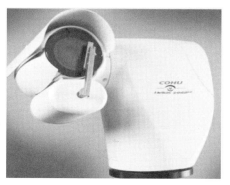

Left: Marine surveillance systems usually have to be stabilized to operate effectively on ships or swaying derricks. This Stabilized High-Accuracy Optical Tracking System (SHOTS) is a long-focal-length telescope that can be combined with other technologies, such as tracking systems. Right: In 2012, SPAWAR announced intent to purchase a Cohu, a 100 mm thermal imaging camera with remote focus and digital zoom capabilities. The surveillance camera operates in two detection resolutions and may be interfaced to a network and storage devices via internet protocol—a capability that is becoming more important to manage growing quantities of image data. [News photos courtesy of U.S. Navy SPAWAR and Cohu Electronics, released.]

Cameras suitable for surveillance often come from military and security industry research labs, but many are developed for scientific research, mapping, computer gaming, and movie animation.

Examples of Improvements in Camera Technology

- lensless cameras that can be focused with electrical impulses rather than relying on hard, curved glass,

- cameras that can focus both far and near at the same time, eliminating the fuzziness associated with objects outside of the current focal range,

- cameras with much broader fields of view without the need for traditional "tiling" of individual images to form a larger picture,

- high-speed cameras for specialized imaging or use on fast-moving conveyances,

- cameras that image in wavelengths other than the visible spectrum at significantly higher resolutions than before,

- cameras that can sense and track motion more effectively and specifically than in the past,

- cameras that interface directly with local networks and the Internet,

- cameras so small, nine could fit on the head of a pin, and

- "smart" cameras that perform image processing not only after the image has been captured, but while it is being photographed

Cameras installed in space vehicles have special requirements. They need to be low power, small, efficient, low maintenance, and capable of imaging objects far away. Thus, many microminiaturization components developed for the space program have found their way into the surveillance industry.

Computer gaming and movie animation development have often sparked new imaging technologies because they are resource-intensive activities. They require fast processors and mass storage, as well as novel camera systems, like motion capture, to create illusory worlds. These can be adapted to surveillance—especially applications that require environment modeling and a rapid assessment of moving phenomena. The movie industry has also spawned many new gimbal-and-crane systems for positioning cameras for quick movement and flexible orientation. Carbon fiber, a material with a good strength-to-weight ratio now commonly used in bicycle frames, has been adapted for camera-and-dolly systems.

There have been many recent breakthroughs in camera technology that are applicable to surveillance. Up to the 1990s, surveillance-capable cameras were mostly limited to spy/detective work and scientific research. Now they are both powerful and inexpensive, and routinely incorporated into cellphones, keychains, pens, sunglasses, and even lapel pins.

Efficient use of cameras isn't always about resolution, color, or night vision, sometimes it is about how effectively they are used. Motion sensors are often used to trigger cameras, so that they activate when something is happening, or to trigger an alert so an operator pays attention to a particular screen (or the correct tile in a split-screen image from several cameras). Motion sensors can also trigger a camera to record at higher resolutions or frame rates. This enables continuous monitoring while conserving archive space and clarity can be increased for situations that might need it.

Another significant advancement is the incorporation of analytics software into the camera (or camera housing) to analyze video data as it is received, in order to flag an alert or prioritize incoming images for storage or later analysis. This is described in more detail later.

Stereoscopic Imaging

Humans have two eyes set a couple of inches apart. This is called *binocular vision* (*bi* two, *ocular* eye) and aids our three-dimensional perception. One way to simulate three-dimensionality in photographs is to model human stereoscopic vision by placing two lenses a short distance apart, capturing two images, and then using a special pair of glasses to perceptually merge the two images into one. This simulates dimensionality similar to human vision.

This spider-like device is a stereoscope originally designed for analyzing military reconnaissance photos. The U.S. Army Corps of Engineers used it for reading aerial maps shot from two slightly different angles. After a significant gain in territory after the U.S.–Mexican War in the late 1800s, the Army Corps was involved in an extensive geographical survey e to 1) gather data for military operations, 2) locate natural resources, and 3) develop transportation infrastructure. Prior to that time, western territories had been occupied by France, Spain, and Mexico. [Photo courtesy of USGS Museum Collection, released.]

Stereoscopic images composed of two pictures taken from slightly different viewpoints were popular after World War II. Now imagine going a step further and using thousands of tiny lenses so that an image gives a sense of multiple depths. This has wide-ranging applications for terrain imaging or biometric surveillance. One person might be mistaken for another by a regular one-lens identification system but if you add hundreds or thousands of tiny lenses so the depth of physical features such as cheekbones, mouth, nose, and ears can be determined, identification becomes easier and the chance of mistaken identity drops.

Left: Stereoscopic images like this photo of NASA's Curiosity rover, an extraterrestrial surveillance vehicle, can be taken with dual (binocular) lenses or simulated with computer graphics software and printed with color shifts to give a sense of three-dimensionality when viewed with 3D glasses. By blocking certain wavelengths with colored filters, different parts of the image are seen by each eye, giving it a dimensional effect—in this case, red/blue glasses. This is not merely an aesthetic enhancement—when achieved with multiple lenses, an image can carry important depth-related information as well. [News photos 2011 courtesy of NASA and NASA/JPL-Caltech, public domain.]

Right: The first live 3D video transmissions from space were captured in August 2011 using the Erasmus Recording Binocular (ERB-2) camera developed by the European Space Agency. Flight engineer Paolo Nespoli is shown floating in space in the International Space Station while handling the ERB-2 camera. Images are later viewed with polarized glasses.

New-Generation Fish-eye Lenses and Scanners

Photographing a large area has always been a challenge. Traditional fish-eye lenses made of plastic or glass can take in a broad area but suffer from fuzziness and distortion the closer you move toward the edges of the image.

A scanner is a specialized imaging device that collects data in sections (often rows) when it is difficult to shoot the entire picture in one shot, either because of bottlenecks in the speed of transmitting large amounts of data or because the size of the area being imaged is too large.

Scanning very large areas is usually accomplished by collecting *tiles* or *swaths* in a systematic way so the images may be combined or "stitched together" to form a composite picture with less distortion.

Scanning in swaths has been very effective in satellite imaging, but there are still technical challenges:

- The trajectory of the vessel carrying the camera (or adjustment mechanisms on the camera) must be as precise and distortion-free as possible or it is difficult to stitch the tiles together.

- Tiles may be imaged in different light or weather conditions making it hard to provide a consistent overall look to the composite image. (To overcome this, different frequencies may be used, or a series of pictures imaged so that the best results may be selected.)

- Significant time delays between the imaging of adjacent tiles may provide confusing data at the seams or in the overall picture.

- Moving objects in the imaging area may appear more than once if they move into another swath as it is being photographed.

Considerable research is going into improved lenses and imaging surfaces.

- Researchers at the University of Alabama have developed new lenses that overcome some of the limitations of tiling. One approach is to develop a *lens array* composed of a large number of lenses trained at the same focal point. This can provide a very broad field of view without the need to tile.

- Another approach taken at Duke University is to invent an entirely new kind of lens where control of the lightwaves passing through the lens is handled differently from ground glass or plastic. Instead of controlling the light waves using a curved surface, individual surfaces within a *lens matrix* provide a controlling mechanism. Since materials can be fabricated in thousands of different patterns, this approach holds great promise for lenses designed for specific applications and those needing to image a broader field.

- At the University of Wisconsin-Madison, researchers tried something different. Rather than modifying the lens, they modified the surface upon which the image is projected. By using flexible materials, they aim at making a curved surface that reduces image distortion.

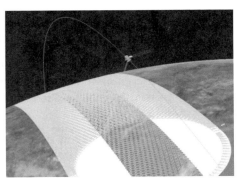
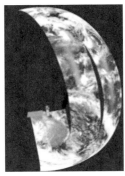

Orbiting satellites like the SMAP satellite (top) and the Aqua satellite (bottom) can't view every part of the globe at the same time, so they image the planet in *swaths* that are tiled together to form a composite picture. This is true of many targets that are too large to fit the field of view of a camera or other imaging sensor. A variety of sensors are used because they yield different kinds of information. Some can image color, some frequencies can penetrate clouds, others can sense heat. Aqua uses a spectroradiometer imaging system that measures 36 light frequencies from an altitude of 705 kilometers. [Photos (top) courtesy of NASA/JPL, and (bottom) the National Oceanic and Atmospheric Administration, released.]

Large documents are also frequently scanned rather than photographed in one shot. It's difficult to fit the entire field into a single picture without sacrificing resolution (or producing fish-eye lens distortion). Instead, the picture is scanned in sections, with a mechanism to control the alignment of seams.

Scanners are very helpful in forensics, especially if they can image a variety of wavelengths. Filtering the light may reveal invisible inks or portions of a surface that have been erased or modified. Broad-spectrum lights can help detect counterfeit money, forgeries, or otherwise imperceptible stains.[3]

Scanners also provide a way to image moving objects, such as products on a conveyor belt or goods that are passing across borders in trucks or trains. High-speed multispectral scanners are suitable for detecting defects or bacterial contamination. Multiple light sources can also be used to increase a scanner's versatility.

---

[3] The Oxford multi-spectral scanner is used by Forensic Document Services for analyzing crime scene samples. The scanner is also suitable for archival and archaeological research.

Scanners can detect hidden objects, such as weapons, or illegal goods being smuggled through transportation terminals or border crossings.

Some scanners, like active X-ray scanners, require the target object to pass between a transmitter and an imaging device. Medical X-rays passing through a body have traditionally been imaged on photographic negatives or other film-like surfaces but digital images projected to monitors are becoming more common. Portable cameras that use microwaves in much the same way are now in development.[4] Portable X-ray machines weighing less than 15 pounds that use regular DC outlets rather than specialized wiring are also commercially available. A portable machine has applications for industrial fault detection and customs and border inspections away from main terminals.

Improvements in Scanning Technologies

Researchers at Manchester University, U.K., have developed an X-ray sensor array that can pick up the diffraction pattern of scattered X-rays to provide more information about the targeted materials, including color interpretation of the signals. Early versions could only image surfaces to a couple of millimeters but even this is sufficient to provide data that can't easily be studied in other ways and more powerful systems are being developed.

Traditionally X-rays were imaged as grayscale photo negatives, but recently researchers at Washington University School of Medicine in St. Louis (WUSTL) have enlisted a wider range of X-radiation to detect the presence of metals and display them in color—a breakthrough called *spectral CT.*[5]

Color X-ray imaging opens up many possibilities. WUSTL is developing ways to inject nanoparticles into an object (or a human's bloodstream) and have them seek out an object of interest, such as a magnetic source or a biological blood clot, to track and reveal information about internal structures.

Color X-ray technology could be used to peer into many kinds of objects, such as a relic suspected of being a fake, or an industrial component suspected of defects or below-standard construction. It could reveal components of an unknown chemical or help analyze evidence from a crime scene. It is the kind of technology scientific researchers and forensics labs would welcome. In production settings, it could help test composition and quality.

Selective Vision

Selective vision is extracting aspects of interest within a scene or image. Humans constantly pick out visual cues of importance. A wolf looking out from behind trees demands immediate attention, as does an oncoming car. Certain colors, things that move, favorite foods, and certain shapes attract our attention

---

[4] A research team led by Dr. Reza Zoughi has developed a portable microwave scanner that can produce images at speeds up to 30 fps.

[5] Jens-Peter Schlomka, et al., "Experimental feasibility of multi-energy photon-counting K-edge imaging in pre-clinical computed tomography," *Physics in Medicine and Biology,* Vol. 53 (15), 2008.

more than others. Many people react instantly to insects or snakes, even if they have never been stung or bitten.

Sometimes selective vision can be a hindrance. Highway advertising signs with flashing lights and moving pictures get our attention but may also take our eyes away from the car ahead of us.

The advertising industry exploits selective vision. Advertisers know what shapes, colors, movements, and sounds will get our attention. Attractive models are widely used in ads because we are instinctively drawn to them.

Selective vision is an important tool in surveillance. A detective looking for an arson suspect needs to scan a sea of faces to pick out people who might fit the description of a suspect who ran from the building shortly before it began to burn. Arsonists and thieves often come back to survey a scene after committing a crime, so watching the area for a few days after the event may yield information on the person responsible.

Selective vision in optical systems is a complex area of research. Humans pick out familiar patterns naturally. We have vast memories and the ability to match up things that are familiar. While a memory for names may be elusive to humans, many are good at recognizing faces. To develop practical vision systems for surveillance, computer scientists, engineers, and biologists have extensively studied pattern recognition, object recognition, and biological neural systems. Now, with better pattern-matching algorithms and large object databases to serve as "memory," there are visions systems that can recognize various objects and events quite well.

Once a vision system has the capability of recognition, the next step is deciding what parts of a scene are important. Thus, separating the *target* from the *ground* (any low-priority items or situations), is the subject of study at a number of labs, including the Electronics and Telecommunications Research Institute (ETRI) in Korea, where researchers are working on the problem of extracting objects of interest from a dynamic scene—a capability particularly relevant to surveillance.

The project involves extracting people, such as sports competitors in a stadium, from the background, which often consists of buildings, bleachers, and crowds. By using a large array of cameras to shoot multiple views of events, 3D information can be extracted from a scene, with emphasis on events in the foreground. The extraction itself depends on a number of cues, including boundary shapes (the edges of objects) and blur that occurs from movement.

Micro-Cameras

Tiny cameras have many surveillance benefits. They are easy to power, transport, hide, and insert into tiny crevices. They are also irresistible as consumer items and are found in cell phones, flashlights, computers, keychains, watches, sunglasses, buttons, and pens. They are also commonly inserted into household items such as wall clocks, clock radios, smoke detectors, stuffed animals and plants.

Tiny cameras have research, forensics, robotics, and security applications, as well as great potential for the visually impaired. Part of the project funding for a micro-camera developed at Cornell University was provided by the Defense Advanced Research Projects Agency (DARPA), which helps maintain national security and the technological superiority of the U.S. military. Those involved in surveillance are interested in micro-cameras—even if resolution is limited.

Tiny cameras have an appeal similar to toys and jewelry and are now carried as necklaces and bracelets. Yet their small size no longer means limited capabilities. For under $50, the palm-sized camera on the left not only looks stylish, but shoots between 20 and 243 still images (depending on resolution) or a few seconds of 1290 × 960-pixel video at a full 30 fps. The camera on the right has similar capabilities and comes in this version and a waterproof version at lower video resolution. In the past, high resolution video was limited to bulky camcorders costing thousands of dollars. [News photos courtesy of Somax China Co., Ltd. SM-SH0203 (left) and JTT Chobi Cam™ (right).]

A team led by Patrick Gill in A. Molnar's lab at Stanford University has developed a tiny lensless camera that can image in places no other camera can go, showing great promise as a medical diagnostic tool. The camera measures only half a millimeter by 1/100th of a millimeter thick.

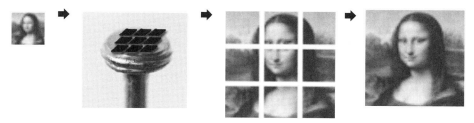

Researchers at Cornell University have invented a lens-free camera so small it is dwarfed by a pinhead yet able to resolve images about 20 pixels across (left). This is not high-resolution, but can still have many applications. Imagine combining nine micro-cameras in a pinhead-sized 3 × 3 array, for example, to create a 60 × 60-pixel image (right) that conveys sufficient information so that familiar or simple objects may be recognized. [Illustration © 2012 Classic Concepts, used with permission.]

DARPA commissioned a company to develop a tiny camera-equipped aerial vehicle small enough to fly through a window for use as a surveillance tool in urban environments. We may think of UAVs as small compared to conventional aircraft, but many are larger than a living room and not ideal for stealth applications that involve maneuvering in small spaces. The new *Nano Air Vehicle* (NAV) or Nano Hummingbird required hundreds of trial-and-error designs and finally emerged as a 6.5-in. beating-wing device capable of vertical ascent, hovering, and basic aerial acrobatics.

The challenge of creating a small aerial vehicle capable of flying through windows inspired the Nano Hummingbird, a palm-sized remotely controlled surveillance 'bird.' With an experienced operator at the controls, it is capable of navigating into a building through a door, flying around, and making its way out again without crashing.

Left: The AeroVironment Nano Hummingbird is a cross between an autonomous drone and a remotely controlled vehicle. Navigation is handled by a remote operator, with pitch and velocity corrected within the Nano bird. The entire bird, including camera and internal mechanisms weighs less than a AA battery. Right: In operational tests, the Nano bird was able to fly through a door, into a building, and out again. In this image, the operator, using a thumb-control remote device, navigates the bird into a 360-degree flip. [News photos courtesy of AeroVironment.com, 2011, used with permission.]

The development of ultracapacitors, energy-storing components that recharge faster than conventional batteries and which may be recharged thousands of times, may have applications for handheld and aerial cameras in the future. Imagine a Nano bird that could poke its beak or tail into a recharging unit as a hummingbird pokes its beak into a flower.

Stanford University has also accepted the challenge of tight-space flight by developing control software that uses data from a small camera to classify local obstacles to help an autonomous aerial vehicle, such as a helicopter, navigate in confined spaces. They see the technology as suitable for intelligence-gathering and search-and-rescue operations

For search-and-rescue, dogs can crawl into spaces too confined for a human, but there are places even dogs can't go and an autonomous or remotely controlled aerial vehicle could.

Robotic Sensors

Sensors are an important aspect of robotics and many of the sensors are optical. Video, infrared, and ultraviolet sensors are common, in addition to tilt and location sensors that help the robot navigate in a variety of environments. A robot lacking sensors is deaf and blind and likely to run into things. With sensors it can find its way around, locate objects, and assess situations that may have significance for reconnaissance and surveillance. It is not uncommon for a surveillance robot to have a dozen or more sensors—optical sensors often face in all directions, or are mounted on something that rotates.

An interesting aspect of robot vision systems that wasn't possible when data storage was limited and expensive is the use of extensive object databases to help the robot recognize the things it sees. The robot can take a picture or select a segment of video and extract various shapes or colors that are distinctive or recognizable and match them against the database to figure out what they are or how to deal with them. Software search "trees" (hierarchical database structures) can help the software find the shapes more quickly.

 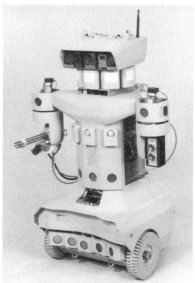

Left: The earliest Robart surveillance robot was only capable of sensing an intruder. Later versions added assessment and autonomous decision-making capabilities. Right: The Robart III's *Vision Computer* is housed in the head. It takes live video feeds from the robot's various cameras and converts them to digital images for storage and later processing and analysis. The robot also includes ladar (laser radar) and near-infrared sensors. [News photos courtesy of U.S. Navy SPAWARS, released.]

The U.S. Navy SPAWARS program has developed robots to support the Army's Mobile Detection Assessment Response System (MDARS) program. With each generation, the robot has become more sophisticated, more capable of working on its own.

- Robart I could detect an intruder, but assessment of the situation had to be followed up by a human.

- Robart II added assessment capabilities, which reduced the possibility of false alarms.

- Robart III is more autonomous than its predecessors and is being developed in a number of directions, including as the leader of a convoy of robots, with the followers recognizing each other using optical barcodes and working together like bees in a hive.

The SPAWAR Systems Center is making efforts to develop software for autonomous communication and intelligent behaviors within a convoy of robots to help them cooperate in unknown environments where they may need to help each other, or may be separated and need to come together again at the end of the mission.

Robart III, the leader, is equipped with more advanced technology than the rest of the pack, including three computers and several kinds of sensors facing in several directions.

Tiny Robots

Researchers at the Harvard Microrobotics Lab went even smaller than the Nano Hummingbird, taking inspiration from the flight capabilities of bees. Their challenge, with funding support from the National Science Foundation, Wyss Institute, and U.S. Army Research Lab, was to develop a swarm of robotic bees.

Why a swarm? Consider the breakthroughs in communication and collision avoidance that may emerge as the project progresses.

The robot resembles an insect not only in scale but in the way it flies. To test the robot, which flaps its wings when electricity is supplied, developers suspended it by fine cables, provided power, and watched as the wing motion propelled the robot upward. Videos of the motion show it is surprisingly similar to biological wing beats.

Storage capacity on small flash memory cards is still improving, and may double or triple without increasing in physical size in the next few years. These are useful in medium- and small-format robots, but the smallest ones may need to rely on radio-wave communications to transmit information from sensors to a controller that can store and analyze the information, rather than saving it on-board.

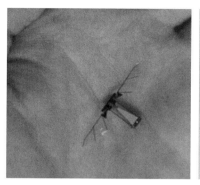

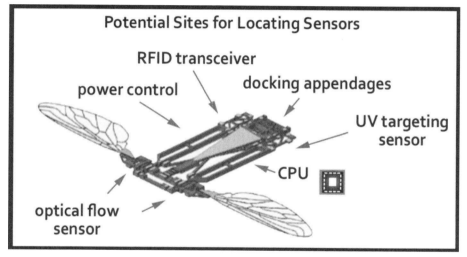

Robert Wood's team at Harvard developed a tiny robot inspired by the flight capabilities of insects. The diminutive Monolithic Bee (or "Mobee") can potentially carry sensors and RF transceivers so it can seek out information and communicate with other "bees" or nearby devices. The fabrication of the robot is novel as well. A scaffold that folds into different configurations like a sheet of origami paper enables the bee to be constructed and assembled like a popup, thus creating a final 3D product. The wings flap when electricity is applied. Optical flow sensors will help the bees avoid collisions. [News photos courtesy of Robert Wood and Harvard Microrobotics, released; labels by Classic Concepts.]

ADVANCED OPTICAL SYSTEMS

Aerial Surveillance

Aerial surveillance is discussed at more length in the following chapter, but the optical aspects will be introduced here.

Aerial cameras have special needs that require a different approach to imaging from many ground-based cameras. Aerial cameras often have to image at great speeds or from great heights. They also may be subject to high levels of buffeting and vibration from winds or the movement of the aircraft on which they are mounted. Aerial cameras are not always level. Most aircraft have to bank, as well as ascend and descend, to contend with winds and clouds. They sometimes have to climb to high altitudes to avoid being seen. Aerial imaging systems must function well under constantly changing conditions. To accomplish this, many are equipped with special gimbals, stabilization systems, location sensors, and shutters.

Visible light, infrared, and optical cameras are all commonly used for aerial surveillance and sometimes the same scene is imaged in more than one sensing mode to provide comparative data.

Biology-Based Robot Vision

Robot vision systems were traditionally based on photographic concepts from the 20th century, but that is changing. Biological vision systems and neural networks are serving as inspiration for many of the more recent optical sensors.

Researchers at the Intelligent Robot Vision lab in Korea are developing robots with autonomous location and navigation systems so they can track and follow people or objects outdoors. Robot vision has special challenges in outdoor applications, such as constantly changing light conditions, motion blur from things that are moving, like cyclists, dogs, waving branches, and obstacles around which they have to maneuver. To deal with these factors, a variety of sensors, including D-GPS (Differential GPS), multiple cameras (both single and stereo), and laser range finders are used. The combination of GPS, which provides global location information, and optical sensors for local assessment, improves the robot's navigation. The researchers call this a *sensor-fusion system.*

An Italian lab has developed an omnidirectional vision system inspired by biological vision. It uses an omnidirectional optical sensor to provide a broader field of view, in combination with a form of sensor that is similar to the *retina* at the back of a human eye. The biological retina is the "projection screen" upon which light is focused. It senses light characteristics such as color and intensity and sends this information to the brain for further interpretation. The aim of the developers is to create vision systems for mobile robots and automated surveillance systems.

Researchers in Spain have developed a robot with a system modeled after the way the neurons in monkey brains work. After studying the way monkey eye/brain/arm coordination works, they created a computer model of the process and incorporated it into the robot's "brain." The robot has a 3D vision system with moving eyes that is synchronized with the use of the arms, and integrated with memory to store images so the robot can recognize its environment and seek out familiar objects in the future.

Super-High-Resolution Imaging

The quest for ever-higher resolution vision systems hasn't ended, and much of it goes beyond what the human eye can resolve. Images displayed as 2000×2000 pixels look good to most people and beyond 4000×4000, on a three-foot-square imaging surface, even trained graphics professionals have trouble distinguishing further detail. Nevertheless, there are situations where you may want to capture more information per square inch for later reference or for enlargement purposes. Researchers at the SOI Research Unit in Italy have developed an imaging sensor with a resolution 1/10th the size of a human hair.

There are many innovative ways in which 3D information may be captured by cameras and visually interpreted.

Imagine bouncing a tennis ball at a garage door from 3 feet away. The ball comes back quickly. If you move back to 10 feet, the ball takes longer to return. Now take a flashlight and illuminate a wall at night. The intensity and size of the bright spot on the garage will be different at different distances. Thus, there are various ways a camera could measure distance and the contours of an object being imaged. Systems that are both high-resolution and 3D-capable have numerous applications in scientific research, biometric identification, robotics, medical imaging, and forensics.

Beyond Binocular Vision

Researchers at Stanford University are working on multiple-aperture 3D systems that can determine depth utilizing the disparity between several apertures. Since the apertures sit side-by-side, each aperture images a slightly different viewpoint in the same focal plane and depth information (and potentially movement information) can be gleaned by comparing the images from the different viewpoints.

One result of 3D imaging research at Stanford is "light-field" devices such as the Lytro™ camera that captures not just a single impression of light hitting a photographic plate or imaging surface, but multiple impressions on a lens array (a series of related lenses) to store not only color, but the direction of the light hitting the imaging surface. This means a person viewing the image can select different parts of the picture to bring into focus after the picture is taken. The Lytro team likens the effect to having multiple light tracks, similar to multiple audio tracks in a musical composition, so you can focus in on a chosen "track" (portion of the image) at any given time. [News illustration and photo courtesy of The Outcast Agency, released.]

One of the results of research at Stanford is a new breed of consumer camera called a *light-field* camera that captures not only the colors and shapes, but the vectors (direction and distance) of lightwaves entering the imaging chamber. This produces a novel effect, a 3D image displayed on a computer monitor, PDA, or light-field all-in-one device that can be focused after the picture is taken, since the 3D information is stored within the digital file and can be accessed and displayed in realtime—click on the part of the picture you want to see more clearly, such as a house in the background or a flower in the foreground, and it comes into focus.

Electronic 3D Imagery

The appearance of three-dimensionality can be created using electronics rather than bicolored glasses. By using visors with liquid-crystal shutters controlled by radio-frequency transmissions, a video image can be synchronized to flip between transparent and opaque to limit which eye sees which frames.

Panoramic technologies and 3D are closely related. If you have a photographic system that takes a wider view, it becomes similar to human peripheral vision or to a field-of-view broadened by turning the head.

NASA and IPIX partnered to create a photographic process called *3D immersion* that creates a 360-degree view of an area, inspired by conditions on space vehicles where working room is cramped, and there is a need to reduce mechanical parts (like pan-and-tilt joints) that might fail. The developers suggest immersive photography can also be used for surveillance, security, and safety monitoring, and may have applications for entertainment-related virtual tours or real estate ventures.

Broad-view cameras reduce the need for panning and tilting and may require fewer cameras to image an area. Some new cameras can perform pan and tilt electronically, rather than physically shifting the lens, a technology that was originally developed for space-robot guidance systems but which is now available in consumer products such as the OmniView™ camera.

Three-dimensional imaging has significant advantages for surveillance. In traditional photography, the camera is focused before the picture is taken. If depth-data is captured when something is imaged, one could focus on a point of interest after the fact. It is not always known what is important in a surveillance image. If you have a picture of a person in the shadows at the back of a convenience store, and one in the foreground near the cash register, how do you know who committed the robbery? With a traditional camera, only one is in focus (or neither). With depth-photography, focal points can be investigated later.

Applications of Robot Vision

Scanners are sometimes used for robot vision, especially if the robot has to navigate in a changing environment. The scanner will "sweep" the surrounding terrain, seeking impediments or changes in altitude to help it move safely. It may even send an image of the terrain to a remote monitoring station. Systems developed for planetary exploration are often applicable to military reconnaissance in deserts or rocky regions.

X-ray-vision robots are also used for security purposes. In recent years, terrorist bombs have been hidden in boxes, handbags, and backpacks—baggage that can be innocuously set down in any public place. If the container or bag is suspected of housing a bomb, it is better to keep humans away from it, so bomb squads sometimes use robots to handle the package. But a robot costs money and it is better to keep it intact for the next project than to have it blown to bits. X-ray vision can be installed on a robot (or carried by a robot) so it can move close to the suspicious object and use the imaging technology to get a better sense of the contents, which are displayed on a remotely viewed video screen.

Robots can be used for surveillance in environments that are deadly to humans, such as nuclear power plants that have been damaged, or industrial environments that are too hot, too cold, too acidic, or too toxic to support human life. They can be used to patrol the premises for security and may include sensors that sound alarms if leaks, cracks, or other problems are discovered.

MICROSCOPES

The microscope is an indispensable tool in scientific research and investigations. It can reveal details not visible any other way, making it ideal for forensics, and can also be a tool for developing new components such as lenses and imaging surfaces.

An optical coherence microscope (OCM) is a device that uses water rather than glass for the lens. It can operate as a camera or as a scanner. To use it as a scanner, electrical impulses shape the lens to focus at different distances to build up a 3D image in layers.

Optics researcher J. Rolland has developed an OCM device that uses infrared light to scan through layers of skin to search for anomalies in tissue, a technology that has the potential to replace invasive surgical biopsies. The same concepts could be applied to imaging artifacts, manuscripts, building materials, or plants, to determine their composition, structural integrity, or authenticity.

Microscopes are precision instruments intended to give the best possible images of minute structures not visible to the unaided eye. As such, they are often sold at prices accessible only to forensic, medical, and research institutions. This is changing, however.

Consumers can now purchase USB digital microscopes for under $40 that can image small items up to 800 times. The data can be viewed in realtime on a computer, or can be broadcast over a network in much the same way as video from a regular cam (see photos on the following page). Remote observation has obvious applications for forensics. Small clues at a crime scene could be digitally recorded or sent to an expert across the country. It is a minimally invasive way to examine objects and transmit information about tiny details.

A USB microscope could also be used to turn a laptop into a viewing device for the visually impaired, aid a student taking courses over the Internet, or enable scientists to immediately examine artifacts at an archaeological dig in a remote location if workers have questions about whether to harvest a find or leave it alone. Instructions on the best way to unearth it could be shared in realtime.

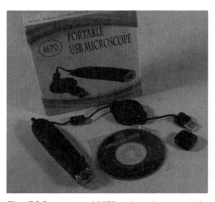 

The FCC-approved M70 color microscope is a good example of an inexpensive device with capabilities that were unavailable and almost unthinkable at this price a few years ago. It uses 8 LEDs to provide light for imaging tiny details through a 2.8-mm lens at speeds up to 30 fps. The images can be transmitted either as still photos or streaming video through the attached USB cable. At higher price-points USB microscopes have improved imaging quality and more features, like image processing and a variety of magnification levels. This kind of technology has many applications in industry and law enforcement but is so economical consumers can buy it and use a digital microscope like a regular videocam. [Photos © 2012 Classic Concepts, used with permission.]

Networked USB or IP microscopes could be used to check the authenticity of currency in banks or at checkout stands. They could help an inspector transmit images of fractures in bridges, rails, or other public transportation infrastructure to a central agency so that rapid response could handle situations that might pose safety hazards, such as damage following an earthquake or large-scale collision.

A USB microscope by itself, or combined with heat-sensing capabilities, would be helpful for arson inspections.

CAMERAS IN TRAFFIC SURVEILLANCE

Cameras are used in many ways to observe and manage domestic and international ground traffic. News crews in helicopters report on traffic conditions from the air, especially when there are accidents or other sources of traffic congestion that impede commuter traffic. Freeway cameras to monitor congestion are common, as well and, in some cities, may be viewable by the public on the Web.

Cameras are also insstalled to detect traffic violations, especially on freeways and at intersections. They can regulate freight vehicles and catch people speeding or running red lights. Red light cameras are usually placed above the traffic signals or mounted at the corners of intersections to give several viewpoints.

Pictures are sometimes combined with data from nonoptical sensors such as radar detectors to identify not only the travel speed, but the make and model of a vehicle, and the person driving it. Sometimes they are interfaced with automated services that issue tickets for violations. In this way they are controversial. Some drivers contend the systems are not there to enhance safety but to increase revenues, an assertion that is denied by local law enforcement who typically respond, "If you don't want us to collect fines, don't break traffic laws."

Traffic surveillance also encompasses pedestrian traffic on city streets, buses, subways, and in train stations. Cameras are intended to deter vandalism and theft, and may monitor ticket checkpoints, vending machines, incidents, accidents, and medical emergencies.

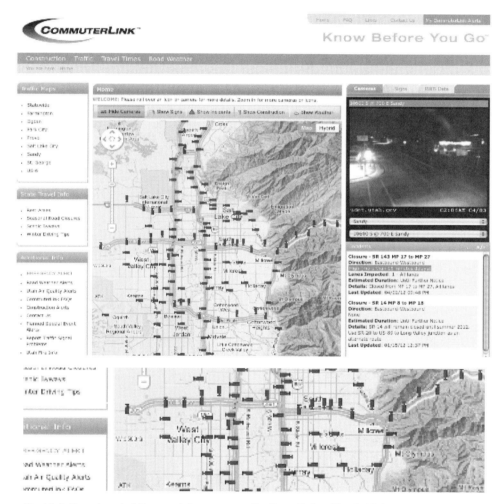

The Utah Department of Transportation sponsors the Know Before You Go™ CommuterLink™ traffic information website to provide extensive information for commuters regarding road conditions and highway services. A map shows the location of each camera and users can click on a camera icon to instantly consult camera feeds at that location to plan their routes. [Traffic conditions Web shot courtesy of Utah DoT (www.commuterlink.gov).]

The Utah Department of Transportation has a system of live traffic information cameras called the Advanced Traffic Management System (ATMS). This Internet-Protocol-based service provides information on road conditions, construction delays, and accidents. Dispatchers relay information to drivers via signs along the roads. The cameras can be combined with remote control of traffic-light timing to adjust varying traffic volumes more efficiently. Information is also posted on a website so drivers can check road conditions before heading out.

Mobile Units

There are now mobile surveillance units that can be towed on the back bumper of a vehicle like a trailer, which are equipped with cameras, other sensing equipment, and sometimes solar panels for power.

A mobile unit can readily be moved where it is needed and used to monitor traffic trouble-spots, construction sites, gravel pits, parking lots, private and public roads, temporary flea markets, gateways, as well as perimeter fences at public events such as parades, circuses, festivals, and sports meets.

For wildlife surveillance, a portable unit is less likely to spook animals than many other kinds of surveillance. Not only does it not throw off a strong human scent, but it is shaped somewhat like an animal. It could be used for deer or bird counts, or for detecting wolves or coyotes. It could monitor livestock to trace their movements, to check their welfare, and to see whether they are stalked by predators or poachers.

Communication for portable units is through cellular services or wireless LAN, and some systems have a satellite uplink option, so the viewer need not be onsite to view incoming video. Some have infrared/thermal imagers for imaging at night or to detect movement or identify thermal signatures.

The Netvision Mobile Solar Trailer is a good example of how many surveillance capabilities can be incorporated into a mobile package. The portable surveillance unit can be readily transported to any vehicle-accessible site and includes high-definition pan/tilt/zoom cameras mounted on a 30-foot mast, with solar panels to provide power. It includes day and infrared night vision, optional thermal imaging (right), online GPS tracking, and wireless cellular and LAN linking capabilities for remote event recording. Portable systems such as this are appropriate for highway and industrial construction sites, public events, temporary building security, seasonal park security, and wildlife monitoring. [News photos courtesy of Skyway Security.]

License-Plate Recognition Systems

The identification and location of stolen vehicles is a common surveillance task and the process of checking plates has been significantly automated in the last decade.

Automated License-Plate Recognition (ALPR) systems are commercial products that take most of the work out of the eyes and hands of police officers. They use optical scanners that read plates at greater distances and speeds than humans. Depending on the system, a plate may be read up to two blocks away.

ALPR systems have three main subsystems:

- hardware to image a license plate and relay it to a computer capable of wirelessly transmitting the information to the central computer that houses the database,

- software (or firmware in the optical scanner) to interpret the image and convert the license plate from a picture to letters and numbers (alphanumeric characters) that can be matched through the central database, and

- a centralized database service that can be queried by the local units to provide matching information, if any is available. The National Vehicle Location Service (NVLS) is one example of such a service, but there are many, including state-maintained databases.

A database is a key component of an ALPR system. Without a way to match the plates to existing information in law enforcement files, a plate number is of limited use.

There are two main aspects to LPR databases. One is gaining access to existing databases, which may consist of police records entered for the last few years or more. The other is the continued contribution of additional information when officers make an arrest or conviction or identify a suspect in an ongoing investigation. At present, it is not clear whether those arrested and released, or arrested and found innocent, are kept in these systems, especially since some are commercially managed rather than by government agencies, but since there are *watch lists* for suspects, it appears that information is archived whether the driver has been arrested or not.

When a law enforcement officer in a patrol car spots a get-away vehicle, a hit-and-run vehicle, or a suspected stolen vehicle, the LPR system enables him or her to image the license plate and, within minutes, add the data to nationwide watch lists like NVLS, which is accessible to other enrolled investigators.

There are a number of vendors of ALPR systems and insufficient space to list them all, but one is used here as an example. Elsag sells the MPH-900,[6] which combines a pair of fender-mounted surveillance cameras with computer hardware and software. There is also additional hardware housed in the trunk. Vehicle license-plate alerts are based on matches with pre-stored hot lists, usually in state-managed databases.

ALPR cameras can image plates from a considerable distance even if the vehicle passes at freeway speeds. They can handle thousands of "hits" per hour, although there may not be that many cars within view of the sensors, especially in smaller towns.

According to a drive-along report in *Car and Driver,* one cruiser alone photographed almost 33,000 plates (more than 1000 a day) in August 2011.[7] This indicates that plate-based surveillance is undergoing the same automation trends as wiretapping—rather than a focused approach on specific targets, a wide net is thrown over everyone to see what emerges from the crowd.

ALPR systems are automated—the officer doesn't have to click any buttons as with a still camera. The system automatically photographs any plates that pass or are passed, even if the driver is outside the car. The system displays the plates on a laptop monitor next to the steering wheel. The cameras are very versatile and sensitive. Misty weather doesn't significantly interfere with the reader and plates with smoke-colored screens to obscure the numbers can be read.

An officer in Salem demonstrates the Elsag ALPR system mounted in a patrol car. The optical sensors capture a license plate, which is converted to letters and numbers and displayed on a laptop inside the police cruiser. The computer and transceiving components in the trunk then send the plate ID to a central database which searches for a match. If it finds one, it sounds an alert and relays the information back to the patrol car with a red border on the image to bring it to the officer's attention. A spoken message explains why the plate was flagged. [Still shots courtesy of Elsag MPH-900 news video, released.]

---

[6] Note that information about specific vendors in this book is intended for educational purposes as examples of common or unique products, not to imply endorsement. Readers interested in purchasing surveillance equipment are encouraged to do due diligence and research a variety of products.

[7] John Philips, "Smile! Your car's on camera: we ride along to learn what the cops know about you," *Car and Driver,* December 2011.

Whether to stop a driver if a plate is flagged depends on local law enforcement policies and philosophies. In some states, petty crimes are not listed for pursuit and plate flagging information is not shared between cruisers. In others, pursuits may be more common, and sharing of information between cruisers posted at intervals may occur. This could be construed as tracking, however—a situation that oversteps legal bounds in terms of privacy.

National Vehicle Location Service (NVLS)

The NVLS is a free service to qualified applicants that was launched in spring 2010 by the National Vehicle Service and Vigilant Video, Inc. It is a centralized commercial database maintained in a secured facility in Northern Virginia designed for law enforcement agencies to access vehicle information. Already the service has logged more than 200 million vehicle location records.

Data for the service is captured via LPR technology through scanners mounted in fleet *spotter cars* that convert the visual image to editable letters and numbers through *optical character recognition* (OCR). The LPR is aimed at license plates and transmits vehicle information in realtime to the database along with time, date, and location information. Thus, civilian spotters are actively enlarging the data storehouse.

Just as Google Streetview™ is mapping city streets, cars equipped with LPRs are mapping licensed vehicles using those streets. The original mandate of the company was to scan vehicles with a greater chance of having forensic intelligence value, such as those that remain in one place for a time, or which might return in the future. Marketing for the commercial NVLS LPR system was clearly aimed at law enforcement and those who accessed the database had to apply and be prequalified. There are positive reports that the system has helped locate felons, stolen vehicles, and missing children but, nevertheless, there are people concerned about privacy, partly because some of the patrol cars equipped with LPRs are using them like radar detectors (to image every vehicle that passes), and partly because surveillance has a way of broadening its reach once the technology is in place.

More recently, it has been alleged that other users besides police officers are accessing the database, such as tow truck drivers and repossession firms.

Controversy Over LPRs

Plans to purchase new surveillance cameras and LPRs in Aurora, Colorado, raised public controversy in March 2012. Police officers felt the automated systems could help them monitor areas that can't be patrolled all the time. Others felt there were already too many cameras in the area.

Most police deny that the LPRs are tracking devices, which may be true if they are used individually, but since data from some LPR systems is transmitted to centralized databases owned by private companies, such as the NVLS system, it cannot be known whether the administrators or operators of the database are charting locations and statistical patterns of vehicles that show up more than once.

If they are, that's a form of tracking. Some police departments have specifically established policy that prevents deliberate or inadvertant LPR tracking.

If LPRs are used as tracking devices, as might happen if one officer scans the plates in one section of town, and another scans them down the road, with the data being coordinated, it could be construed as illegal search, according to a January 2012 U.S. Supreme Court ruling.

Around the time of the court ruling, CBS San Francisco published an article clarifying that police agencies have restrictions on how long they can keep surveillance data such as video records, but private firms do not operate under the same constraints. They may be storing them indefinitely and could theoretically sell the information to other firms.

Licence-Plate Recognition Systems in Canada

License-plate surveillance is not limited to the U.S. The concept of automated license-plate scanning was originally developed for political surveillance by the British government.

There was both surprise and opposition when it was discovered that Canada had a system similar to American LPRs.

The RCMP and B.C. government tested ALPR in 2006 and, as of Feb. 2012, more than 40 patrol cars were equipped with the system. As with American LPRs, the units can scan thousands of cars per hour as they pass on local highways, even if they are traveling at more than 100 mph.

The police and provincial auto insurance agency have been using the system to locate stolen vehicles, missing persons, and unlicensed or uninsured drivers. This kind of surveillance was not the main subject of the controversy, however. The bigger concern was that it might be used for profiling travellers, people near protest sites, or simply logging law-abiding drivers. Other targets, according to the official Privacy Impact Assessment, include people who have been involved in incidents due to mental health problems or who have been to court to establish legal custody of a dependent. It also included people who have been linked to someone else who is under investigation—making a law-abiding citizen appear guilty by association.

There were also questions about whether the system was as effective or more effective than other forms of law enforcement.

"Since 2007, about 100–150 stolen vehicles in BC have been recovered thanks to ALPR (in Victoria: zero). By comparison, Vancouver volunteers doing weekend patrols boast an annual stolen car recovery rate ten times that.

Primarily, ALPR hits have led to 7,191 charges for no driver's licence, 2,215 for no insurance, and 1,199 for driving while prohibited or suspended. But Nelson couldn't show if crime or charge rates have been trending up or down due to ALPR, or are above or below areas in BC not using ALPR."

Rob Wipond, "Hidden Surveillance," *Focus Online*, Feb. 2012

More information on traffic surveillance and Intelligence Vehicle Systems is included in the *Effectiveness of Surveillance Systems* chapter.

SPECIALIZED VISION SYSTEMS

Infrared Motion Detectors

Most motion sensors are based on optical detectors using pyroelectric infrared (PIR) that can respond to changes in the surrounding environment by sensing thermal emissions emanated from nearby humans and animals (or equipment with warm engines). PIR detectors are passive technologies. They read incoming infrared.

Active infrared devices, such as remote controls, shine an invisible infrared beam and trigger an event such as turning on another device (e.g., a DVD player or car alarm). The range of passive and active devices varies, but many consumer systems can sense or trigger up to about 15 or 20 feet. The angle can sometimes be changed by bouncing active infrared off of reflective services.

PIR systems are based on line-of-sight. It is necessary to aim the PIR detector in the direction of the doorway or other area to be secured so it activates appropriately. If the angle or sensing range is too long, the system may trigger too many false alarms.

For building security, passive infrared sensors are commonly bundled with floodlights that switch on if motion is detected. A lens in front of the pyroelectric component can increase the breadth of the sensing range. PIRs may be connected to solar panels to save electricity and selectively control power to the light.

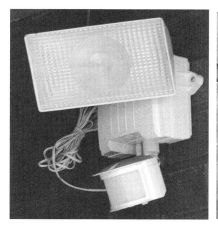 

Left: Motion sensors are common security devices that are usually combined with lights and sometimes with cameras. The curved component below the light houses the motion sensor. Devices of this kind use passive infrared pyroelectric sensors to detect heat radiating from human bodies, animals, or warm vehicles. Right: A satellite image created with data from infrared sensors is coded so that weather forecasters can visualize the relative temperature of cloud patterns. [Photo © 2012 by Classic Concepts, used with permission. Weather map courtesy of NOAA/NESDIS, 2012, released.]

Motion detectors are sold as a low-cost deterrents to trespassing, graffiti vandalism, or theft. If a light comes on when someone sneaks into a yard or building, the idea is that they will retreat to avoid being seen in the glow of the light, or will back off on the chance that security cameras have been triggered by the motion sensor.

Night Vision

Many animals see well at night, but we are not one of them. Human eyes are optimized for daylight and seeing color. We have some night vision, but it is not highly developed and tends to diminish as we age.

There are structures in the human eye called *cones* and *rods* that help us sense different wavelengths and light intensities. The cones aid in color vision and the rods help our night vision.

Seeing at night is an important aspect of surveillance. Much surveillance occurs at night because the watcher can observe from the shadows or creep up to a window with less chance of being noticed. Many crimes, such as arson and theft, are committed at night since fewer witnesses are around and criminals have a better chance of fleeing a crime scene if it is hard to be seen.

Night vision is also an important aspect of industrial site security, where machines may be working twenty-four hours a day and must be monitored for proper operations and safety.

In recent years, consumer cameras have been enhanced with night vision features that enable picture-taking in very low light. The images appear a bit ghostly because we are used to seeing in color, but they can clearly reveal basic shapes in a scene by using infrared and thermal frequencies that are not dependent on bright daylight.

We may not think of heat as light, but light in the far-infrared region, called the thermal region, consists of heat emanations that can be felt (put your hand near a fire or hot stove) and seen if converted to visual frequencies. Various night and thermal vision devices operate in different regions of the infrared spectrum. In the past *image intensifiers* were used to amplify the signal for devices that worked in very low light, but some of the more recent products can function without this component.

Night vision has long been used by the military for field operations carried out on foot, and for operating vehicles and aircraft at night.

More recently, automakers are adding infrared imaging to cars. The sensor fits under the hood, behind the grill, and sends the data to a regular video monitor inside the car. This can extend night vision by as much as three or four times— enough to avoid a deer on the road or a pedestrian about to cross in the dark. It can also be used to patrol military zones, or international borders where fences are minimal and people might be tempted to cross at night.

Thermal imaging has applications for industrial inspections, as well, as it can show up heat stress or other faults that might not be visually apparent. It can also be used to detect areas of a building that are poorly insulated. Heat leaking through walls and roofs show up as glowing areas in infrared photos.

X-Ray Vision

Almost everyone has seen ads for "X-ray vision"—glasses that supposedly help you to see through walls and clothing. Most of these ads are for glasses that slightly distort edges and make things appear fuzzier and thinner, with imagination filling in the rest, but X-ray vision isn't entirely fantasy. X-rays can penetrate objects that block visible light rays. The problem is that high-energy X-radiation can be hazardous and is typically used under carefully controlled conditions.

Some X-ray imaging systems are sold in smaller sizes for applications such as checking a container for bombs or checking luggage at an airport, but they contain shielding and must be used carefully according to instructions. Other electromagnetic spectra can, to some extent, penetrate materials, without the dangers inherent in X-radiation. Often a combination of sensors is used.

In military reconnaissance, in situations where people in a building might pose a threat, soldiers using sensors to peer through walls are thinking more about their own safety than what X-radiation might do to the health of the people inside.

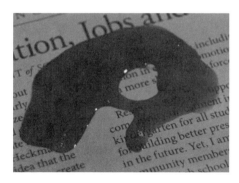 

Backlighting (top) and imaging in different wavelengths (top and bottom) can reveal features that aren't otherwise visible. Specialized cameras, such as those with infrared-sensitive imaging surfaces and specialized filters, can penetrate materials like paper, fabric, and certain liquids that are opaque in the human-visible spectrum but which allow longer wavelengths to pass through and be reflected back to the imaging device. [Photos © 2012 Classic Concepts, used with permission.]

Seeing through things is an invaluable forensic tool. If something has been blotted out (or obscured by blood or other dark liquids), sometimes an infrared-imaging camera or microscope reveals the item or information underneath. Seeing through things also creates opportunities for criminals to locate goods and currency they may wish to steal or alter.

Check washing is a crime in which chemicals are used to remove parts or all of the ink that is used to fill in checks. The information can then be reapplied to change the date, amount of the check, or the recipient. Envelopes with bill payments are sometimes stolen from flagged mailboxes, "washed," and reused. A camera capable of imaging the contents of an envelope before opening would make it easier to identify those with documents that could potentially be stolen or used in felonious schemes.

Looking through clothing is a popular theme for voyeurs and they have discovered that off-the-shelf camcorders with infrared sensitivity, coupled with specialized filters, can image through some materials like clothing, revealing the more reflective surfaces behind them, which means piercings, weapons, jewelry, and images on T-shirts may be visible. It is possible to image through sunglasses as well, revealing hidden eyes.

Seeing Around Objects and Through Walls

Surveillance would be easier if we could see around corners and through walls, but that's not possible, is it?

We see when light hits our retinas and our optic nerves send signals to the brain. If it were possible to inject electromagnetic energy through walls or bounce light off various objects to get it around a corner and capture the returning rays with an imaging device, then many new technologies may be possible.

At the MIT Media Lab, researchers have a theoretical prototype for seeing around walls, based on ultra-fast femtosecond lasers.[8] A pulsed light is reflected off surfaces that we don't usually think of as reflective, such as doors and walls. Enough light is reflected back to create a recognizable image and when the light vectors are taken into consideration, as in light-field photography, 3D depth information is available as well.

With more study and software algorithms to deal with a diversity of scenes, such as those with light, or whose geometry may be unknown in advance, practical applications may be possible.

We know that X-rays can penetrate walls and that infrared light can pass through certain materials like cloth or liquid, but can far-infrared, also known as thermal imaging, pass through walls or other solid objects?

Seeing through walls can be accomplished by analyzing electromagnetic patterns within a space and interpreting changes or interference patterns. Many offices are interconnected by wireless computers and printers creating, in a sense,

---

[8] Ahmed Kirmani, et al, "Looking around the corner using ultrafast transient imaging," International *Journal of Computer Vision*, June 23, 2011.

a radio-wave fog that will show disturbances if the emanations are converted to visual form and someone walks through the area. Radar has also been used to look through walls. These are described in more detail in the *Radio-Wave Surveillance* chapter.

The capability to see through walls and around corners has always been a goal of surveillance. In the 1920s, this Tresoroscope was developed in Germany to enable people to peer through walls without being seen from the other side. The example shows a security guard checking on a bank vault through a wall periscope. [Photo courtesy of Modern Mechanix, Nov. 1929, copyright expired.]

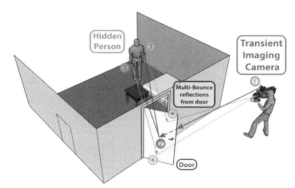

Left: In the past, a periscope was the most common way to see around corners or, in the case of submarines, above the waterline. It works by reflecting and thus directing light through a series of lenses or reflective surfaces. It is now possible to see around corners with fiber optic periscopes called endoscopes. Endoscopes are mainly used in diagnostic medical imaging but there are also palm-sized endoscopes optimized for surveillance purposes that interface with cameras. They can be used for viewing in tight crevices, such as locks, or for seeing through holes to image something on the other side of an obstruction. [Photo courtesy U.S. Navy, 1945, released.]

Right: The MIT Media Lab is working on cameras that might some day see around corners for terrestrial applications. The developers suggest this could be used for peering into hazardous sites, like burning or collapsed buildings, or around corners before entering a potentially unsafe area in a police raid. It may someday also make vehicle navigation safer. [News illustration courtesy of Ahmed Kirmani, 2011.]

Another project at MIT for seeing through walls takes a different approach. Instead of bouncing light around corners, researchers at the Lincoln Lab have developed a transceiving array of antennas, mounted in two rows, that uses S-band radar frequencies to penetrate concrete walls. Most of the radiant energy is bounced off the concrete, but a small amount makes it through. Even less makes it back, but with amplification of the ingoing signal and sensitive devices to intercept the outgoing signal, it is possible to get a sense of what is going on behind the barrier at a rate of almost 11 fps—enough to image movement.

Stationary objects are not imaged because video processing is used to compare each image with the previous one to register changes. This may be an advantage, however, if the intention is to detect people—furniture and other objects won't clutter the view.

CLOAKING DEVICES

Spatial Cloaking

Most visual surveillance technologies are designed to help us see farther, smaller, and better, but some stealth technologies are intended to hide, rather than reveal, physical details. To this end, optical "cloaking" systems are in development.

Cloaking devices, once limited to science fiction stories, may soon be a reality. Since light can be manipulated, and reflected light is what brings objects to our visual attention, the possibility of hiding objects exists.

Generally, light travels in a straight line, but its course can be changed by bodies with sufficient gravitational influence and, closer to home, it can be bounced in different directions off reflective surfaces. If the reflectors are spaced very closely together, the path of the reflected light, in a sense, curves around an object.

Metamaterials that can "bend" light, by incrementally directing it along a curved path rather than bouncing it back to the source, could make a vessel virtually invisible to many detection devices. Taking this idea a step farther, phenomena other than electromagnetic waves might be directed around objects as well. Imagine a boat hull with a redirection cloak that facilitates movement of objects around the hull's surface to protect it from damage by driftwood or severe waves.

Some of these ideas are still development but, by using materials that can be fabricated in nanometer ranges, certain telecommunications micro-wavelengths and even some visible wavelengths can be cloaked.[9]

A significant limitation of cloaking technology is that it cloaks in both directions—if you can't see in, you can't see out. Researchers in China have proposed a way to overcome this limitation with an anticloak mechanism that would provide

---

[9] Tolga Ergin, Nicolas Stenger, Martin Wegener, "3D invisibility cloaking device at optical wavelengths," *2Physics,* April 2011.

someone within the cloaked region to have a "window" through which to see out.[10]

Cloaking isn't limited to visible light. In military applications, vehicles may need to be hidden from sensors outside the visible spectrum. Heat-seaking missiles and infrared motion detectors are sometimes used to detect enemy combatants.

Taking this a step farther, an object could be manipulated to look like something else to an infrared detector by configuring its temperature print. The concept itself is not new. For decades, military units have hidden tents, equipment, and people from visual surveillance with camouflage nets and tarps that simulate the colors and shapes of the surrounding terrain, but this form of visual camouflage doesn't deter infrared and thermal detectors that can pass through nets and fabrics and sense heat patterns.

The U.S. military has experimented with products that change the *heat print* of an object so it appears as something else.

Imagine the heat print of a truck—the muffler and engine are hottest, the bumpers are cooler. Occupants inside are somewhere in between—not as hot as the engine or as cool as the dashboard or seats. Now imagine screening the vehicle with a dynamic matrix of materials whose parts are individually programmed to emit different heat signals, like those that are characteristic of a hot dog stand. Parts of it are cooler, like the wheels touching the sidewalk or the top of the canopy. Parts are hotter, like the grill and the area around it. A hot dog stand can readily be recognized by its thermal patterns. If you digitize that pattern and then send signals to small thermal units to replicate the pattern, just as visual patterns are sent to a computer monitor to represent a scene, the truck could appear as a hot dog stand to nearby infrared sensors.

This system may be heavy and take some time to construct around the objects, but it doesn't suffer from the two-way cloaking problem. People behind the infrared false-picture could monitor the surrounding environment through small viewing holes.

Temporal Cloaking

Most cloaking is based on spatial cloaking, changing what we see in terms of shapes, colors, or heat emanations. Scientists at Cornell have experimented with the idea of *temporal cloaking,* interrupting a beam of light for a fraction of a second too quick for human perception to create a time gap to make it appear as though some event hadn't occurred or some object wasn't there. It is not yet a practical reality, but exploring the concept might lead to novel applications.

---

[10]Huanyang Chen at Shanghai Jiao Tong University, reported by American Institute of Physics, "Invisibility undone: Chinese scientists demonstrate how to uncloak an invisible object," *ScienceDaily®,* Sept. 4, 2008.

VIDEO ANALYTICS

There was a time when cameras simply recorded events. Now they can be equipped to make decisions on when, where, who, and how to record, through a process called *video analytics* or *video content analysis*.

Video analytics is a way to interpret, analyze, and evaluate video data—captured visual data of a temporal nature. Live streaming video or recorded data can be analyzed by software or human experts. Video content analysis (VCA) is a subset of video analytics that focuses on analyzing the content of video streams or recordings.

Video analytics provides a new level of utility and automation that promises to revolutionize the visual surveillance industry. Some examples of applied video analytics include

- motion detection and tracking,
- backward flow and congestion recognition (such as a car going the wrong way, products backing up in a production line, or a growing crowd),
- environment-sensing and alerting for changes or anomalies in temperature, smoke, humidity, stability, light levels, color, or noise levels,
- counting objects or identifying specific people or groups,
- boundary crossing (doors, fences, walls, bodies of water, borders, etc.),
- facial, body, animal, or object recognition,
- license plate or other identification recognition,
- currency detection and authentication (as in ATMs or vending machines),
- package and other delivery monitoring,
- production line monitoring,
- perimeter or tripline security,
- weapons detection,
- vandalism detection,
- medical patient, geriatric, and baby monitoring,
- body scanning for hidden object detection,
- monitoring and analysis of traffic patterns and transportation infrastructure, and
- determination or location of mechanical malfunction.

Terminology

The term *video analytics* is sometimes loosely used to describe a broad range of technologies in confusing ways. To prevent ambiguity, this text distinguishes the terms as follows:

- *video analytics*—analysis of the content and integrity of video streams or recordings.

- *video content analytics*—analytical routines that focus on the content of the video stream or recording.

- *video integrity analytics*–analysis of streaming or recorded video, or the cameras that provide the video, for signs of tampering or alteration.

- *video-viewing analytics*—statistics-gathering and demographical analysis of user video-viewing patterns.

- *video-targeting analytics*—analysis of personal profiles or habits, with the aim of showing individual users or groups selected video content.

Video analytics is probably the most significant development in visual surveillance since the invention of hardware to capture visual events. While imaging systems continue to develop, the data are only as good as the information that can be derived from them, and video analytics aids in detecting, tracking, or making sense of important data or events.

A camera to detect motion of an intruder while ignoring trees in the background, or which distinguishes a boat from the surrounding waves, is a video analytics device. Sometimes the analytical software is separate from the camera, sometimes some or all of the analytics are built into the camera housing. A system for picking out a specific individual in a crowd with facial recognition software or tracking marine traffic in a crowded harbor usually requires a separate computer to process the information, but some of the preliminary analysis may begin with the camera.

Analytics can also be used to optimize video quality. By analyzing ambient conditions, such as light, moisture (fog, steam), or motion, the camera settings or angle can be adjusted on systems with multiple sensor modes, swivel capabilities, or electronic pan-and-tilt.

Increasingly, cameras are equipped with Ethernet connections to link them to a network, and some have network server software built into the camera.

Benefits of Analytics

For those seeking to study or secure an area, automated analysis has many advantages over human surveillance. Cameras are inexpensive enough to blanket an area, including places where humans can't go, such as toxic or hazardous areas, tiny crevices, precarious building structures (e.g., roof eaves or fire escapes), treetops, docks, or ice floes, and they can monitor 24/7. Thus, they can provide both cost savings and better safety for scientific, industrial, or commercial applications.

Cost is not always the main issue. While cameras usually cost less than wages, redeployment rather than reduction of staff is sometimes the aim. By providing eyes and analysis capabilities, an analytics system can free employees from concentrating on detection, so that they can spend more time evaluating data (as in scientific applications) or devote their energy to other tasks that benefit the organization.

Video Quality

Analytical systems cannot provide optimal results with suboptimal video. A common programmer saying is *garbage in, garbage out*—better cameras and better installation increase the chance of gleaning accurate or relevant information.

Basic cameras may not capture optimal video. Sometimes something is better than nothing, but to get the best video, many cameras are now equipped with image stabilization, zoom, and swivel capabilities, as well as light metering and lenses that adjust for current conditions or that include night vision. Macro lenses are appropriate for many applications, such as forensics or laboratory research.

New light-field cameras capture not only colors and shapes, but also light vector information to provide three-dimensional data. Low-end models may exhibit some distortion in shapes, but nevertheless come with a significant benefit—the capability to choose the point of focus after the image is taken. If light-field cameras were combined with analytics software, the software could evaluate images, select some of the more meaningful areas on which to focus, and prioritize them for later viewing by humans, thus saving operator time.

Video Analytics Implementation

Like speech recognition software, most video-analytic software doesn't work effectively out of the box. It needs to be configured for a task and may need to be trained to recognize target phenomena or patterns. Once it is in use, it must be associated with recorded footage or live cameras installed at appropriate angles and light levels, and which are protected from projectiles and weather.

Video analytics may be paired with audio analytics, but it depends on the application and the location. A simple motion-detection system rarely requires audio data, but an employee-monitoring system may. Legal issues prevent some forms of audio recording under eavesdropping laws. In the workplace, this may not be an issue. On a public street, audio-equipped cameras may be illegal.

Video can generate massive quantities of data. Depending on resolution, video format, and the number of cameras, a single hour of video from one to five cameras can range from 10 to 100 gigabytes. If the cameras are imagining 24/7, that adds up to almost a million gigabytes per year. And these are small systems—some buildings have hundreds of cameras. While multi-camera systems often use aggressive compression to reduce the amount of data, and may even selectively control whether cameras are on or off, it still adds up.

An analytics system not only comprises the original video footage, but also the data and analysis files that go with it. Sufficient working space, storage, and backup routines are essential. Some analytical programs are designed to check the footage for tampering or data corruption in addition to evaluating specific traits or events.

Systems where storage and analysis are performed near the camera, rather than at a remote location, are called *edge* systems. This can create substantial savings in storage and bandwidth, but may also become a security risk. If the data is stored in the camera housing and the camera is stolen, the data is gone. Remote storage involves greater cost for bandwidth and installation of the connections (e.g., fiber optic cable) but has the advantage of creating *data redundancy*—a data repository in more than one storage device or more than one remote location. Even this isn't a perfect solution, since the data might be snooped or intercepted en route. All these factors should be considered when selecting a system.

Analytics systems are more effective if the data can be cataloged. Many are integrated with databases to store and retrieve information or to search existing archives for matches to certain criteria. As examples, a staff identification program might search employee records for specific face profile or a traffic recording device might search a database of license plates for stolen vehicles or vehicular patterns at border crossings.

It is important to have good data-searching capabilities for analytical systems to be effective. Not all analytical software takes care of archiving and recall of relevant information, partly because it can't be known in advance what might be important, and also because the software may be sold as a modular component, so buyers can choose databases that best serve their needs.

Many video-analytics implementations are site specific—it is difficult for vendors to anticipate exactly how they may be used or what kinds of targets are relevant. To accommodate a wide range of applications, hardware that is designed to be programmed after manufacturing has advantages. A field-programmable gate array (FPGA) is a semiconductor that can be configured or reconfigured for specific applications even after product installation. Logic elements (LEs) within the device can be physically connected to perform simple or complex logic functions for applications such as signal processing or sending and receiving. This makes it an ideal technology for video analytics.

Another trend in analytics is to make systems modular, so imaging units can be removed or added as needs change.

Increasingly, digital cameras are used to capture data. With video formats such as H.264, which exceeds MP4 in quality and efficiency, digital data (which can be replicated without the data loss common to analog systems), are used in analytics. Analog cameras are still supported by many, however, especially in cases where it would be costly to replace legacy systems.

While the process is never perfect (some loss occurs), it is possible to convert analog data to digital, if desired. If the camera system is intended only for live streaming and not for archiving, then analog cameras may be as effective as digital.

Algorithms to ensure the integrity of the data may be included. The system may provide an alert if a camera is stolen, damaged, blacked out, or pointed away from a target incident.

Optimizing Resource Allocation

It is not always practical or cost-effective to glean relevant information from large quantities of data. Sometimes analytics are used to determine when a camera should be on or off in order to stream or record only incidents or phenomena that have a higher probability of being useful. While there is always the risk of missing something, there is a considerable savings in data backup and the time it takes to sort through video to find information of consequence, as well as more efficient allocation of staff.

Analytics and Airport Security

Analytics can be used to distinguish intruders from background noise such as moving branches or traffic, but they may also be used to analyze people's faces to determine their intentions.

---

facecrime *noun*

"He did not know how long she had been looking at him, but perhaps for as much as five minutes, and it was possible that his features had not been perfectly under control. It was terribly dangerous to let your thoughts wander when you were in any public place or within range of a telescreen. The smallest thing could give you away. A nervous tic, an unconscious look of anxiety, a habit of muttering to yourself—anything that carried with it the suggestion of abnormality, of having something to hide. In any case, to wear an improper expression on your face (to look incredulous when a victory was announced, for example) was itself a punishable offence. There was even a word for it in Newspeak: facecrime, it was called."

1984, George Orwell, 1949

As diabolical as such systems may seem, they are being developed in the European Union and may be installed in airplanes to monitor passengers as part of the Security of Aircraft in the Future European Environment (SAFEE).

The overall plan is to broaden transportation security to effectively sense both human and electronic intruders, but one of the more specific goals is to monitor planes in flight as part of an Onboard Threat Detection System (OTDS).

Some of the proposals include explosives sniffers in the lavatories and cameras and microphones to monitor the aisles and to be installed on airline seats so that the faces and conversations of each passenger can be analytically monitored. Personnel might be alerted if terrorist intentions are sensed by the analytics. Those involved in crowd control at large events have also expressed an interest in software designed to sense perceived intentions.

Analytical software can be programmed to look for loitering, both in the terminal or on a plane. Activities that might trigger an alert include running in the cabin (watch out, 5-year-olds), movement against the normal flow of traffic, unattended bags, several people entering a lavatory together (the Mile High Club may be disappointed), or excessive perspiration in combination with other signs of nervousness. A prototype was tested in January 2008 in a mockup Airbus fuselage.

The cost of installing such systems will probably be passed onto passengers as higher ticket prices. The general idea is to rebuild traveler confidence in transportation systems in the wake of terrorist attacks targeting aircraft and trains. It may backfire, however, if security systems become too tedious or too invasive—travelers may find other ways to recreate or travel to avoid long security lines or personally intrusive surveillance.

Video Analytics Progress and Limitations

Two of the biggest challenges facing the field are the complexity of developing reliable systems that don't generate misleading data or false alarms, and the high cost of maintaining systems that incorporate transmission, cataloging, and archiving of large data storehouses.

One of the primary goals of those developing video analytics software is to reduce the amount of time it takes to locate or recognize significant events.

Misleading data or false alarms can lead to high staffing costs if they happen too frequently. While analytic systems aim to be as autonomous as possible, human interpretation of events or alarms is still a necessary part of most systems. Frequent alarms can also result in complacency. Like the tale of the boy who cried wolf—if staff members get used to false alarms, they may disregard a real one.

The design and implementation of good analytics software is technologically challenging and the field is still developing, but specific applications are now commercially viable.

As an example, the U.S. Security Transportation Administration (TSA) evaluated perimeter security systems, based on commercial off-the-shelf products, at selected airports. The system at Buffalo Niagara International Airport, for example, operated around the clock and included automated detection and target-tracking capabilities. National Safe Skies Alliance enacted intrusion scenarios and attempted to bypass the detection system to approach vulnerable areas such as the runway, and concluded that airport security analytics "had a positive effect" on perimeter security.[11]

Another challenge facing analytics developers is adapting the systems to different environments without having to individually construct them. Progress has been made in incorporating artificial intelligence into systems so they may learn the surveilled environment without users having to train them or feed them a set of rules by which to operate. There are now systems that not only adapt to detect objects in different locations (zoos, parking lots, hydro generators, gates, building entries), but which can distinguish them from background noise (e.g., moving leaves and branches) without operator intervention.

Hypocepts™ is a term coined by BRS Labs to describe systems that build and store memories or *hypo*thetical con*cepts* similar to models of human memories and cognitive relationships. They are essentially digital neural networks. According to BRS, "Hypocepts enable the creation, storage, and decay of observed behavior patterns" including short- and long-term memory and the process of forgetting. Given the power and autonomy of such systems, it is expected to be a commercially fertile area of development.

There has been significant growth in video monitoring and availability of massive storage devices, increasing the need for video analytics to deal with the data. In December 2011, ABI Research released a report that the video analytics market was expected to reach almost $1 billion dollars by 2016.

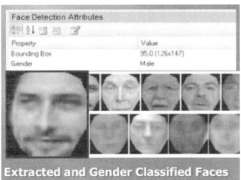

VideoRecall™ is an example of a commercial video analytics product designed to detect and distinguish relevant objects or events. It can extract data regarding people, vehicles, and activity levels and generate reports from the metadata. The example on the right shows extracted faces that are further classified for gender to facilitate recognition or matching. [intuVision VideoRecall News Release product sheet, 2011.]

---

[11] Final Report, Buffalo Niagara International Airport, Office of Security Technology, Airport Perimeter Security Projects for FY08–09, Transportation Security Administration, U.S. Department of Homeland Security, copyright National Safe Skies Alliance, Inc., 2011.

System Intercommunication

One of the problems of integrating different components that make up a surveillance system is the myriad data and physical standards that govern communication, and the size or shape of interconnecting parts. Some standardization has occurred—many cameras now have USB, Firewire, or Ethernet connectors and, increasingly, they are being equipped with wireless capabilities—but the process is not yet seamless or secure.

Consumers usually prefer a single standard—it is easier. But manufacturers favor a number of standards, partly because different formats have different strengths and weaknesses, and partly because it creates niche markets for accessories and add-on components for proprietary systems. One compromise is to develop routers and converters that facilitate communication between differing components.

Ad hoc systems, ones that configure themselves as a need arises, also have potential for surveillance. The *Reconfigurable Ubiquitous Networked Embedded Systems* (RUNES) project seeks to make it easier for devices to intercommunicate and to do so in a way appropriate for the task at hand, such as an emergency situation where rescue workers or technical crews can assess the area and locate entries, exits, and problems.

Wireless is a convenient way to transmit data to a remote source, but encryption is important if the data is of a sensitive nature.

Privacy and Potential Exploitation

Video analytics can be a great boon to research and security applications but also may be used to gather potentially intrusive information. Not everyone is aware that cameras are increasingly equipped with video analytics software.

When people are recorded in stores, hotels, casinos, transportation terminals, and workplaces, many assume a basic VCR is creating a traditional tape. They don't know that the data is often stored in digital databases where it can be cross-referenced with other information or sold to third parties.

Video data may also be subject to legal subpoena, which means it must be turned over to law enforcement agencies upon request, regardless of how many innocent parties are included in the recording. In most cases, those incidentally included are not informed that the video is at the police station. This may not seem like a concern if the people are innocent of any wrongdoing, but what if the tape caught someone leaving a bar shoulder-to-shoulder with a suspected serial killer? Tapes might suggest an association between people, even if it was nothing but a coincidence.

Using video analytics, facial recognition data from one location can be cross-matched with that from another, forming an ever-growing profile of a person's movements, habits, and interests, again without the individual being aware that video footage is analyzed and stored, perhaps indefinitely. By creating an electronic travel itinerary of a person's projected whereabouts, combined with a profile of his or her personal tastes and interests, ads and face-to-face encounters with marketing personnel can be orchestrated to entice or manipulate a person's activities or buying habits.

If a man is approached by a pretty brunette in high heels and fishnet stockings, he may not remember he mentioned on a forum the year before that he liked brunettes in fishnet stockings. He could then be unwittingly enticed into a relationship where he is coerced into buying something he didn't need, or be more readily defrauded by someone posing as his girl-friend. Such things have already happened on the Web even before analyt-ics became capable of streamlining data collection and profiling. There are predators, both male and female, who befriend victims through video chats or forums with the express purpose of cheating them out of money or prop-erty. Surveillance data makes it easier to identify "marks."

This leads to some privacy concerns that can be generalized to most surveillance technologies.

- It is difficult for people to protect themselves from tech-nology that is so sophisticated that they don't immediately (or ever) understand what it does and how it could poten-tially manipulate or harm them.

- It is not always easy to recognize that you are being exploited. Children and adults are easily led by imposters, ads, or other enticements such as attention and affection from another person, or free goods or generous sweepstakes offers from commercial vendors.

It is almost impossible for lawmakers to anticipate harmful applications of potentially helpful technologies and to craft proactive laws to protect victims. This is partly due to busy schedules and lack of technical knowledge (most are lawyers, not hardware/software professionals), and partly due to the speed at which technology is evolving.

How do individuals protect themselves where access to technology and analyt-ical products is selectively available to those with power and money? At present, many surveillance technologies are within reach of household consumers, but as the technologies evolve, will this continue, or will some become less accessible?

Societies have always struggled with inequities, but technology can increase or decrease gaps in ways never before imagined.

The biggest inequity at the moment doesn't appear to be the technology itself, but the resources to gather, store, and process information obtained from these gadgets. Well-funded institutions can afford high-bandwidth Internet investments, for example, and are building huge repositories of personal information that indi-viduals cannot see, access, or update even if the information or analysis of that data is inaccurate. Many of these repositories are in foreign countries operating under different laws and, since many are in private hands, there are no freedom of access laws, as with some government documents, for checking their presence or accuracy. There are currently few restrictions on the resale of the information.

Imagine a manager applying for a directorship at a new branch office. He or she knows several subordinates who might also apply. Employees pass through video checkpoints in the course of their workday and their faces can be cross-referenced with information in employee files. The software may harvest further information from data aggregation sites on the Internet about the person's economic status, relatives, hobbies, former brushes with the law, and academic performance. In addition, the manager might patronize previous employers who are selling information on past employees.

Armed with this arsenal, the manager can gather ammunition to shoot down rivals vying for the directorship he or she covets. In hierarchical institutions, those in higher positions frequently have privileged access or knowledge that is unavailable (or unknown) to subordinates. This inequity creates a digital glass ceiling or "silicon ceiling" reducing opportunities for promotions or mobility.

Many video analytics systems are developed and marketed by multinational corporations that have no allegiance to local communities or even specific nations and may not care how use of their video surveillance systems affects individuals. Protecting a specific institution with state-of-the-art security or providing marketing tools to eager businesses always results in incidental video recording of people who pose no threat or who don't realize they are a commercial or social target.

Convenience Versus Targeted Subcultures

There are companies that sell video-viewing and video-targeting analytics that enable social networking and entertainment-video viewing sites to automatically gather and track detailed information on their viewers' gender, locations, ethnicity, and interests. They are then showed selected advertising or short lists of entertainment videos specific to their tastes.

Sometimes it is convenient to see only sites of interest, given the mass of content on the Web, but software may make false assumptions about people or may segregate them into digital cliques—separate areas of the cyberspace honeycomb—through manipulative content presentation. This is sometimes known as "bubbling." This may not seem like a problem now, but given increasing efforts to regulate the Internet, both by governments and large corporations, it may emerge as a problem later.

AERIAL SURVEILLANCE

KEY CONCEPTS

- *common forms of aerial surveillance*
- *how aerial surveillance is used*
- *specialized surveillance for military and human rights purposes*

Introduction

Aerial surveillance is used for many purposes such as detecting forest fires, search and rescue, aerial mapping, geological research, weather forecasting and storm tracking, wildfire detection, border security, and resource management.

Aerial surveillance encompasses a variety of technologies mounted on towers and buildings, and carried aboard piloted and uninhabited aircraft and orbiting satellites. Airborne devices that sense in visual spectra and radar are the most common, but there are many that use infrared and ultraviolet frequencies, as well.

Several hundred miles above Earth, satellite imaging systems now sweep the planet in about a day, and are gathering imagery that is increasingly precise. Improved-resolution cameras are coupled with advanced software algorithms to combine data from separate images to yield a composite with better information content than either image alone.[12]

Multispectral bands that capture different kinds of terrain, and software programs for differentiating between water, vegetation, and minerals are improving, as well.[13]

Aerial imagery is used for mapping, urban planning, weather forecasting, climatology, environmental studies, agricultural assessment, archaeological investigations, forensics, military planning and analysis, wildlife monitoring, and tracking of human rights violations. It is difficult to think of disciplines that don't use satellite imagery in one way or another.

---

[12]H. Demirel and G. Anbarjafari, "Satellite image resolution enhancement using complex wavelet transform, "*Geoscience and Remote Sensing Letters*, IEEE, 7 (10), Jan. 2010.

[13]C.K. Toth, J.H. Oh, and D.A. Grejner-Brzezinska, "Airborne hyperspectral image georeferencing aided by high-resolution satelllite images," *ISPRS TC VII Symposium*, Vienna, Austria, July 2010.

Aircraft and weather balloons used to be the primary source of aerial sensing data, but orbiting satellites are increasingly important providers of data and imagery, and remotely controlled aerial vehicles are being used in war zones and, more recently, by local law enforcement agencies.

BALLOON SURVEILLANCE

Balloons were one of the first aerial vehicles to carry cameras aloft. Balloons may seem primitive compared to remotely controlled vehicles or satellites, but they are viable surveillance vehicles and can be outfitted with a variety of sensors, including optical, infrared, and thermal imaging devices.

Balloons, also known as *lighter-than-air surveillance systems* (LTASS) or *tactical aerostats*, are relatively inexpensive and quiet, making them suitable for wildlife observation, weather-watching, and stealth applications. A medium-sized system can hold up to 200 pounds of sensing equipment.

In the mid-2000s, the U.S. Army and Marines expanded their balloon programs. Some were deployed to Iraq, and some complemented existing surveillance balloons along U.S. borders.

Balloons have limitations. They can only operate within certain temperature ranges, and may need to be tethered, depending on their design and application. High winds can impede their use (particularly if they are untethered), and they can be shot down if used in war zones.

Left: Aerostats or "blimps" have long been used by military forces and researchers as wind indicators and carriers of surveillance tools. Larger ones may carry staff members, while smaller ones are often tethered and may send signals through wireless transmitters or optical cables that reach the ground. Right: Balloons are not limited to atmospheric applications. They can also be sent into space, as were the Echo and PAGEOS orbiting satellites. [NASA/Ames Research Center, 1996; NASA/Langley Research Center, 1965, public domain.]

Balloons also have many advantages. They are economical to transport, conserving of space when deflated, and don't take much time to launch. They are economical to operate, since balloon operators require minimal training and balloons don't need fuel to stay aloft. They can remain aloft longer than most aerial vehicles (with the exception of orbiting satellites).

Some military balloons or "blimps" are larger and more costly, and carry more sophisticated communications gear, but are still economical to build and operate compared to high-tech stealth planes.

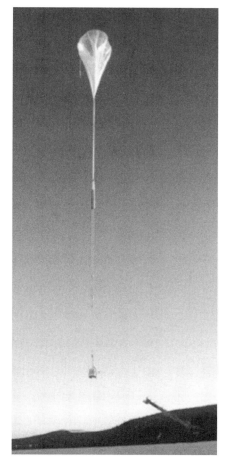

Left: High-altitude helium-filled balloons have been used to study the stratosphere. While it may look like modest technology, this balloon is able to lift approximately 1000 pounds of equipment to an altitude of more than 30,000 meters. Top right: The Echo I was America's first passive radio relay satellite shown here in an inflation test that ballooned it to 100 feet in diameter, in 1961. Bottom right: A sailplane is, in a sense, a reverse balloon. It is carried aloft by a high-altitude balloon and then released and remotely piloted as it drifts back to Earth, gathering sensing data on its way down. [Photos courtesy of NASA/JPL, 2000, NASA/Langley, 1961, and NASA Dryden Flight Research Center, 1995, public domain.]

Balloon communications can be through wireless transmissions or fiber optic cables that are sometimes as long as 300 meters.

Balloons have also been sent out into space. The Echo I satellite was an over-sized 100-foot mylar balloon—a giant version of a party balloon—that served as the first American passive communications satellite for reflecting radio waves over distances much greater than typical for ground-based relay stations.

Balloon surveillance isn't always for covert purposes or national security. In *Red-Eyed Sky Walkers,* a 2011 art installation at the Kumu Art Museum, Estonia, a cloud of weather balloons, some of which were outfitted with cameras, surveilled the museum visitors in the courtyard and wirelessly transmitted images to an indoor video projection room. In the follow-up silver series, the images were combined with media from sources such as YouTube™, WiLeaks™, Twitter™, and Flicker™. Images were further combined with public video from the Internet depicting violence, crowd control scenarios, and political upheaval, thus creating social commentary on our surveillance society.

Uninhabited Aerial Vehicles

There are two common kinds of uninhabited aerial vehicles (UAVs)—those that are controlled remotely, and those that are mostly or wholly autonomous, controlled by smart software algorithms and onboard computers rather than by pilots communicating through radio-wave controllers.

Some UAVs can take off from the ground like a commercial airliner, others are launched by a catapult or towed and launched from the air.

Radio-controlled UAVs used to be a hobbyist passtime. Enthusiasts would convene at local parks and schoolgrounds to fly planes and helicopters ranging in size from about a foot to about four or five feet in wingspan. Now UAVs from the size of a fly to the size of a piloted plane are either in development or in active use for many military and surveillance applications.

UAVs have been actively used over Afghanistan and neighboring areas for a number of years. Most are equipped with cameras and other sensors and may also include tracking mechanisms. The military uses both remotely controlled and autonomous vehicles. One of the craft now used by the Marines was originally designed to help fishing boats find schools of tuna.

To increase stealth, specialized materials are applied as "skins" on UAVs to deflect or confuse radar signals. In the case of ScanEagles, mufflers were installed facing skyward to make them harder to hear from the ground.

One of the most significant changes in regulating UAVs is a mandate by the government for the Federal Aviation Administration (FAA) to open up airways so that UAVs can share space with commercial aircraft. This development was foreshadowed in a UAV "Roadmap Report" in 2003 in which Dyke Weatherton commented on integrating UAVs into civilian airspace to enable "the full capabilities that UAVs promise to provide," their primary purpose being intelligence gathering, reconnaissance, and surveillance. In 2010, for the first time, a Customs and Border Protection aircraft landed at a civilian airport, for an air show event.

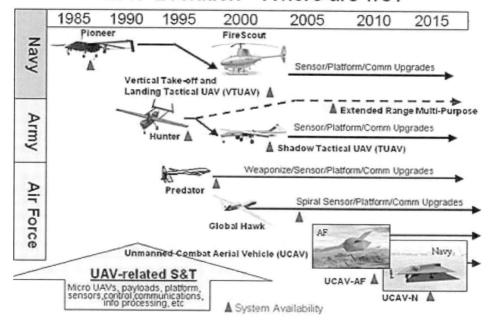

As described on the U.S. Department of Defense information site, "Historically the greatest use of UAVs have [sic] been in the areas of intelligence surveillance and reconnaissance." This illustration shows how multi-sensor aircraft have become an important aspect of military reconnaissance and surveillance. [DoD news photo courtesy of DoD News Briefing.]

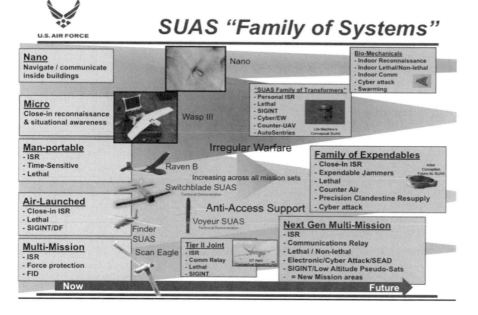

This U.S. Air Force diagram categorizes some of the common forms of UAVs that are used in reconnaissance missions and for irregular warfare support. UAVs can be built as both recoverable and expendable craft, depending upon the mission. [Educational chart courtesy of U.S. Air Force, released.]

Thus, legislation was enacted in early 2012 to integrate UAVs into airspace so that they could be more broadly deployed. Other than small hobbyist remotely controlled aircraft, UAVs are almost entirely intended for surveillance and most are equipped with multiple sensors. The FAA mandate is discussed in greater detail in later chapters.

UAVs are not always small. This U.S. Air Force Global Hawk, with a wingspan of over 100 feet, was deployed in the Middle East for three years, logging almost 5,000 flight hours and capturing tens of thousands of images in all kinds of weather with sensors that can penetrate cloud cover. The Global Hawk can stay airborne for more than 24 hours. The Global Hawk program is managed by the Air Force Reconnaissance Systems Wing. In 2010, procurements included two additional Global Hawks and a variety of radar, in-line, and other sensors. [News photo courtesy of U.S. DoD, 2006, released.]

Left: The ScanEagle UAV, a craft equipped with cameras for both day and night, is explained by a Boeing contractor. It is a smaller, lower-cost alternative to large systems (like the Global Hawk) that flies autonomously after launch, and can stay aloft almost as long as the Global Hawk (about 19 hours) while consuming very little fuel. Right: Sgt. Chad John, of the Marine Corps, is a UAV pilot deployed to Camp Leatherneck Afghanistan to fly remotely controlled aerial vehicles over Afghanistan up to 12 hours a day. While one operator pilots the vehicle, the other controls the onboard surveillance camera that watches over troops below. [DoD news photos courtesy of the American Forces Press Service, 2004 and 2011, released.]

Left: The Predator B UAV is remotely controlled via C-band radio signals for line-of-sight and via Ku-band radio signals relayed through orbiting satellites for longer distances. It can stay aloft up to 20 operational hours and fly at altitudes up to 50,000 feet. Right, a CBP Predator B comes in for a landing at the Oshkosh airport in Wisconsin, representing a milestone UAV landing in a civilian airport. [News photos courtesy of CBP, 2010, released.]

The Predator B is a versatile, long-endurance UAV used by the military and Customs and Border Protection for aerial surveillance. It is a larger version of the RQ-1A Predator reconnaissance plane used for intelligence-gathering in Afghanistan and Iraq, with a wingspan of over 60 feet.

The Predator B is equipped with radio transceiving equipment so it can be controlled from the ground via satellite links, and includes electro-optical/infrared sensors. One version of the Predator B, which began operational testing in 2010, is fitted for maritime operations for patrolling the southeast maritime-traffic corridor.

The Altair/Predator is a modified Predator B with even longer endurance, up to 32 operational hours. It can carry up to 750 pounds of radar equipment and other sensing and imaging components in the forward compartment. As it is designed to fly remotely, it is equipped with collision-avoidance systems.

Predator B aircraft are typically deployed in teams, with four aircraft working together. Multiple sensors including radar, infrared, and color video cameras aid in navigation and the gathering of surveillance data.

UAV Disaster Surveillance

UAVs can help monitor avalanches, floods, storms, and volcanic eruptions. They enable firefighters to follow the course of a fire and track whether it is nearing human habitations.

In 2007, a Predator B system that included an infrared-imaging modification from NASA was used to image through smoke and haze from the wildfires in California, with the data transmitted through satellites to a NASA/Ames center near San Jose. There, the imagery was overlaid on maps from the Google Earth™ 3D terrain-imaging application that is freely downloadable on the Internet. In this way, realtime aerial images of the fire could be relayed by a Boise, Idaho, fire center to California firefighters to deploy crews and allocate resources. This level of realtime long-distance interagency cooperation through aerial imaging and satellite transmissions was unthinkable 20 years ago.

The Ikhana is a Predator B UAV modified for civilian uses. The control console for remotely flying the aircraft is shown on the left, with research pilot Mark Pestana at the "con" (command console). On the right is an infrared image taken by the aircraft that has been colorized to reveal hot spots from wildfires ablaze in San Diego county in 2007. The areas that have already burned are coded in green, for example. Aerial imagery can help firefighters locate trouble spots and send out appropriate response teams. Image data from the Ikhana was correlated with satellite imagery to provide data from different altitudes and different sensing instruments. [Predator image and news photo by Tom Tschida courtesy of NASA, released.]

Predator B aircraft provide streaming video and synthetic-aperture radar (SAR) mapping data to the U.S. Geological Survey and the U.S. Army Corps of Engineers to help them make assessment/response decisions in areas devastated by floods. The realtime video stream (called Big Pipe) is also available to government agencies to help them develop policy and disaster responses.

Aerial Real Estate Photos

UAVs are used by commercial interests for surveillance as well. Helicopters and remotely controlled planes have been contracted by journalists and traffic reporters to observe crowds or transportation routes, and by real estate firms to shoot aerial images of properties for sale, especially high-end multi-million-dollar castles and mansions with large spreads.

There are companies specializing in providing aerial photos to commercial, residential, and rural brokers of real estate. The images are similar Google Maps™ satellite images at their highest magnification, but are even more bright and clear because they are shot from altitudes of about 30 to 80 feet. UAVs images are not limited to overhead views. They can also be taken from the side.

The FAA and police are concerned about safety issues, however, especially when they noticed UAV images from higher altitudes displayed on public venues like the Internet. High-flying UAVs could interfere with police and commercial aircraft. In 2011, the FAA asked UAV firms to stop flying their vehicles until new FAA regulations for integrating UAV into civilian airspace are drafted.

In January 2012, the Los Angeles Police Department issued a warning to real estate agents that UAVs might pose safety hazards and requested that realtors not use images taken from UAVs.

AIRCRAFT SURVEILLANCE

One of the most common forms of aircraft surveillance is traffic surveillance by news teams and local governments seeking to report on congestion, accidents, and other events affecting local streets and highways.

Helicopters are also commonly used in search-and-rescue operations, particularly in areas that are difficult to access by planes, such as mountain passes, glaciers, and forests with clearings that are too small for planes to land.

Urban Surveillance

Aerial surveillance is used in many aspects of urban planning and security.

Urban planning can benefit from surveillance by providing data to planners that help them assess environmental impact, population density, human and vehicular traffic patterns, growth, sprawl, inbuilding, and energy usage.

Aerial images taken with infrared sensors can identify buildings that are leaking heat, so that better roofing and insulation can be installed. Industrial plants can be monitored for heat or chemical leaks and safety compliance.

Aerial surveillance can also help assess construction safety and locate construction projects that lack proper permits.

Aerial surveillance is also widely used in law enforcement. Many police departments, including the Los Angeles Police Department (LAPD), use aerial technology as part of their surveillance operations. LAPD's Air Support Division is considered one of the most extensive in the world.

At least one LAPD helicopter is equipped with a high-resolution stabilized camera that can covertly image events on the ground from thousands of feet in the air. This is sufficient to distinguish a gun from a mobile phone. Data from the camera can be transmitted through encrypted radio-wave signals to police command centers, mobile ground units, or other aerial vehicles more than 100 miles away. Infrared sensors and laser range-finders are also available as sensor options.

Many police officers welcome technological tools to help them do their jobs, but some citizens have countered that high-technology increasingly distances law enforcement from the communities they serve. Others see the adoption of high-tech devices by police departments as a trend toward a military model of community policing.

Border Surveillance

Aircraft have been used for decades to patrol borders and coastlines, including the monitoring of offshore oil spills.

The Dash-8 (DHC-8), known to many commuters as a "puddle-jumper," is flown on short hops on many commercial airline routes, but is also flown as a surveillance plane by Customs and Border Protection (CBP).

In March 2012, CBP announced that a Dash-8 had been patrolling off the coast of Florida when it spotted a vessel matching a boat described in an FBI bulletin related to suspected parental kidnapping. The name had been changed, but the boat was the same.

Once a tentative ID of the boat was made, aircraft crews directed a Coast Guard vessel to the location of the suspicious boat. Police checks showed there were active warrants out for the father, so he was apprehended and the three children on board were returned to their mother.

Drug smuggling occurs constantly along U.S. borders and enforcement agents can barely keep up with locating and apprehending smugglers. Aircraft sitings of unusual vessels, or vessels sailing in restricted areas or in unusual ways, may alert surveillance teams to criminal intentions.

In 2009, U.S. and Mexican forces were in the same area conducting patrols. The American aircraft was scanning the area with radar when two "vessels of interest" were noticed below. Often drugs are cleverly disguised inside hulls and other containers on larger boats and ships, but these were smaller craft, under 50 feet in length, with no canopy to hide the cargo, which consisted of thick cubic bales in one, and what appeared to be plastic fuel containers in the other.

Radio communications were established (via the Joint Interagency Task Force) with a Mexican surveillance aircraft. Pilots in the CBP plane guided the Mexican plane to the site. Surveillance photos were analyzed to determine which vessel was carrying contraband. Then a Mexican frigate launched a helicopter and the helicopter hovered overhead, guiding the frigate to the boat.

In 2009, a pair of open boats loaded with powder-packed bales in one (left) and apparent fuel containers in the other were spotted by CBP surveillance aircraft off the southeast coast of North America. The go-fast boat, powered by four engines, was apprehended by Mexican authorities using location information from the American pilots in coordination with Mexican surveillance aircraft, including a helicopter launched from a Mexican frigate. Orion P-3 high-endurance surveillance aircraft (right) are equipped with long-range sensors that aid in spotting suspicious marine vessels in coastal waters. Sometimes they work in tandem, one to watch multiple targets, the other to zero in on suspicious vessels. [News photos courtesy of CBP, 2009 and 2012, released.]

This demonstrates how multinational surveillance aircraft are coordinated to handle surveillance, analysis, and apprehension of illegal drugs. The contraband cocaine, in this instance, was estimated as being worth $150 million.

While drug-runners packing drug bales in an open vessel might seem like sitting ducks from the air, it hasn't stopped others from trying the same thing many times. In 2012, two separate go-fast boats were spotted off the coast of Colombia and Costa Rica within hours of each other. One was loaded with numerous packages, the other with square bales. Both vessels were apprehended and found to be carrying more than 10,000 pounds of cocaine. During that week, more than $800 million in drug seizures were made as a result of counter-narcotics surveillance by the crew of Orion P-3 aircraft.

Not all marine smugglers are easy to spot. A semi-submersible vessel, one with much of the hull under the waterline, was observed by the crew of a CBP P-3 aircraft off the coast of Colombia. A P-3 is equipped with long-range tracking capabilities, which are essential for locating small vessels in the Gulf Coast and vast Atlantic sea. Sometimes P-3s work in tandem, one to observe multiple targets, another longer-range craft to identify and track targeted suspects.

Since it can be difficult to determine what a semi-submersible might be carrying, the crew were tipped off to the intent of the smugglers by their behavior. When the four people in the semi-submersible realized they were being pursued, they took to an inflatable raft and scuttled the semi-submersible. It was later found to be loaded with 100,000 pounds of cocaine.

CBP officials point out they aren't only looking for drugs. Coastal patrols prevent and deter the illegal import of explosives and other destructive weapons to the U.S.

Large events, like concerts and sports events, are sometimes monitored by aircraft such as helicopters and smaller planes, as well.

Military Surveillance

The military makes extensive use of aerial surveillance, using balloons, dirigibles, UAVs, jets, and specialized stealth planes. One could fill many books with information about just a small part of these operations, so this section summarizes just a few examples of how the military conducts surveillance using planes.

The U.S. Air Force employs a variety of aircraft for national security and combat-related missions. Aerial missions are an important aspect of recent events in the Middle East and also enabled air crews to assess the effects of the 2011 tsunami in Japan.

Surveillance planes sometimes are dedicated to reconnaissance and surveillance activities without having any weapons capability. Others use surveillance technologies to specifically assist in targeting, aiming, and deploying weapons. Modern inhabited aircraft have cockpits that look like command consoles on science fiction shows, with readouts from many sensors and high-resolution display devices similar to video game systems to image and zoom in on targets.

The U-2 "spy plane" is a high-altitude aircraft developed in the 1950s specifically for providing reconnaissance and surveillance to intelligence agencies. It came to public attention when one was shot down for trespassing in the western portion of Soviet airspace in 1960. Authorities assumed the pilot would go down with the plane rather than give away national security secrets, but the pilot survived and was taken as a political prisoner while the aircraft was examined, dismantled, and destroyed. The secret was out.

The public is often unaware of the existence of specific surveillance aircraft until five or ten years after they begin flying missions. With the Internet, and greater transparency of armed forces contracts and missions, the operations of spy craft are not as covert as they used to be.

Top: The long wings of the U-2 surveillance aircraft help it fly at high altitudes above 70,000 feet but the flying properties of the plane also make it hard to land. Pilots need special life-preserving equipment, resembling space suits, to fly the plane, since humans are designed for terrestrial levels of oxygen and different air pressure levels from those found at high altitudes. Bottom: A wet-film, super-high-resolution *optical bar camera* was installed in a U-2 aircraft in South Korea prior to flying a mission over the tsunami disaster zone in nearby Japan. [News photos courtesy of the U.S. Air Force, 2012 and 2011, released.]

Top: The A-12 stealth plane was developed soon after the U-2 but didn't last long, as it evolved into the SR-71. Bottom: The SR-71 stealth plane was also known as the *Lady in Black* or *Blackbird*, due to special fabrication materials to reduce detection as well as the dark color of the hull and the covert nature of its missions. It was very fast, designed to outpace attacking missiles, and could perform surveillance from high altitudes like the U-2. The SR-71 was in active service for a quarter century. [News photos courtesy of the U.S. Air Force, 2010 (Lance Cheung) and 2008, released.]

Other planes were later developed, such as the A-12, which quickly evolved into the SR-71.

The U-2 has actively flown missions over Iraq and other parts of the Middle East, using Saudi Arabia as a landing spot. Reaction to U-2 planes in the Middle East is mixed. Some welcome American air surveillance, others don't want American planes in their airspace. Even within a specific country, an initial welcome may change to prohibitions and attempts to shoot down the planes, as happened via Iraq's air defense system in 2001.

The U-2 is not limited to military reconnaissance. It is also used to assess natural disasters and their consequences.

U-2s have long lifespans compared to surveillance craft that are designed to be expendable within a few hours or months. U-2s capture thousands of feet of high-resolution terrestrial imagery to film, with the resulting data, weighing hundreds of pounds, brought down from each mission.

The SR-71 was developed soon after the U-2. The SR-71 represented a new approach to stealth vehicles, using materials specifically designed to prevent detection. The quest for materials to reduce the chance of detection is still an active area of research that applies to other vessels, as well, such as submarines and reconnaissance ships.

Stealth technologies include materials that are colored or otherwise fabricated to be less visible, and which reflect or absorb energy from electromagnetic radiation from other sources such as radar or infrared probes.

The F-22 and F-35 fighter jets are stealth vessels designed for air defense and reconnaissance. They are installed with a wide array of surveillance sensors including radar and infrared. In the past, stealth coatings were typically applied to the outer surface of reconnaissance vessels. Now, with planes like the F-35, stealth fabrication is built into the structural fuselage.

This may look like a communications ground station, but these banks of terminals are managed from far above ground on an E-3 Sentry (top), with surveillance data relayed by the crew (bottom) to U.S. and NATO commanders. Occasionally the crew will take a break from battlefield surveillance at Christmas to track the progress of Santa on his annual ride. [News photos courtesy of the U.S. Air Force, 2008, released.]

Beginning in the early 1970s, a wide range of surveillance activities has been carried out by airborne warning and control system (AWACS) planes such as the E-3 Sentry. The Sentry looks similar to commercial jetliners because it was based on the design of a Boeing 707. With upgrades, the design is expected to have an active life of about 60 years. The main outward difference is the addition of a 30-foot dome that houses a rotating radar with a range of more than 200 miles (depending on the direction being monitored and the nature of the targets).

Location and tracking information on suspicious aircraft can be processed in realtime, with the crew relaying information to commanders in the U.S. and North Atlantic Treaty Organization (NATO).

The E-3 has remarkable data collection and storage capabilities, with the ability to record entire incidents from beginning to end.

Customs and Border Patrol have their own aircraft, as described earlier, but CBP agents sometimes ride in the E-3 Sentry to find and apprehend smugglers.

Transitions in Aerial Surveillance

The SR-71 program was retired in the late 1990s, but U-2 planes are still operated by the U.S. Air Force more than 50 years later—a remarkable lifespan for a specialized plane.

This may change, however. According to the U.S. Air Force, UAVs like the Global Hawk are projected to replace the U-2 by around 2015, and the film camera will likely be superseded by digital imaging.

The FAA mandate to integrate UAVs into civilian airspace over the next few years could dramatically alter the way aerial surveillance is conducted, including opening up far more places in which spy craft can fly. Operational costs must be justified, however. UAVs aren't always less expensive to fly than traditional planes.

Despite operational concerns, however, the days of the U-2 are numbered. In the future, pilots may operate from ground units rather than cockpits.

SATELLITE SURVEILLANCE

Satellites are "eyes in the sky" that communicate with base stations to relay information around the world. The first satellites were launched by the Soviet, Canadian, and U.S. governments, but commercial satellites now share orbital space with government vessels. In fact, it has become very cluttered in space, not only with space vehicles and the orbiting space station, but "space junk" that is building up in large quantities. The debris can collide with satellites and do damage.

Satellites handle communications, location services, and many imaging tasks. They also serve as relays for other kinds of communications systems. Satellites can send information from one kind of surveillance sensor to another.

Many people are familiar with satellite maps that are displayed through services like Google Maps. In the 1990s, a single aerial map of a small section of land could cost $20 to $100. Now millions of maps that tile together to form a picture of the Earth's surface are freely accessible on the Web. High-resolution satellite printouts can still be purchased for a fee.

Left: By 1960, commercial communications companies were urging NASA to develop a satellite that could occupy a high orbit, synchronous with Earth's rotation, thus remaining at approximately the same spot in relation to ground-based transceivers in order to facilitate transmission between the satellite and broadcast stations. The result was a series of Syncom (synchronous communications) satellites that demonstrated that the concept was sound. The third Syncom was used to transmit live broadcasts of the Tokyo Olympic games to the western world. Right: Ground-based stations weren't always on solid ground; sometimes the best location for receiving a satellite transmissions was on open water. The first ship built for satellite transceiving was the USNS Kingsport. Within the large plastic dome was a 30-foot parabolic antenna that could communicate with the Syncom satellite. [Photos courtesy of NASA/Goddard, 1963, public domain.]

Left: A composite image of the Earth as seen from the high-orbit GOES-7 and GOES-8 environment-monitoring satellites. GOES satellites, owned and operated by the National Oceanic and Atmospheric Administration (NOAA) are geostationary satellites similar to the earlier Syncom series in that they remain at about the same location relative to the Earth. North and South America are clearly visible in this GOES image, as are weather patterns reflected in cloud patterns. Right: Satellites used for imaging closer to the ground, such as this TRMM satellite, circumnavigate the globe to map its contours, a swath at a time, with the imagery being processed and assembled or tiled in various ways, depending on the application. [Images courtesy of NASA/Goddard & NASA/Goddard Space Flight Center Scientific Visualization Studio, 1994 and 1998, public domain.]

Satellite dishes, designed to capture satellite transmissions, used to be large, expensive, and unappealing to the eye. With recent developments, and transmission at higher frequencies, the size of dishes has become smaller, bringing down the cost and also the variety of places in which they can be installed. In fact, smaller dishes have become almost as common in some places as television antennas were ten years ago. Building owners in some areas have told tenants to remove them, mainly for aesthetic reasons or to prevent damage to buildings.

Courts in some countries have supported owners' right to retain the dishes, based on the freedom to receive religious broadcasts or because the judicial system has ruled that access to global information through television broadcasts and Internet connectivity is a human right.

In the past, most consumer satellite communications were receiver-only, but demand for Internet services has created a growing market for satellite transceiver services, especially in areas where it is impractical to install optical cables.

Early Warning Systems

The Defense Support Program operates a constellation of satellites that collectively scan the surface of the Earth in less than a minute. One of the functions of this system is to serve as an early-warning system, using infrared/thermal sensors to seek out anything with a significant heat signature that could indicate explosions, missile plumes, oil fires, or nuclear detonations.

A system with these capabilities could also be used to provide alerts for other events that emit heat, such as volcanic eruptions, explosions, or wildfires.

Geosynchronous surveillance satellites in the Defense Support Program provide early warnings by using thousands of infrared sensors to sense signs of heat emissions from explosions, missile launches, and nuclear detonations. If events are detected, reports are relayed to the North American Aerospace Defense Command and other strategic commands. The system has been in place since the 1970s and has gone through a number of upgrades to improve the number and quality of the sensors. [News photo courtesy of the U.S. Air Force, 2010, released.]

There are three main components to most satellite systems: the orbiting satellites, one or more ground-based antennas for facilitating radio communications, and a command center that enables staff to interpret data from various satellites, including combat satellites, weather satellites, communications satellites, and mapping satellites.

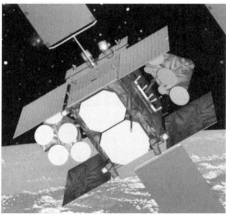

Top: SATCOM is a wide-band communications satellite that can provide combat-zone tactical communications with ground-based defense information systems. Bottom left: This helical structure is a radio antenna to provide increased range for in-the-field satellite communications, being set up by radio maintainer Master Sgt. J. Shearer. Bottom right: Image consoles show data from satellites. These are monitored for weather patterns by meteorologist Shawn Dahl at the Scott Air Force Base. [News photos courtesy of the U.S. Air Force Space Command, 2012 and 2011, released.]

U.S. Government Aerial Imagery

Aerial imagery is now common on the Net, but there was a time when it was difficult to access and much of it was classified. Some of the government agencies involved in geological study and mapping that collected aerial data from planes and satellites have gradually made the data available. These are often used for surveillance projects by local and regional governments, as well as by individuals, private detectives, and commercial firms.

Satellite images reveal a great deal about urban construction, agriculture, warfare, marine environments, ice flows and melts, coastline fluctuations, and other aspects of Earth's topography.

When many images are collected over a period of time, they can be studied for trends and changes. Logged forests, areas altered by floods or dams, the results of nuclear incidents or bomb explosions, or the craters left by volcanic eruptions can be monitored and studied.

Aerial imagery is also useful for planning, whether it is urban planning, military planning, or deciding when to harvest certain fish or crops.

The U.S. Geological Survey (U.S.G.S.) has collected aerial images in both visible and infrared spectra for many decades, many of them through the Earth-Explorer satellite system, making it possible to look back through the photos and see how things have changed and evolved. The U.S.G.S. also provides access to images taken from the International Space Station and from Lear jets operated by NASA.

A NASA satellite captured an extensive oil spill in April 2010 in the Gulf of Mexico, southeast of the Mississippi delta, following an explosion on the Deepwater Horizon oil rig. A series of time-lapse shots were taken with the MODIS imaging system that is used on NASA Terra and Aqua Satellites. They show the shape and continual spread of the spill. Explosions and oil spills can endanger nearby marine craft and significantly damage the local ecology and resources and thus are carefully monitored. Surveillance imagery may also be used as evidence in ensuing lawsuits. [News photo courtesy of NASA/Goddard.]

NASA captures and archives satellite imagery, as well, through their Terra and Aqua satellites, and NASA is a key player in launching government and commercial satellites into space.

A significant improvement, in recent years, in both government and commercial aerial imagery is the greater use of true color images, those in which the color is obtained through the camera rather than being calculated and added later to grayscale images by computer software.

Global Positioning System Surveillance

The U.S. Global Positioning System may not seem like an aerial surveillance technology, since many people use it on the ground, but it is based on satellites equipped with high-precision atomic clocks that coordinate with each other to send signals to ground stations. It is also an essential navigation component in many surveillance aircraft.

GPS is described in more detail in the *GPS Surveillance* chapter.

Satellite Surveillance of Human Rights Violations

Aerial surveillance has become a significant source of information on human rights. It has been used to surveil border skirmishes and refugees fleeing from war zones. It also adds perspective about local conflicts the disrupt village life.

In the past, governments have been reluctant to get involved in foreign human rights issues unless the voting population is aware of them and calls for action. Satellite imagery brings foreign conflicts to people's desktops via the World Wide Web and has greatly broadened public awareness of brutal actions against both armed and unarmed civilians.

In 2005, the American Association for the Advancement of Science (AAAS) became involved in human rights satellite projects by documenting the demolition of residences in Zimbabwe, Africa.

Amnesty International, an organization that monitors human rights worldwide, launched its first "satellite watch" project, with technical support from AAAS, in 2007, called *Eyes on Darfur,* using high-resolution satellite imagery to document human rights violations in the Sudan. *Satellite Sentinel* is a similar project to digitally document humanitarian crimes in the region.

Then, in 2009, geographers studying digital imagery that was part of the Science and Human Rights Program at the AAAS, noted craters and damage to buildings in a section of Sri Lanka that had previously sheltered refugees. Nearby, increases in the number of graves was seen in subsequent images.

Sometimes even seasoned analysts are surprised at what satellite photos reveal. Parts of the world that seem distant to us, with less access to technology, understand they are being viewed from above. In a moving demonstration of both despair and hope, people in a region of Kyrgyszstan, where homes had been torched during ethnic violence, had crafted huge SOS signs in fields and byways. In this way, news reaches the rest of the world in days rather than weeks or months.

Eyes on Darfur was followed by *Eyes on Pakistan,* a project that utilizes a spatial database of human rights incidents. *Eyes on Nigeria* was similarly put into action in 2011.

Satellite imagery can help identify areas of conflict where villagers may have been evicted or had their homes bombed or demolished by fire. It also reveals gas flares that contribute to air pollution, acid rain, and cause damage to nearby crops. Gas flares continue to occur in Nigeria despite a 2008 government moratorium.

In January 2012, an op-ed article in the New York Times suggested that drones similar to those used in the military could monitor events from a vantage point nearer the ground.

While satellite photos are increasingly detailed, they are imaged from far above. UAVs could fill out the picture by imaging damage from nearer to the ground and to the sides of structures as well.

Destruction of residential areas by military surface weapons in Syria were observed in satellite photos in February 2012. By comparing a series of satellite images, it is possible to track the progress and extent of the damage.

Reports of activities like these are no longer restricted to small circles. They are now routinely uploaded to Internet news sites, as well as social networking sites such as Facebook and Twitter.

MICRO-SCALE AERIAL SURVEILLANCE

Insects make a surprising contribution to aerial surveillance. The pollen of bees can be studied for its contents to see where the bees have been and what kind of plants or toxins are in the vicinity. A number of insects have been tagged with tiny radio transmitters to see where they go. But insects, like humans, like to set their own agendas, rather than following someone else's, so they're not suitable for every kind of mission. As a result, inventors have looked for ways to design tiny remotely controlled or autonomous aerial vehicles.

Tiny aerial surveillance devices have advantages over large and medium-scale UAVs. They can fly at lower altitudes without violating FAA guidelines. They fit in places other things can't. With intercommunication abilities and collision-avoidance, they can team or swarm, and they are easy to transport. They are also easier to fabricate and more expendable than large million-dollar craft.

When objects are very small, their flight dynamics change. Paper airplanes fly quite well, with a bit of wrist action, but below a certain size, the wings don't provide enough lift and the plane quickly plummets.

The Nano Hummingbird has been mentioned in the chapter on optical surveillance. The tiny remotely controlled UAV has proven itself capable of maneuvering through doorways and windows and even doing a vertical backflip. It is one of the most interesting designs so far in small UAVs.

When UAVs are scaled down even smaller, they run into some practical hurdles. It becomes harder to install mechanical parts that move, or to include radio transceivers and other sensors used for surveillance. Stability and the ability to maneuver are more difficult to achieve. Gusts of wind can send them in erratic paths or quickly ground them.

Significant progress in tiny flying machines is being made, however, at a number of research labs, based on studies of the characteristics of insects.

Insects as Models for Micro-Aerial Vehicles

"Winged insects were solving obstacle avoidance, odometry and navigation problems long before there were any humans on Earth, and we have good reason to believe that their guidance principles could provide roboticists with innovative ideas."

Franck Ruffier and Nicolas Franceschini, IEEE International Conference on Robotics & Automation, April 2000

Imagine a tiny robot that can mimic the flight of insects.

In France, the Biorobotic Research Group has developed a vision-based auto-pilot system that enables a Micro-Aerial Vehicle (MAV) to take off, cruise, and land using a motion sensor as a regulator for the optic flow. The idea was modeled after the photoreceptor cells of the eyes of houseflies.

U.S. defense agencies and the National Science Foundation are interested in the reconnaissance and environmental-monitoring capabilities of tiny aerial robots, and have been funding a number of projects to further MAV technologies.

Research at the Harvard Microbiotics Laboratory has resulted in the Mobee, a tiny robot made of carbon fiber, a material that is both light and strong. It can stay aloft by beating its titanium wings like a bee, up to 30 beats per second.[14] The next step is to provide a way for MAVs to maintain stability and steer.

To this end, Harvard researchers have created an automotive-style differential the size of a pinhead that could enable MAVs to asymmetrically move their wings or other joints. There may be many design possibilities that enable a robot to fly effectively.

At Johns Hopkins University, researchers have been using high-speed cameras to shoot series of high-resolution images that can be compared frame-by-frame, to study the movements of butterflies and fruit flies, to discover how they perform their aerial acrobatics. The research on the aerodynamics of different biological structures and how they maneuver could be invaluable in improving MAV stability and maneuverability.

---

[14] See the *Optical Surveillance* chapter for more information on the Mobee.

At Johns Hopkins, Tiras Lin has been using a suite of high-speed video cameras and mathematical analysis to study the flight dynamics of insects to discover how they use their bodies to maneuver in the air, information that could potentially contribute to better stability and navigation systems in micro-aerial vehicles. [News videos clip courtesy of Johns Hopkins University, released Feb. 2012.]

At the University of Pennsylvania, students have developed tiny heli-robots that can intercommunicate and fly together without colliding. The robots look like tiny helicopters with four routers, with the routers providing both loft and control. By varying the speed of individual rotors or pairs of rotors, the motion of the UAV changes. When it is flying, it buzzes like a bee and looks like a dragonfly doing aerobatics. It can fly through doorways, windows, and hoops, and do backflips and other aerobatic maneuvers.

The Quadroters can autonomously built a structure based upon a blueprint, rather than relying on piloted remote control, and play music while navigating a room full of instruments.

Since MAVs are very small, there's a limit to what they can carry. If they can cooperate, they could potentially work together, just as children can cooperate in carrying something heavy by putting it on a blanket and each lifting a corner.

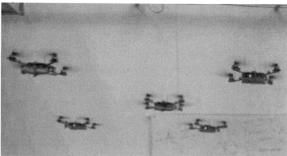

"Quadrotor" MAVs developed by Alex Kushleyev and Daniel Mellinger, with Vijay Kumar as project leader at the GRASP Lab, University of Pennsylvania, can perform aerobatics, cooperative tasks, fly in formation, avoid obstacles while in formation, lift things together, build structures, and play music using preprogrammed instructions that can be interpreted dynamically as the robots fly. [Clips from news video, courtesy of the GRASP Lab, University of Pennsylvania, released January 2012.]

The developers at Penn Engineering envision teams of robots facilitating search-and-rescue missions, but it doesn't take much imagination to realize that flying robots with these capabilities could transform our world in many other ways.

As can be seen by these examples, research and development in tiny flying robots is booming. There are many reasons why MAVs are suddenly in the news and a hot area for research.

- Improvements in microelectronics make it possible to build smaller, lighter modules, and to equip MAVs with more sophisticated radio-wave transceivers and computer smarts.

- Improvements in battery technology make it possible to keep a small vehicle in the air longer. Operational time used to be ten or fifteen minutes. Now some fully working systems can stay airborne for more than an hour.

- GPS can improve location and navigation systems and help operators prechart a flight path on a ground control computer before launching the MAV. With some systems, it is possible to make adjustments to the flight plan during flight. Improved radio-frequency components aid, as well.

- Development of more sophisticated software algorithms enable MAVs to fly autonomously or to make location-specific independent decisions such as collision-avoidance. They can cooperate and communicate with other nearby MAVs, even if they are being remotely piloted.

Commercial MAVs

While research continues in labs, companies are selling MAVs for a variety of purposes.

There are commercial systems similar to the University of Pennsylvania Quadrotor that can hover and maneuver by individually controlling the four routers. In 2007, *microdrone, GmbH*, released video of a Quadrocopter MAV weighing just over two pounds that can fly to altitudes of about 500 yards and remain airborne for up to 20 minutes. The United Nations UNOSAT has used the MAVs for environmental and disaster assessments, and to more efficiently deploy emergency response resources. Newer versions can fly for more than an hour.

In 2011, microdrone announced that the quadrotor or Quadrocopter had helped archaeologists capture images of an ancient burial mound, almost 3000 years old, in northeast Russia. Since MAVs can maneuver near the ground and capture imagery from many angles, the photos can be used to construct a 3D model of the burial area.

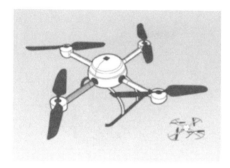 

Microdrone Quadrocopters have been adopted by UNOSAT for gathering surveillance images for humanitarian missions. They have also have been tested by Thyssengas as an adjunct or replacement for traditional helicopters for monitoring high-pressure gas pipelines. The Quadrocopters are, for the most part, independent, once they are airborne, and are equipped with optical imaging devices. [News illustration and photo courtesy of microdrones, GmbH, 2011, released.]

Quadrocopters have also been tested for police work and a decision to purchase the systems was announced by the German Minister of the Interior. The German authorities envision the MAVs filling in a gap between ground-based cameras and traditional helicopters for gathering search-and-rescue, reconnaissance, and forensics intelligence at lower altitudes.

The developers of the Quadrocopter have a sister company in the U.K. that provides a system based on the four-rotor design of the Quadrocopter except that the U.K. system is remotely controlled. The MAVs are utilized by police and fire departments and other government agencies.

The Quadrocopter is based on the model of the helicopter. There are also commercial MAVs based on the design of winged aircraft.

In additional to fire and police surveillance, the Microdrones (UK) MAVs can be used for pipeline inspection, emergency services, documentation/preservation of historic and archaeological sites, and military reconnaissance. [News photos courtesy of Microdrones (UK), Ltd., released.]

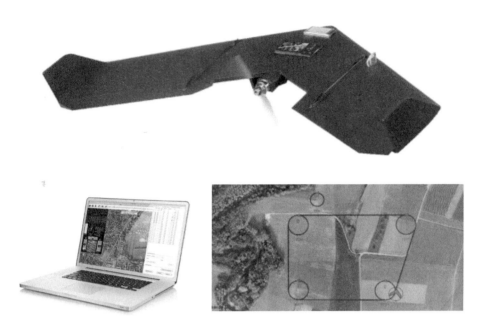

The senseFly, LLC, swinglet CAM looks like a tiny stealth plane equipped with a camera. It weighs only 500 grams, can fly for up to half an hour, and links through a USB radio-wave modem. The flight plan (bottom right) is programmed via computer by providing points on a map that correspond to Global Positioning System (GPS) waypoints or by entering the GPS points directly. The photos can be tagged with geo-points (bottom right) as long as the UAV is within communications range of the ground control computer. [News photos and flight plan video clip courtesy of senseFly, LLC, 2010 and 2012, released.]

In February 2012, the United Nations used a swinglet CAM with infrared and visible sensors to image settlements in earthquake recovery projects in Haiti, reporting that it provided the authorities with "the most detailed aerial imagery ever acquired over the city."

The United Nations also provides photos of other humanitarian projects, including satellite images of refugee camps to track the numbers of tents increasing and decreasing, to assess the local political climate and situation of the refugees in the area.

Maps taken over a period of time aid in damage assessment from battles and natural disasters. Many of the UNOSAT surveillance images and maps are available for download on the Internet.

Military MAVs

The U.S. military has deployed UAVs since the 1950s, in a variety of sizes. In 2002, the U.S. Marine Corps evaluated the Dragon Eye, a UAV that fits in a backpack as a mission take-along.

The Dragon Eye weighs about five pounds, has a wingspan of less than four feet, and is assembled from five pieces. It can stay aloft for about an hour and is launched via bungee-catapult. It can monitor whatever is up ahead, or over the hills in rugged terrain, including potential hazards rigged with explosives.

Top: Cpl. R. Broome, a forward scout with a U.S. Marine Corps reconnaissance battalion, holds the Dragon Eye during a pre-flight check. The Dragon Eye is equipped with two realtime video cameras that are capable of imaging in grayscale, color, and infrared night-mode. A three-person crew operates the craft from the ground using a laptop and radio controller. A flight plan can be preprogrammed and adjusted in-flight, if necessary, by the ground crew. Right: During a training exercise in Jordan, in 2006 (bottom right), Col. R. Johnson observes the Dragon Eye in flight at Camp Al Qatranah. [News photos courtesy of U.S. Marines, released 2000 (photographer M. Hacker), 2008 and 2006.]

The Dragon Eye is equipped with two realtime video cameras in the nose, one pointing forward, the other to the side. Images are transmitted for viewing to the laptop and also to a head-worn monitor from heights of up to 1500 feet or more. Its small size makes it difficult to intercept.

> *"During testing of the Dragon Eye, they had an entire company shoot at it in flight for two days. It only took four hits, but was never shot down."*
>
> Cpl. Joshua L. Britner, as reported by Sgt. Robert M. Storm, Sept. 2005.

In 2005, the Marines tested another aerial vehicle called the Silver Fox. It uses a lightweight catapult to launch and fly to altitudes of 12,000 feet, with autonomous flight time of about 10 hours. Imagery is displayed on touch screens, along with flight plan information, and stored on a removable hard drive. The current imaging system is capable of a full 360-degree view in flight. The system is compact, fitting into a bench-sized case.

The Silver Fox is a high-altitude, high-endurance compact UAV designed to fly autonomously and carry a variety of sensors for day or night reconnaissance. Images are transmitted to the ground control station or relayed to other locations via satellite. [News photo courtesy of U.S. Marines, released 2008. News photos courtesy of Bae Systems, released.]

The Silver Fox has also been tested by the New York Police Department and used by the Canadian and Colombian military services.

Members of the military using various small and mid-sized aerial vehicles have described them as invaluable to reconnaissance missions.

MAV Autonomy

Another reason for the increased popularity of MAVs is improvements in auto-pilot systems. In the past, it took practice to fly remotely piloted vehicles—often they would crash. It was particularly difficult to fly remotely piloted helicopters, since helicopters don't have wings to give them lift if the routers fail or, if there are multirouters, if one or more of the routers is improperly maneuvered.

With autopilot software and hardware, which enables UAVs to steer themselves, more rotors can be added to a MAV-copter to make it more stable, and collision-detection and GPS navigation can help both copter-style and plane-style UAVs avoid the ground or follow a charted course.

Sensors can also be added to UAVs of any size to help determine tilt/acceleration, and other aspects of the aircraft that may not be discernible to a human operator. UAVs are not always in perfect view at all times. They may be briefly obscured by cloud, brush, branches, chimneys, or other impediments. If gusts of wind change their orientation, the operator may not note the wind effect until the craft is unstable. Thus, crashes are more likely to happen. Sensors can reduce accidents by tracking attitude and vicinity changes while the pilot handles general instructions.

Sharing Private and Commercial Airspace

Countries all over the world are scrambling to redefine commercial airspace in a way that will regulate and incorporate UAVs. This is partly because of concerns of the proliferation of UAVs that may cause safety hazards or fly in restricted airspace, and partly because there are many eager to exploit the commercial potential of UAVs who want access to civilian airspace.

It is not known how changes in FAA regulations will affect residential privacy. If permissions for flying UAVs are lenient, commercial companies will soon be mapping neighborhoods in minute detail using small aerial surveillance products to supplement street-view maps with bird's-eye maps. While it will give us an unprecedented high-resolution view of the world from low altitudes, it may become impossible to enjoy a hot tub or full-body tan in the back yard without being watched.

.

Audio Surveillance

Key Concepts

- *bugs and other common audio surveillance devices*
- *logging and recording of calls*
- *controversies over wiretapping and other forms of audio surveillance*

Introduction

Spy movie "bugs" are the classic audio-surveillance archetype—special agent Bond peering under a lamp to see if anyone is listening, or a stock investor tapping into a telegraph cable in the 1800s.

Bugging and wiretapping used to be technically simple—contact was made with the wire carrying the transmission, or a small microphone was placed near the source of the sound. Most wiretapping was by individuals eavesdropping on gossip or business news, or by law enforcement agents investigating crimes. It was mostly a local phenomenon.

Now wiretapping, digital eavesdropping, and other forms of audio surveillance monitor cell phones, landlines, online audio conferences, and other forms of domestic and foreign communications, building up information storehouses related to possible future crimes or future entrepreneurial opportunities.

Audio surveillance is no longer a local phenomenon, either physically or temporally. Videocams and cell phone microphones can be activated remotely, even if the owner isn't actively using the device at the time. In this way, common consumer items can be used as room monitors since most gadgets are now multifunction devices. With features such as Global Positioning System (GPS), it may also be possible to pinpoint the location of the device and the person carrying it, without him being aware he is being tracked.

Audio surveillance goes beyond the monitoring of rooms and phones. We are rapidly reaching the point where every conversation, whether in chat rooms, through online services like iChat™, AIM™, Skype™, or via landlines or cellular phones, is either logged for time/date and caller, or recorded for its content.

Basics of Audio Surveillance

The Nature of Sound

Audio surveillance encompasses sounds humans can hear and sounds outside the range of human hearing. Room bugs operate within human hearing. Sonar, which is widely used in marine applications, is mostly outside human hearing.

Sound waves propagate in materials that have the ability to compress and decompress. Since materials vary in density and flexibility, some transmit sound better than others.

Sometimes electricity is used to promote the transmission of sound or to amplify it. In other cases, the sound waves are converted to electrical or optical signals, transmitted, and then turned back into sound waves so they may be readily understood at the receiving end.

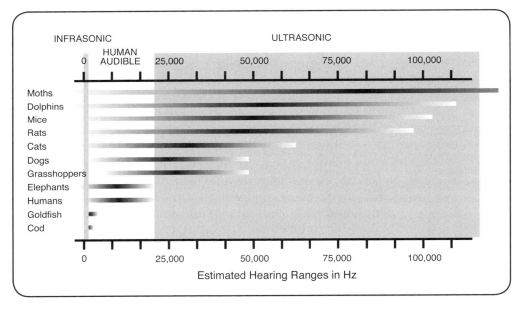

Sounds outside human hearing include infrasonic (below human ranges) and ultrasonic (above human ranges). Elephants can sense infrasonic sounds, and many mammals and birds can hear ultrasonic frequencies. Birds can also distinguish sounds played closer together than humans. Dolphins use a form of biological sonar to echo-locate and to communicate and to "see" through certain obstacles and visualize what is behind them—somewhat analogous to a fetal ultrasound scan. [Illustration courtesy of Classic Concepts © 2000, used with permission.]

Human hearing ranges are quite narrow compared to insects and other animals, but we are specialized for generating and understanding speech within that narrow range and we make good use of it—spoken communication is a powerful tool.

Aspects of Sound Useful for Surveillance

The simplest form of audio surveillance is human eavesdropping—cupping an ear, listening through a peephole, sidling up behind conversants, listening in on a second telephone, or putting an ear to the floor. Technological eavesdropping is popular, however, because a bug is small and easy to hide and can be used to record a conversation any time of day or night. A recorded conversation is also more accurate than human memory and is better evidence in legal suits than hearsay.

Wearing a wire is another form of audio surveillance used by investigators or journalists. It involves wearing a microphone that is interfaced either wirelessly or by wires to a transmitting device, and sometimes also a recording device. Magicians and entertainment "psychics" often utilize wires (tiny radio transceivers) to glean information about an audience from associates who are mingling with the crowd.

In the movies, someone wearing a wire often transmits to associates across the street in a van equipped with sophisticated receiving instruments. In real life, the sound is often recorded on a tiny tape recorder or flash memory stick. Investigative journalists sometimes wear audio transmitters or recorders when investigating organizations that are suspected of illegal activities or of exploiting their members.

Audio Content

Audio surveillance doesn't always mean listening to words. Sometimes it means transmitting a sound and listening for an echo to see if something is there, or bouncing a sound and timing how long it takes to return to estimate distance. Since sound travels well through water, it is an invaluable tool for marine applications and is the basis for sonar sensing.

Sound is sometimes studied for its character rather than its content. For example, sound can be used to recognize an individual, or to determine if someone is lying. Stress affects the way humans speak in subtle ways that can be picked up and magnified by instruments. While stress detectors based on sound may not be completely accurate, the information may be useful when combined with other observations.

Sound may also be used to detect changes in an environment such as a lab, parking lot, building, or wildlife refuge. It serves as an auditory environment sensor to trigger an alarm or other events such as turning on a camera.

Sometimes sound is converted to text. In broad-scale surveillance, there may not be enough staff or time to listen to the character or content of monitored sounds. To save resources, software may be configured to detect changes in sound dynamics (e.g., a quiet area becoming noisy), pitch or expression, or specific keywords if conversation ensues. A computer may record sound characteristics as statistical graphs or may transpose speech into text that can be logged or emailed to an agent who will follow up with more direct involvement in the communication.

Keyword monitoring may also be done directly from text communications such as email or text-chat sites and may be logged as text or converted to sound so it can be heard by staff working on other things. If agents hear something of consequence, they can turn their full attention to the conversation.

Common Ways to Apply Audio Surveillance

- *To indicate changes in an environment,* and optionally to trigger an alarm or other devices.

- *As an indirect source of information* by converting sound characteristics into graphs or alerts, and text into speech so that it may be more accessible to certain people or get the attention of someone near a computer who isn't always watching the monitor (or the software window in which the activity is occurring).

- *As a source of keywords, or other information* that draws attention to the sound, or activities associated with the sound and, optionally, conversion of the speech into text.

- *As a direct source of information*, as from eavesdropping, room bugs, wiretaps, prior recordings, or direct access to online audio chats or phone conversations.

- As an imaging technology to see through or behind something. Ultrasound can penetrate certain materials and make other structures more visible.

Sound can also be used to start other surveillance devices in the field. Many computer operating systems are equipped with speech recognition that enables hands-free operation for certain applications. The same can be applied to small electronic devices like cameras or sensors.

As an example, imagine a wireless bug planted in a room. If it is broadcasting all the time, there's a higher chance it will be detected. If it recognizes significant words, it can start and stop based on certain commands, or can be programmed to respond to certain keywords, record for a specified length of time, and then turn off again until the next triggering event.

Applications of Audio Surveillance

How is audio surveillance commonly used?

The most common application is eavesdropping, either by getting physically close to a conversation or using a microphone to transmit to a remote receiver.

Private detectives and professional wiretappers, such as the infamous Bernie Spindel (an associate of Jimmie Hoffa in the 1950s), have stated that the majority of bugging contracts are to glean business intelligence or information regarding personal relationships. Since 9/11, we can also add more aggressive surveillance of terrorist communications by the Intelligence Community.

Law enforcement agents use audio recordings to monitor conversations such as crimes in the planning stages or crimes in progress. Political candidates often try to monitor the competition. People in business listen in on meetings of their competitors, and spouses and exes often look and listen for evidence of cheating, gambling, or expenditures that aren't part of a couple's shared agreement. Even after they part, some people will monitor an ex to wreak havoc on his or her life and future relationships.

There is a brisk commercial business for tiny audio recording devices that are easily hidden in clothing, briefcases, or accessories such as pens or portfolios. Clearly many people are buying them. How many are using them is not known.

Cell phones have been used to spy on people, and have sometimes been turned on by kidnap victims in the hopes that someone would hear the conversation and rescue the victim. Many PDAs and tiny computers can be used the same way and small devices are increasingly equipped with location targeting systems so they not only can be heard, but can be found.

Cameras have long been a security staple at large-scale events like concerts and sports meets but now sound monitoring is used as well.

Controversy erupted when it was announced that law enforcement would use closed-circuit TV, and microphones that could pick up conversations up to a distance of 100 yards, to monitor crowds at the 2012 Olympic Games in London. The "smart" microphones were interfaced with a system designed to pick up changes in tempo or pitch that might signal emotions such as aggression.

In November 2011, discussions were held about using remotely controlled aerial vehicles equipped with high-definition cameras and thermal imaging devices. The public accepted that a certain amount of security for the games was necessary, but the continual extension of surveillance in a way that increasingly intrudes into personal conversations is causing a stir.

Unlike analog recordings, where each copy degrades the quality of the recording, digital recordings are clones and include the same information as the original as long as no lossy compression algorithms are used.

Telecom Surveillance

The possible abuse of audio surveillance is not a uniquely western problem.

Surveillance technology was purchased from China by Iran's largest telecom, in 2010. People have expressed concern that the technology could potentially enable broadscale monitoring of national communications.

Iran has ancient ties with China, including the intermingling of royalty in past centuries, so it is not surprising China would provide commercial support to Iranian industry despite sanctions against Iran by other countries.

The U.S. only permits the sale of humanitarian products to Iran at the present time. Critics are concerned that telecom surveillance could be used to suppress democratic ideals in Iran. Noteworthy, however, is that the system provided by China includes many components by well-known American companies. A Chinese spokesperson commented that the only motivation for the deal was to help Iranian operators upgrade their networks.

Audio to Supplement Other Surveillance Methods

Audio devices are sometimes used for associates to communicate while monitoring a situation in other ways. Detectives or patrol cars posted at different locations might relay their observations to one another or coordinate a chase. Sportscasters and referees might coordinate observations of a controversial play while overseeing a game. Search-and-rescue workers might provide audio reports of visual or sensor observations to someone on a rock face, in a cave, or attempting an avalanche rescue some distance from the other team members.

Surveillance is always used for less noble tasks. For example, at a past political convention, one man visually monitored people from the rafters, looking for potential voters who might be drunk. If he spotted one, he wirelessly contacted an accomplice on the floor whose job it was to sidle up to the "mark" and see if he could influence or buy the vote.

Such tactics are illegal, but they happen.

Audio Analytics

Audio analytics is digital processing of sound to improve its quality or informational content or to select pertinent sections to emphasize, catalog, or store. It may also be used to test the integrity of audio streams or recordings and to selectively suppress or enhance various aspects of the sound.

Just as video analytics is dramatically changing visual surveillance, audio analytics is transforming sound surveillance.

In the past, most audio surveillance involved a human covertly eavesdropping on a conversation or recording it on tape. The quality might be poor and environmental noises could make it difficult to interpret the sounds.

Digital surveillance provides a significant set of tools to capture better sound quality, as well as making it possible to adjust, analyze, or repeat sections in realtime.

Here are some of the ways in which audio analytics can be applied to audio surveillance:

- Software can pre-learn an environment so that if something of interest occurs, it will recognize the change in sound dynamics and create an alert or initiate a recording.

- Audio filters can analyze the environment and selectively apply noise reduction to sounds such as traffic or wind.

- Memory buffers can enable something important to be repeated while the device or computer continues recording.

- Multiple tracks can save different versions of the captured sound, including raw audio, filtered audio, flagged audio, and audio otherwise processed to make it more meaningful.

- Sound characteristics can be represented visually with graphs.

- Targeting or recognition software can match a sound with known patterns (e.g., birds or vehicles) or with a voiceprint from a particular individual and may optionally call up and display a picture of the speaker.

- Intelligent algorithms can make informed guesses about the content of a marginal recording, offering options as to what might have been said in a weak spot or a section where other noises intruded. While it cannot be known whether the guesses are correct, the information might aid the listener in interpreting information.

- Alerts can flag a sound of particular interest based on previous specifications or a match with sound characteristics in a database.

- Sections that might be more important than others can be flagged by the listener so they can be more readily found later.

- Stress level monitors can assess voice characteristics that might indicate a speaker is lying.

- Multiple microphones can be monitored, with priority given to ones that are selected by the listener or by the software as more significant or more likely to yield useful information.

- Multiple microphones can also be used, in conjunction with parameters such as loudness, to geographically estimate the source of a sound.

Speech-to-Text Software

One of the tools useful in surveillance is the conversion of speech to text. Often transcripts are needed of audio conversations, and speech-to-text can automate the process. Even if the conversation isn't transcribed perfectly, converting and having a human clean up the text is often faster than typing it from scratch.

The more affordable consumer software falls somewhat short of what is needed for good conversion, however. Even with hours of training, most off-the-shelf programs are only about 40–60% accurate, despite manufacturer's claims.

It is a significant programming challenge to correctly interpret speech, considering the vast differences in the ways people enunciate words, in addition to speaking at different tempos, different pitches, and with different accents.

Simple words can be recognized, if spoken clearly, and can be understood by automated telephone menu systems, but surveillance is rarely simple or spoken clearly. Often plots are whispered or private conversations are held at a distance, or in noisy street settings or cafés. Having the speakers train the software is usually out of the question.

Using software to first screen out the clutter, then to amplify, and then to make sense of the words is a technical challenge, yet developers claim they are getting closer to practical systems.

In 2011, Microsoft announced that they had significantly improved recognition of the building blocks of speech by taking a "neural network" approach to interpretation.

SOCIAL EVOLUTION OF AUDIO SURVEILLANCE

Eavesdropping

The tendency to eavesdrop is common among humans. Curiosity about what others are saying, doing, or thinking, motivates people to listen in on their conversations.

One of the reasons audio bugging or wiretapping is considered unlawful, even if one of the parties being monitored is suspected of a crime, is because other people involved in conversations may be innocent and thus entitled to their privacy. Even the suspect is presumed innocent until proved otherwise.

Technical Standards for Wiretapping

"We must ensure that law enforcement has the tools to fight crime and protect our citizens in the digital era, while at the same time safeguarding the public's legitimate concern for privacy."

FCC Chairman William E. Kennard, March 27, 1998

Thus, federal statutes were implemented in the 1930s and, more stringently, in the 1960s to prevent tapping of phone lines and to require law enforcement agents to get warrants or "super-warrants" as they were sometimes called, to engage in wiretapping. Britain didn't implement a wiretap law until two decades later.

At first, only the most serious crimes were considered for court-approved taps. Over the years, however, the list of qualified crimes has become broader (just as the list of qualified persons subject to collection and storage of DNA evidence has become broader).

Regulation of wiretaps, the tapping of communications lines, has a long history extending back almost a century. When it began, most taps were through wired lines, although radio enthusiasts were turning to the airwaves to intercept conversations. Since then, wireless communications have grown, and digitally based Voice-over-IP (VoIP) has become an increasingly important source of audio communications intelligence.

Left: Before software-automated wiretapping was possible, there were many allegations of manually installed illegal taps. In 1938, Louis R. Glavis described how wiretapping of department telephones was carried out, but denied that White House wires had been tapped.

Right: Bernard Spindel (professional wiretap expert commonly seen going from job to job with a black briefcase, screwdrivers, and various phone components rigged with bugs) and James R. Hoffa were acquitted in 1958 of charges of tapping the phones of union subordinates. Spindel was strictly a freelancer not interested in working for law enforcement but carrying out a brisk business nonetheless, in days when tapping was slow, primitive, and required physical installation. Now technology makes widespread tapping from remote locations not only possible, but relatively easy. It is hard to imagine how much eavesdropping happens now given that technology facilitates the process by orders of magnitude. [Photos courtesy of the Library of Congress, Hoffa photo by World Telegram & Sun photographer Roger Higgins, 1957, public domain.]

One of the earliest efforts to regulate communications and protect privacy was the *Federal Communications Act (FCA) of 1934*. This prohibited interception and divulgence of wired or radio communications. Although it was mainly intended to protect broadcast conversations, the courts interpreted it broadly such that it prevented unapproved wiretapped conversations from being admissible as evidence.

During the war years, J. Edgar Hoover argued for broader powers to tap, and allegations of improper tapping emerged.

The pendulum swung back again with *Title III* of the *Omnibus Crime Control and Safe Streets Act of 1968*, often called the *Wiretap Act. Title III* set out regulations for obtaining and administering taps. The act was passed into law as a result of allegations of improper tapping by both government agencies and private citizens.

Some aspects of the *Wiretap Act* have since been amended through the following regulations[15]:

- the *Electronic Communications Privacy Act (ECPA) of 1986,*
- the *Communications Assistance to Law Enforcement Act (CALEA),*
- the *USA Patriot* reauthorization Act, and
- the *Foreign Intelligence Surveillance Act (FISA) Amendments Act of 2008.*

One of the most important consequences of the amendments is a broadening of the terms of a tap to require telecommunications service providers to build their systems so that law enforcement can remotely access communications to perform a tap. Physical access to a premises or to the lines leading to the premises to install a tap is no longer necessary for many forms of taps.

Legislative Efforts to Safeguard Privacy

"Mr. Speaker, I have recently introduced H.R. 2551, the *Privacy Assurance Act of 1989*. This legislation is necessary because the tremendous scientific and technological developments during the last several decades has made possible today the widespread use and abuse of electronic surveillance techniques, rendering much of the current law governing individual privacy ineffective, and largely unenforceable."

Ronald V. Dellums, House of Representatives, Nov. 1989

There was an attempt to ban the sale of devices that enable communications interception, but the regulations did nothing to stem the tide of small listening devices available to both law enforcement and the public. Sale of these devices is a multi-million-dollar industry.

---

[15] These acts are described more fully in Section 5.

Regulation of the public's use of surveillance devices tends to focus on capturing eavesdroppers after the fact rather than preventing commercial interests from selling them, partly because many devices are sold under euphemistic names that purport they are for purposes other than spying.

As a result of the proliferation of spy devices, a number of bills addressing privacy, not only from police taps, but from general spying by people with illicit devices, were proposed through Congress.

In the case of law enforcement use of taps, court approval is required, however. Installing legal taps requires a warrant and warrants are typically only granted after other investigative tools have failed to provide enough evidence to make an arrest or conviction. The warrant process is not easy and taps can only be in place for a limited time. Given these restrictions, law enforcement agents are always lobbying for an easier process and broader powers to tap.

The federal government is not the only governmental body seeking to facilitate law enforcement use of taps; local politicians are interested in streamlining the process as well. In 2012, the mayor of Muskogee, OK, was seeking to simplify the process of getting a warrant to wiretap without having to get approval through Oklahoma City, and to increase the kinds of crimes that qualified for a warrant. He assured the public he was not advocating warrantless wiretaps.

This did not appease everyone, however. Some felt that reducing oversight would weaken the checks and balances that keep wiretapping from spilling over its legal bounds.

INTERNET-BASED VOICE COMMUNICATIONS

Digital Voice

Phone conversations can now be transmitted over the Internet. VoIP is a process of encoding spoken voice into digital data so that it can be readily sent over computers that utilize *Internet Protocol* (IP).

This means it can be received in digital format in an audio file (e.g., MP3 or any other common audio format) and stored on a hard drive. VoIP calls can also be re-coded, if necessary, to broadcast as spoken voice on a variety of traditional landlines. Since phone communications are largely digital these days, very little translation is needed and there is the possibility that VoIP may supersede traditional services.

At the present time, many VoIP services cost only half of their traditional counterparts, and the quality of the communications, and ease of use, continually improve. It is now possible to buy a telephone that plugs into an Internet router and works like a normal wireless phone, letting the user accept or dial calls, through subscription services. The telephone interface to the Internet handles the negotiation of the call over the VoIP service.

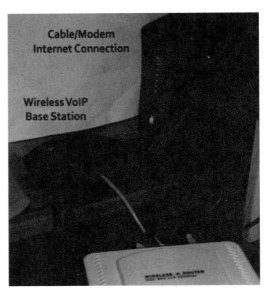

VoIP doesn't have to be communicated through a computer microphone or USB headset. It can also be accessed like a traditional phone. This GE handset is Skype-certified to communicate through a base station and Internet connection in a transparent manner. The base station, right, communicates wirelessly with the phone and interfaces via Ethernet to an Internet DSL or cable modem to send conversations digitally over the Internet. [Photos © 2012 Classic Concepts, used with permission.]

VoIP also provides voicemail capabilities—if the recipient of the call isn't online, the callee can set preferences to store the call, the same as if it were a telephone answering machine or phone company voicemail service. Unlike most answering machines, the user doesn't have to permanently erase calls to make room for more—they can be downloaded and stored on a hard drive or DVD.

VoIP is often more flexible than traditional services, allowing users to discontinue or continue their subscriptions with few or no penalties, and users can choose whether they want incoming or outgoing calls to landlines. If they don't need those services, VoIP is a bargain—it is often free to call from one computer to another. Peer-to-peer links are part of what make the system affordable—users contribute bandwidth.

This has made VoIP very attractive to consumers and has increased the desire of law enforcement to be able to monitor VoIP services.

Why is this significant to surveillance?

VoIP is digital and digital files are easy to store and reproduce. Someone who uses VoIP services often has an archive of voicemail messages on the service provider's hard drive. To access the files, which are not much different from other kinds of data files, law enforcement officers may be able to subpoena the service provider for information about the digital files rather than going through the process of applying for a warrant to tap. A tap would still be necessary for live conversations and content, but might not be needed for archived date/time/sender logs. A telephone tap of a traditional phone can provide useful information even if the conversation has ended. The same is true of VoIP messages.

VoIP and Privacy

VoIP users became concerned that their privacy might be at risk when Microsoft announced, in May 2011, that they had acquired the popular Skype™ VoIP service. Since much of the communication on Skype is peer-to-peer and encrypted, this might seem alarmist, until you consider that in June 2011, Microsoft was awarded a patent for "silently recording conversations" on VoIP networks. It is not yet known how the company will use the patented technology, but the two together might add up to a ready-made surveillance system.

Skype subscribers aware of both the buyout and the patent fear that their communications might be used for testing "silent" VoIP recording technology or for covertly logging their conversations. Not all Skype calls are of the same nature. Some are sent or received on landlines, which involves going through other service providers. The encrypted peer-to-peer model doesn't necessarily extend to every destination point.

With digital communications, users have no way to know if silent recording is engaged. The service provider/subscriber relationship is based on trust and all the control is in the hands of the service provider.

There was also concern that law enforcement agents might look the other way if Microsoft overstepped its legal bounds with regard to recording because law enforcement has a vested interest in getting access to conversations if persons under suspicion are using VoIP services.

Microsoft Patent for A Silent "Recording Agent" for VoIP

"As used hereafter, the term VoIP is used to refer to standard VoIP as well as any other form of packet-based communication that may be used to transmit audio over a wireless and/or wired network. For example, VoIP may include audio messages transmitted via gaming systems, instant messaging protocols that transmit audio, Skype and Skype-like applications, meeting software, video conferencing software, and the like."

U.S. Patent # 20,110,153,809 granted June 23, 2011

Not all the controversy was about surveillance, however. Other companies, such as Cisco Systems, were concerned that the purchase of Skype could result in the service being combined with Microsoft's Lync system to move toward a monopoly in videoconferencing services, if the services were developed further with proprietary rather than open standards. Cisco appealed a European Commission decision regarding the merger.

Proprietary development and a possible monopoly may have consequences for users, as well. If the service becomes popular enough, pushing out the competition, then Microsoft is in a position to steer users to their own operating system and could conceivably drop support for Linux and OS X. If enough people on other operating systems are paid subscribers, they might not want to, but since many Skype users take advantage of the free peer-to-peer computer communications, it is conceivable that Microsoft could stop support for competing operating systems.

Since the acquisition of Skype is recent and much of the concern about monitoring of user calls is speculation, let's pretend for a moment that Skype is owned by a humanitarian organization, concerned about user privacy and maintaining above-board relationships with law enforcement agencies, with no intention of pushing out other operating systems, then the next question is whether the system will be secure against outside surveillance.

Skype is a multinational service and each country has different privacy and call-interception laws. Callers outside the U.S. have many of the same concerns as U.S. citizens do about privacy, but security and democracy are not stable in many of those areas. Skype users are concerned that governments might use Skype to monitor their calls.

Videoconferencing for Audio/Video Surveillance

Skype supports videoconferencing and computer-to-computer calls are free. It is also possible to have multiple accounts so one can be used with a video camera and the other for accessing audio/video images remotely.

Given this easy accessibility, many people have set up laptops with video cams or IP-compatible dedicated video devices in nurseries, businesses, and other locations, to look in on the premises through Skype. Skype has an option to automatically answer incoming calls, so that the remote camera can be accessed whenever desired. There are even firms that bundle all the components needed to create a Skype-based audio/video surveillance system.

VoIP surveillance systems can be used indoors with both audio and video. However, due to eavesdropping laws, cameras with audio may not be permitted in public places without first obtaining permission from those being recorded.

Phones as Multifunction Devices

A phone is no longer just a phone. It is an entire communications arsenal, incorporating voice, text, scheduling software, databases, music players, GPS capabilities, and games.

Many of today's phones are ultra-portable computers.

Thus, users can "talk" to each other in different ways, through text chats, forums, microblogs like Twitter™, audio chat and video chat. If someone tried to monitor a person's stream of thought from a few feet away, it is no longer as simple as eavesdropping on oral phone conversations. Five minutes of voice might be followed by a quick video chat, followed again by the interchange of some still pictures, followed by a few lines of text.

Tracing this meandering communications path from a distance would be difficult unless the actual datastream from the phone could be wholly captured and stored. Capturing data is possible, if an ISP (or law enforcement officer with remote access to the communication) captured all the data and used software to reconstruct the communication as it was displayed on the user's handset.

It is also possible to capture the packets if they are transmitted via unencrypted radio-wave signals, as from a minimally secured wireless "hot spot" provided as a free service. Someone nearby with packet-sniffing software that rebuilds the various media formats, such as text and graphics could see the datastream in much the same way it would display on the user's phone.

Mobile Phones for Audio Surveillance

Smartphones are multifunction devices, so there are many ways to use them to spy. The internal microphone can be left on near a conversation of interest, or paired with software to record—or selectively record if certain keywords are spoken.

Software developers have been quick to develop new surveillance applications that exploit microphones and other components built into the phones. In fact, phones are such powerful devices, they have replaced many traditional forms of covert surveillance, such as room bugs. With spyware installed on the phone, the person carrying it may be broadcasting everything he or she is doing, including location (through GPS), conversations, Web surfing, and Internet chats.

The person who installed the spyware can then monitor the information remotely through the Net. Spyware can even be used to remotely turn on the microphone so the phone becomes a "room monitor" for transmitting audio.

Software to accomplish surveillance is readily available online and can be installed manually by someone who has access to the phone, or remotely, if the user clicks on an email link without realizing it activates the download and installation of the spyware.

Mobile Phones that Spy on Computers

Smartphones can be set down next to someone on a bench or at a table as an audio eavesdropping device and the conversation recorded as a digital file. Smartphones can also sense and interpret vibrations from a nearby keyboard, which is described later in the *Sensor Surveillance* chapter.

RADIO-WAVE SURVEILLANCE

Introduction

Radio surveillance or, more correctly, radio-wave surveillance, is one of the largest categories of surveillance next to visual surveillance and is strongly associated with audio surveillance through the broadcasting and interception of sounds transmitted through radio waves.

Radio waves travel invisibly through the air all around us and may pass through objects such as walls and furniture. They are frequently used in commercial and residential security systems and they enable computers to communicate with wireless routers. They can be bounced off natural satellites like the Moon as well as manufactured satellites orbiting the Earth. Radio waves are like light in that they can travel through a vacuum. Radio signals can be sent far into space.

Radio waves carry news and entertainment broadcasts as well as cell phone and wireless computer communications. Radio waves are used for direction finding and ranging (radar) and for a variety of imaging applications such as weather images. Radio transmitters are being incorporated into cards and clothing and are increasingly used in tracking and tracing people, animals, cargo, and inventory.

Radio communication is a recent phenomenon. Broadcasting as we know it didn't begin until the early 1900s. Since then, however, it has become the most common form of distance communication on the planet and is thus of great interest to those engaged in surveillance.

Regulation of Radio Waves

Radio waves constitute a broad section of the electromagnetic spectrum, but that doesn't mean radio waves are limitless. Since a strong signal can overcome a weak one on a similar frequency, it has been necessary to regulate radio transmissions and assign certain frequencies for particular tasks or to particular groups. The Federal Communications Commission (FCC) is tasked with allocating radio-wave frequencies. Licenses are necessary for sending long-range radio signals. Short-range devices can be operated without a license, but the device itself may require FCC approval before it can be sold.

Often catastrophic events, such as the attack on the World Trade Center or the sinking of the Titanic, will motivate changes in procedures or regulations. The sinking of the Titanic provoked stronger requirements for marine vessels to have wireless equipment and resulted in the *International Radio Convention* and the *Radio Act of 1912* to regulate use of airwaves from that point forward.

Many other significant events happened around the same time, from the first military use of wireless radios to the laying of wired communications cables to distant outposts such as Alaska.

Character of Radio Waves

Radio waves typically travel in all directions from the transmitting source and are not difficult to intercept, so privacy has always been a concern for radio communications in war zones or even on cordless phones used in the home. To protect communications, most radio broadcasts other than public broadcasts are now encrypted and may also be spread through a variety of wavelengths to make the entire message difficult to intercept.

For us to hear radio transmissions as speech, music, or alarms, they must be converted to audio signals in a frequency range audible to humans.

Microwaves, the part of the radio-wave spectrum most distant from visible light, were once considered too short to be useful for commercial communications. They are now important for cellular phone transmissions and for cooking food.

Radio beacons are used to track terrestrial vehicles, aircraft, wildlife, surface-level marine vessels, and sometimes prisoners on parole.

Most people think of television as a visual technology, but the sound and images rely on invisible radio waves to travel from a wireless broadcast station to a local antenna. Optical cables transmitting light have replaced radio waves for TV transmission in many places, but in rural areas where it is difficult or uneconomical to lay cables, television is still transmitted through radio relay towers or satellites.

Adding Communication to Radio Waves

By itself, a radio wave doesn't carry communications. Communications data is "added" to the wave by modulating its form and then interpreting the modulated wave at the receiving end. Amplitude modulation (AM) and frequency modulation (FM) are the two most common modulation schemes.

Eavesdropping

Individual inventors and amateur radio operators have had a significant effect on the development of technology and shaping of the radio industry. They are also well-known for listening in on other people's conversations through home-based systems and mobile scanners. Listening to police communications is a particular favorite of eavesdropping radio fans and some have been known to intercept and relay military communications. Intelligence Community members have also long listened to radio broadcasts around the world.

Eavesdropping on communications has become both more difficult and easier. Many forms of radio communications are encrypted, which makes it difficult to access the content, even if the raw signal is detected and intercepted. In some ways it has also become easier. In the past, usually only a single line or two was physically tapped and the conversations that could be heard were limited. Recent laws require communications service providers to set up their infrastructure in a such a way that law enforcement personnel can remotely access hardware to turn on tapping capabilities and listen to multiple streams of conversations.

Restrictions on Communications

During periods of unrest, governments often restrict radio broadcasting, sometimes turning communications control over to the military.

In the U.S., the use of radio receivers by private citizens was prohibited during World War I, a ban that was lifted in April 1919. Remaining wartime restrictions were lifted in February 1920.

Surveillance of radio-wave communications also increases during times of unrest.

Human Rights and Covert Communications

For decades, amateur operators have been working behind the scenes on human rights issues. In repressive regimes, sometimes the only way to get a message out of the country is through radio broadcasting. When unrest was boiling up in Yugoslavia, a few amateur operators put themselves and their families at risk by broadcasting covert news and distress messages to the outside world through radio. They would sign on, broadcast for perhaps 20 seconds and then sign off again, then randomly sign on again at a different frequency a few hours later, and hope that no enemies intercepted the signals.

North Korea is strictly monitored by government and military forces—getting information in or out is very difficult and punishable with severe penalties. Cell phones and other electronic devices must be checked at the border. Cameras are sometimes permitted, but their use is strictly monitored. Thus, North Koreans get very little information about the rest of the world. A few North Korean amateur operators manage to tune into broadcasts from South Korea, but given the political climate in North Korea, the people listening are sometimes unsure if it is real news or more political propaganda. In dictatorial regimes, it is hard to know who to believe.

Human rights watchers have been listening for radio distress signals around the world for years, but not everyone has access to technology. In many parts of the world, the only fax machine in a 200-mile radius is one at the post office and radio transmitters are almost nonexistent. In the absence of radio signals and sometimes in addition to them, human rights watchers can monitor low-tech areas by analyzing satellite imagery that may indicate village burnings or fleeing refugees.

Community Radio

Until recently, in the U.S., low-power radio stations were restricted, due to claims by commercial broadcasters that the signals might interfere with their programming, but after the *MITRE Report*, a study that contradicts these claims, the FCC has been working to open up airwaves for small stations in local communities in accordance with the *Local Community Radio Act* that was signed into law in January 2011. That's not to say all commercial stations are larger than community stations. Sometimes the situation is reversed.

RFID CHIPS

RFID stands for *radio-frequency identification*. RFID devices are small storage and transmitting components that emit identification information through a tiny antenna using radio waves. They fall into three general categories.

- *Passive tags* can transmit from a few inches to a few feet, sometimes up to 20 feet under ideal conditions. Those used by the U.S. military have a range of about 12 feet. Passive tags are dormant most of the time, but respond if they are queried by an outside device. Passive tags can often function without a battery. Passive tags are used as *proximity* tags.

- *Active tags* transmit continuously or at frequent intervals and, with the aid of a battery, can transmit up to a few hundred feet or farther if more than one relay repeater retransmits the signal. Active tags can be used as proximity tags but, with their longer range, are often used as *vicinity* tags.

- *Enhanced* or *programmable tags* are smart RFID devices that have additional components that can store information, include sensors, or which may be updated remotely to include logs such as shelving information or a short history of recent use.

Basic elements of an RFID access system include an RFID card, tag, chip, or key fob that includes the RFID identification transceiver and a small antenna. Active RFIDs may also include a battery. The reader consists of a sensor/transceiver to query the RFID and interpret the returned ID signal to see if it is valid. The reader can then trigger other devices such as locks, cameras, messages, alarms, alerts, or whatever is appropriate once the tag-user (or tagged object) has been identified.

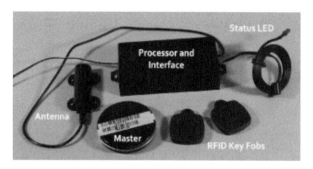

 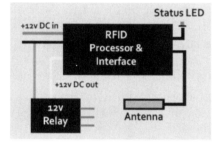

The example shown here is a basic RFID reader and key fob ID system that has a transmission range of a few inches, but there are systems with longer transmissions distances. This RFID system is intended as a simple access authorization device, for a residence or small office, but RFID can also be embedded in other kinds of objects to detect unauthorized activities, such as someone trying to steal a library book or a shoplifter taking a tagged object out of a store without paying and having it deactivated at the checkout. [Photo and illustration © 2012 Classic Concepts, used with permission.]

RFID transmissions are read by radio transceivers, called *RFID readers,* that interpret the ID code and match it up with information in a database.

RFID tags are most commonly used for three purposes:

- *access control in secured areas,* usually by being embedded in an ID tag that is worn by an employee, or attached to objects or vehicles that pass through checkpoints,

- cargo, *inventory, or supply control, and tracking,* and

- *asset identification.* RFID can be used in a way similar to a serial number or barcode, or as an ownership mark (just as silverware used to be monogrammed to aid identification or deter theft).

These waterproof consumer-priced key fobs and access cards are embedded with RFID chips that broadcast a unique 10-digit number at 125 KHz to a nearby RFID reader. This can be used similarly to a password, but without the inconvenience of remembering it or having to key it in. Since a keyring is easily lost or stolen, some systems combine RFID with biometric ID such as a voice print or palm scan. RFID access cards and keychains are also incorporated into some video door phones to provide visual identification together with selective RFID access. Some systems include a master key that can be used to reprogram the individual keys in case one in a set is stolen or lost. [Photos copyright © 2012 Classic Concepts, used with permission.]

In its purest form, an RFID chip doesn't include personal information, medical information, or other data. It is simply an identifier that needs to be matched up with information from a database. It is like a wireless serial number or barcode. As components become tinier and more powerful, however, the potential for broadcasting additional information from smart RFIDs exists.

When combined with a database that logs each use of the tag, or interrogates it on a regular basis to determine its location, RFID can also serve as a track-and-trace mechanism.

RFID isn't limited to object identification and inventory control. It can be used to identify employees with RFID cards or tagged domestic animals or wildlife. There are also some novel applications in the medical field.

Consumer Products and Inventory Control

RFID is incorporated into a number of identification cards and attached to many retail items to track inventory and deter shoplifting. It is not yet inexpensive enough to hide in every retail product, but that day may come.

Since RFID works by transmitting radio waves, it is open to surveillance not only by the original provider or purchaser, but also to anyone nearby with a radio receiver tuned to the right frequency. Even if the signal is encrypted or phased through a number of frequencies, there may be people with access to the manufacturer's receiving equipment who can read the signal covertly.

For the retail industry, RFID has some advantages over barcodes. A barcode lets a retailer know what you've bought when you go through the checkout, but RFID can potentially tell the store much more. In conjunction with information from ceiling surveillance cameras that are now common in many supermarkets and department stores, RFID can reveal what items people pick up and put back on the shelves, or can trigger an alert if someone tries to carry a shoplifted item out of the store. It can also make the checkout process faster. With barcodes, each item needs to be sent through a scanner. With RFID, theoretically the entire shopping basket could be identified in one scan—a convenience that could put most cashiers out of work.

Another way in which RFID differs from barcode tags is persistence. A barcode is usually removed and thrown away and even if it remains on the product, it cannot be identified from a distance. A tiny RFID insert may remain indefinitely, especially in products like shoes, jackets, purses, and hats. Then, as a consumer walks through a mall, RFID-equipped stores (all of which may have bought the technology from the same manufacturer) can use receivers and computer software to build a profile of the consumer's spending habits for as long as the device is able to transmit and could place targeted ads for similar products in the window or doorway of a store as the consumer passes by.

RFID has some practical applications as an inventory control mechanism. A barcode enables stores to track inventory, but RFID takes it one step further by making it easier to locate something that has been incorrectly shelved or boxed, or which has fallen behind other products.

The U.S. military also has a need for organizing and tracking a complex supply chain to military bases and personnel in the field. RFID can help distribute rations, uniforms, weapons, ammunition, and medications, replacing older, manual systems of inventory maintenance and distribution. RFID implementation is part of the Naval Inventory Control Point (NAVICP) effort providing supply support to the U.S. Navy, Marine Corps, and Joint and Allied Forces through the Department of Navy Automatic Identification Technnology (AIT) program.

The main limitation of RFID, other than transmission range, is sorting out the signals if there are many tags in the same place broadcasting on similar frequencies. Radio-wave frequencies, like highway lanes, can result in collisions.

Firearms Tracking

RFID has been proposed as a way to identify and track firearms within prison facilities or other secured areas. An Italian gun manufacturer is adding RFID chips to their firearms to aid in identification or tracking if the firearm is stolen. Maintaining a history of the firearm in a database is intended to deter theft. Presumably this is in addition to serial numbers or other unique identifiers that are already impressed into most guns.

Cargo and Luggage Tracking

In 2009, FreightWatch issued a report that claimed cargo thefts in the U.S. had increased 12% in the last year. Analysts suggested that the dollar value of the thefts may have increased as much as 67%. To reduce or prevent theft, some companies have implemented RFID systems to more closely monitor the location and allocation of goods and cargo. RFID can help locate specific items, and also make it possible to verify whether the stated items are within designated boxes or containers without opening them. It is a less invasive way to scan goods.

By combining local cellular technology (with satellite communications capability) with RFID, Siemens IT announced in 2008 that monitoring of cargo at sea was possible. Similar combinations of local RFID readers and long-distance relay systems for continuous monitoring will probably be offered by other companies.

The same principles can be used to manage luggage in transportation systems and to facilitate their transport to the correct destinations.

RFID is not a magic bullet against crime, however. RFIDs can be stolen, swapped, disabled, or used by thieves to identify packaged targets. They are only helpful as long as their transmissions can be protected from unauthorized scans and that's difficult with radio-wave technology. Encryption can help secure the data so that generic scanners can't read the signals, but if the thief also steals the RFID reader associated with the system, encryption doesn't protect the goods.

RFID also has some limitations. Certain materials can interfere with radio waves, especially very dense materials or those that are similar to chaff (long and stringy reflective materials) that scatter radio waves. There are also limits on how many objects, packed close together, that can be scanned at one time. Sometimes tagging the container works better than tagging individual items.

RFID systems that are integrated with satellite-based GPS also have limitations. Many cargo containers are stacked, or pass through tunnels or mountain passes that might interfere with line-of-sight transmissions.

Despite limitations, RFID can help manage and track the movement of goods in many industries and, with biometric identification as a backup, restrict access to only those authorized to handle them.

Product Protection

Not all RFID chips are for inventory tracking. Some are to identify original manufacturer products so that counterfeits or cheap substitutes are not sold to consumers. Oki® printer cartridges, for example, have recently been chipped with RFID tags to validate installed cartridges, monitor toner, and reset messages to the user. The RFID chips can be reprogrammed one time with an "empty" message as the toner runs out. The idea is to encourage the user to replace the cartridge rather than refill it. Using reconditioned cartridges can damage printers, since part of the drum is often incorporated into the cartridge, and drums have a limited duty cycle. Nevertheless, as soon as printer companies began chipping cartridges, third-party vendors started selling replacement chips so that users could refill their cartridges.

Border Protection

U.S. Customs and Border Protection (CBP) monitors about 300 entry/exit points to the U.S. using a wide variety of technologies. In July 2008, the Department of Homeland Security (DHS) announced that CBP had begun using RFID to expedite frequent crossers in El Paso, Texas.

To streamline border crossings, CBP has developed the NEXUS and SENTRI programs to expedite frequent crossers of Canadian and U.S. borders. After background checks, preapproved travellers are issued photo ID that includes an RFID tag. RFID readers are installed at entry points so that the card holder can wave the card at the device to quickly call up information stored on computers for evaluation by border security for authorization to cross. RFID is complemented with a License Plate Recognition system along the Mexican border. RFID tags attached to vehicle windshields can be read if a vehicle traveling up to 100 mph passes within range of the reader.

In fact, port directors are encouraging travelers to apply for "enhanced" passport cards and drivers' licenses to facilitate border crossings via embedded RFID chips. An Enhanced Driver's License (EDL) incorporates RFID and "multiple layers of overt, covert and forensic security features." Apparently travelers want the convenience, as more than 21,000 were issued in Washington State alone in the first half year of the program.

In January 2012, CBP announced that Enhanced Tribal Card (ETC) would be issued to eligible U.S. and Canadian residents in the Kootenai tribe, which occupies a region that crosses the Canadian/U.S. border in Idaho. The ETC is an RFID-tagged security document that enables travel across marine and land points of entry.

Content Alerts and Consumer Interactivity

Providers of communications devices such as phones and other multifunction gadgets have announced plans to put RFID readers in the next generation of their products. Thus, a phone approaching a movie theater with an RFID transmitter might trigger a trailer for the next film to be shown, or an RFID tag in a zoo might trigger a video presentation about the animal in the closest exhibit. One would also expect advertisers to jump on the opportunity to trigger infomercials on a monitor in a nearby window.

Enhanced RFID and Beyond

Enhanced or smart RFIDs include additional electronics so the tiny components can do more than transmit an identifier code. If memory is added to the system, short-term logging and tracking is possible. Applications for smart RFID cards include media tracking in libraries, copy center services, toll services, or subway access. They can be combined with financial system access to deduct usage points or currencies, and make it possible to develop more sophisticated employee location and inventory-and-supply management systems.

Dot™ platform technology is an open architecture that goes beyond smart RFID tags by including additional components such as a microprocessor and sufficient memory to store about three pages of information. It is, in essence, a miniature computer. Developers can use Dot as the basis of information-enhanced tracking and logging components or systems. In 2010, the technology was contracted for monitoring oil platform workers, with RFID badges relaying on-the-job status and location information, and sounding alarms if someone is injured on the job.

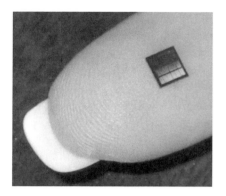 

Left: The Dot™ platform is an FCC-approved open architecture component that has many of the features of a computer, including a processor and memory, combined with radio-frequency communications similar to enhanced RFID tags. This makes it capable of RFID applications that require storage and transmission of information beyond an identification code. Right: A smart wireless sticker can be easily attached to a card or other surface as a logging, tracking, and access identifier tool capable of sensing items up to 1000 feet away. [News photos courtesy of Axcess International, Inc., released]

Variations on RFID using other ways to encode or send a transmission show promise in increasing the range and versatility of traditional RFID chips. One such innovation, developed at the NASA/Johnson Space Center is the SAW ID, based on conversion to acoustic waves. SAW ID can operate in a wider range of temperatures and withstand certain forms of radiation that interfere with traditional RFID.

By combining SAW ID with spread spectrum technology (broadcasting over a range of frequencies), it is possible to read a larger number of closely-packed chips without their signals interfering with one other.

SAW RF tags

Researchers at the NASA/Johnson Space Center have been looking for ways to extend the signaling distance of RFIDs and have created a cousin to RFID in the form of SAW (surface acoustic-wave) ID.

SAW ID uses a pair of SAW RF tags to convert a radio signal into an acoustic wave to "imprint" the ID. The modulated signal is then converted back to radio waves to respond back in the direction of the original interrogation. It is robust, more efficient in terms of power-to-distance ratio, and can sense pressure and temperature changes, making it more versatile, as well. As SAW RF technology develops, it may replace RFID for some applications and find its way into devices that are beyond RFID capabilities.

RFID Implants

RFID technology is so small, it's possible to implant a rice-shaped glass RFID container in a human body. A few years ago, the Federal Drug Administration approved them as safe for this purpose. "Safe" is a relative term, however. Drugs that are generally safe are not safe for everyone—some people can have adverse reactions—and some people may react to RFID implantation, which is an invasive surgical procedure. There is also the possibility of small implants migrating to other sites in the body. The body is a dynamic system, always making adjustments, and people who have stepped on pins or nails or items that are injected under the skin have sometimes had them show up years later far from the original puncture site, or even in the heart or brain. Anything small implanted in the body has to be monitored to be sure it doesn't travel elsewhere where it may be life-threatening.

Depending on the application, implanted RFIDs don't necessarily need batteries. It depends, in part, on the type of implant and the distance between the RFID device and the radio receiver. RFIDs have been proposed as a way to monitor heart health for those at risk for heart disease. In this case, passive transmitters are used, as they only have to transmit as far as a doctor's diagnostic receiver.

For longer transmissions (e.g., as a GPS location device), either a battery is required with the implant, or the person has to carry a bulkier relay receiver on or near the body to amplify and rebroadcast the signal to GPS satellites—which negates some of the purpose of the implant. The battery would have to be recharged periodically, as well, similar to a cell phone battery. Safety devices such as smoke detectors or RFID relays are useless if the battery isn't regularly monitored and maintained.

A less invasive alternative to implanted RFID is a wristband, as is worn in hospitals to identify patients, or even a piece of jewelry made of materials that don't interfere with the transmission, such as resin or decorative plastics.

More information about RFID is included in the chapter on legislative issues.

GPS Surveillance

Key Concepts

- *the evolution of geolocation systems*
- *spread of GPS to consumer devices*
- *GPS tracking*

Introduction   In the past, humans used constellations to help them navigate—both on land and sea. They had to understand the movement of the Earth in relation to the Moon and stars, which changed with the time of day and the seasons and, further, to use instruments such as a staff of a consistent length or, later, a quadrant or astrolabe with marks and measures, to help chart a course.

Now we use manufactured "stars" in the form of orbiting satellites to give us more precise information about our location on Earth and we don't have to look up to do it—we can look down onto any handheld device equipped with a Global Positioning System (GPS) geolocation capability and know the latitude, longitude, and altitude to a few meters, and sometimes closer. The Russian Federation has a similar system called GLONASS and the EU is launching the Galileo constellation to provide more precise positioning than current systems.

The development of GPS was technically demanding. It involved learning to use radio waves to track a constantly moving body. Radio waves are affected by fluctuations in the atmosphere and ionosphere and reflect off a variety of objects and layers around the Earth. Radio waves are also subject to degradation and scattering when sent long distances and they undergo changes in frequency as they travel through different environments. Radio waves exhibit changes similar to changes in pitch when sound passes by—a phenomenon called the Doppler effect. This can be both a hindrance and a help. Once understood, it can reveal characteristics of a transmission.

The origins of GPS have already been described in the brief history, so this section will summarize the basics of geolocation and how it is used in surveillance activities.

How GPS Works

Components

A basic GPS system consists of orbiting satellites, antenna-equipped transceiving ground systems, and a GPS-compatible device capable of communicating directly with satellites or with the ground stations. The U.S. Air Force controls and maintains the system.

GPS satellites comprise a constellation of almost three dozen orbiting satellites equipped with atomic clocks for precise timing. They are positioned around the Earth at an altitude of almost 13,000 miles so they can coordinate with one another and transmit to ground stations with as few obstructions as possible.

GPS was originally a military system but is now open to commercial use and is incorporated into many handheld devices and automotive products. These products have antennas and radio transceivers that can send and receive signals from local relays that communicate with GPS satellites.

There are currently more than 2000 land-based GPS receiving stations worldwide—a number that is expected to grow. Hundreds of thousands of individual handheld or vehicle-mounted devices communicate with these stations.

Location Information

To perform geolocation, several GPS satellites synchronize to transmit a signal to the same point on the ground. The signals arrive at slightly different points in time because each satellite is a different distance from that target point. This discrepancy makes it possible to estimate distance and, with some calculations relative to the shape and position of the Earth, to determine the location of the GPS receiver. U.S. Air Force ground stations track the position of each satellite and the satellites themselves transmit positional information.

GPS can be used anywhere there is an unobstructed means to transmit information to relay stations. Solid barriers, such as metal building materials, cave walls, and other heavy obstructions may interfere with radio signals.

Common Uses

GPS can be used as a location system or as a tracking device to monitor the whereabouts of commercial fleets, marine vessels, and military vehicles in the field. It is also incorporated into guidance systems in aircraft.

GPS is commonly used in navigation and is a basic navigation-control system in piloted aircraft and uninhabited aerial vehicles (UAVs). Many companies use GPS tracking to monitor their trucks or sales reps as they traverse the country.

Since GPS depends on radio waves to make calculations and broadcast the information to users, it works within certain regulated radio-wave frequencies. The popularity of the service is creating challenges since radio-wave bandwidth is not unlimited, and the demand on satellites, which require technical expertise and funding to maintain and upgrade, is increasing.

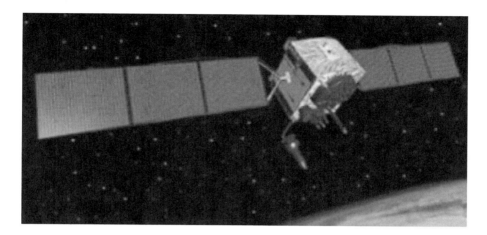

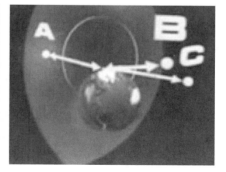

Top: Global Positioning System IIF satellite manufactured by Boeing. Middle: Satellites communicate with ground stations, with the master control station being located in Colorado. Bottom left: A constellation of almost three dozen satellites, some of which are spares, constantly orbit Earth and maintain communication with each other and more than 2000 receiving stations on the ground. Bottom right: Three satellites, in relation to the position of the Earth, are sufficient for a rough estimate of geographical position. By using four or more satellites equipped with atomic clocks, more precise positioning can be calculated. [Boeing GPS illustration courtesy of the U.S. Air Force, 2010. Orbit illustrations courtesy of NASA, 2005, released.]

Today navigation requires no understanding of the heavens for the general consumer of tech gadgets. GPS sensors are built into many devices, such as cell phones and vehicle navigation systems. They can display location as coordinates or, more simply, as a point on a map. Most consumer GPS receivers are accurate to about 15 meters. They can be combined with local wireless systems to provide more precise positioning, as in a worksite or busy shopping mall. This is called Differential GPS and may improve accuracy up to about a meter.

How GPS is Used

The most common uses of GPS are for navigation, stationary geolocation, and tracking of movement by people, animals, or objects. GPS is also an invaluable aid to surveyors and anyone studying terrain such as molten lava or moving glaciers or icebergs.

GPS is used by hikers, boaters, and servicemembers in combat zones for determining location. It also displays navigation on dynamic mapping systems within vehicles and is now standard equipment on commercial airlines.

Many uses are practical, but some are just fun, and developers are always coming up with novel ideas, such as a software application to automatically broadcast a person's location to a website mapping system, or utilizing the internal GPS sensors in a handheld device to turn it into a pointing device for playing games.

Phones as Tracking Devices

GPS is now common on many handheld devices and also in newer cars. The FCC has permitted GPS on mobile phones since 1999, but automotive GPS didn't become widespread until recently.

GPS now enables a phone to tell the user his or her position and may also broadcast that information to others. This positions it as a natural tracking device, especially when consumers load software in their iPhone™, Blackberry™, or Android™ to help them find nearby friends in realtime. If they are sending wireless broadcast query signals (and receiving signals in return) they aren't just broadcasting to their friends, they are broadcasting to the whole neighborhood. This opens the door to someone spying on the entire exchange.

Many people are eager to use GPS services. Teenagers can find their friends in the mall, parents can find their teenagers. Co-workers can find one another and rendezvous for lunch. Even though they could simply phone one another, there's something enticing about probing for nearby buddies, especially those you might not have known were in the area, and seeing pinpoints on a map showing their location.

Analysts estimate hundreds of millions of people will be using this technology by the end of 2012 and dozens of companies have already come out with applications to help people find one other.

Some companies are charging individuals reasonable rates to access their tracking services, but this may change now that websites are offering the service free.

InstaMapper™ is one example of a free service where the user signs up for an account and downloads an application that broadcasts the user's location to InstaMapper servers. The server, in turn, plots the user's location on a Google map with the current time and additional information about altitude, speed, and heading. The map can then be embedded in a Web page, blog, or social networking profile for friends and family (and sometimes complete strangers) to see.

This kind of software will be embraced by friends and family who want to keep track of each other, but is potentially dangerous to youngsters who see it as cool without realizing that stalkers may use information to ambush or kidnap a child. Parents are often busy working and dealing with other children and don't always know about the latest technology (some don't even use computers) and wouldn't necessarily know that their child signed up for a service that tells the whole world where he or she is located.

Search and Rescue

Surveillance isn't always about chasing criminals or spying on ex-spouses. Many surveillance activities are focused on scientific research, mapping, weather forecasting, and public safety.

Top: The MicroPLB™ Personal Locator Beacon (left) works in coordination with SARSAT geolocation data to find a person or vessel broadcasting a distress signal (right). Bottom: Communication with other satellites can help narrow the beacon's location to even more precise coordinates and information is relayed to ground stations so that rescue operations can locate the distressed victim. [Photos courtesy of NASA/ Goddard Space Flight Center, Rebecca Roth, and NASA/Goddard educational video, released.]

Search-and-rescue use of GPS location devices can help find a hiker, downed pilot, earthquake victim, or sailor in distress. If a person needing assistance has a personal locator beacon (PLB), a signal can be sent to the international *Search and Rescue Satellite-Aided Tracking* (COSPAS-SARSAT) system—a constellation of geosynchronous satellites in low Earth orbit. Since its establishment in 1979 by international cooperation among Canada, the U.S., France, and the former Soviet Union, the program has rescued more than 28,000 people.

SARSAT is managed by the National Oceanic and Atmospheric Administration (NOAA) and utilizes technology developed internationally and by the NASA/Goddard space program. Distress beacons from the ground can be picked up by satellites and used to locate a person in need within about an hour.

Improved Systems

Research is ongoing to improve the system further. NOAA, NASA, the U.S. Air Force, and the Coast Guard are developing and testing an updated search-and-rescue system called the *Distress Alerting Satellite System* (DASS) that will reduce the time needed to locate distress beacons to less than five minutes. It is expected to be operational around 2014.

DASS is the U.S. portion of a global effort since 2000 to improve satellite search-and-rescue systems. *Medium Earth Orbit Search and Rescue* (MEOSAR) will add search-and-rescue transponders to newer GPS, GLONASS, and Galileo satellites over the next few years.

One of the advantages of MEOSAR is that large portions of the Earth are visible to each satellite and there will be many satellites, making it possible to provide continuous monitoring of the entire planet.

Satellites in medium Earth orbit (about 21,000 km above Earth) require larger antenna transceiving systems, but researchers are looking for ways to reduce the size of radar-aperture antennas without degrading their performance.

SENSORS

Introduction

Sensors are devices that detect motion, temperature changes, wind velocity, and other indoor and outdoor environmental changes in order to sound an alert, log an event, or trigger a device. The Global Positioning System (GPS) is a form of sensing system, in that it polls several signal sources to pinpoint the location of a device. But there are now sensors that can provide location information that don't rely on satellites for the positioning data.

Cameras are sensors that detect light intensities and sometimes color frequencies, but since they are covered earlier, this chapter focuses on sensors in the nonvisible optical spectrum, such as infrared, as well as nonoptical sensors that detect changes in the environment.

Motion sensors are one of the most common forms of sensors. They are found near doorways, outbuildings, greenhouses, warehouses, and property boundaries, and are often powered by solar panels so that it is unnecessary to replace batteries frequently or run wires out to the device (which may not be near any electrical outlets, especially if the device is on a farm or industrial site). Traffic intersections may be equipped with sensors, either on the light poles or under the asphalt, to detect the presence of cars so that traffic lights can be timed more efficiently.

Wind, temperature, and barometric sensors are used in weather monitoring and forecasting activities. Seismic sensors monitor the Earth's coastlines and fault lines for shifts or shaking that might indicate ensuing earthquakes, tsunamis, or volcanic eruptions.

Sensors are an intrinsic aspect of robotics as well, enabling a robot to interact with its environment to avoid obstacles or pick up or deliver objects to their intended places. Robot "vision" isn't always accomplished with cameras; it is sometimes achieved in other ways.

ALCOHOL SENSORS

DWI Sensors

More than a third of all traffic deaths in the U.S. are related to alcohol consumption, making drunkenness a serious concern to law enforcement.

The American Medical Association has stated that impairment occurs at about a 0.05 blood-alcohol level, that is, approximately 0.05 grams per 100 ml of blood. Most states set the standard for impaired driving at about 0.08 or 0.1 blood alcohol, which is higher than the level at which a driver is considered to be impaired.

Sensors for determining DWI (driving while intoxicated/impaired) and DUI (driving under the influence) are chemical detectors that sense molecules of drugs (the most common being alcohol) when they are blown across the sensing surface. The concept is similar to the chemical sensor "portals" used in airports to sense explosives. Thus, a suspected alcohol/drug user is asked to blow into the device with sufficient volume to get lung air, rather than just mouth air, into the wind channel. This is more practical and less invasive than taking blood or urine samples.

Most breath samplers work on two basic principles, by detecting a chemical reaction or by discerning certain patterns using infrared spectroscopy. Thus, both chemical and optical testing can yield information about substances in a person's lungs.

The devices to test for DUI are not perfect. If they are old, incorrectly calibrated, poorly handled or maintained, or if the person being tested doesn't blow sufficiently hard, they may give inaccurate readings.

A minority of people would be incriminated by a faulty sensor. If it is faulty, it is more likely to give a false negative rather than a false positive. It may be broken and not give any readings at all. Even if it is partially working, a suspect who doesn't blow hard enough will pass less air through and will under-stimulate rather than overstimulate the sensors. However, there may be situations where the device reports insufficient flow (an uncooperative suspect) when in fact, the suspect did blow hard enough—this defect can cause problems for the accused.

DWI sensors are never the sole indicator of whether a person is impaired. Other tests, including behavior, expression, balance, and eye movements, are taken into consideration as a whole.

Nevertheless, faulty sensors, if the defects or misuse can be proven, may render evidence inadmissible in court and undermine public faith in the process of giving a sample to law enforcement.

Bar Sensors

Bar owners have long sought to keep track of how much beer or hard liquor has been dispensed to customers. Now it is possible, with small sensors that fit in liquor spouts and taps on draft beer kegs. When liquid is poured, a wireless radio-frequency signal is sent to a computer and the data can be logged in realtime, giving statistical images of what liquors are being dispensed and in what quantities.

The magic is accomplished with a smart RFID tag that not only transmits the wireless signal, but can monitor other factors such as temperature, acceleration, and the degree of tilt. Tilt sensors are also present in handheld gaming devices and pinball machines.

Tilt sensors can be used to detect tampering. If someone tries to pry or pick up something they shouldn't, an alert can be triggered that the object was moved.

Tilt sensors can also be integrated with GPS to determine which way an object is pointed. If combined with mapping software, a handheld device could tell you which way to go to find the nearest club or café.

ENVIRONMENTAL SENSORS

Sensors are widely used in environmental monitoring. Some examples include:

- avalanche-prediction to protect skiers and snowmobilers,

- weather prediction,

- industrial sensors to warn of toxic leaks or spills,

- information for fishing boats to locate fish stocks and stay away from impending storms,

- warning systems for natural disasters,

- surveillance information for patrol boats and submarines,

- generation of data used by fish and wildlife managers to protect resources and endangered species, and

- security systems for labs, clean rooms, zoos, museums, and other institutions that need to maintain certain environmental parameters to protect the contents from theft or damage due to heat, humidity, or other factors than can harm artifacts, works of art, or industrial components.

Sensors aren't always land or vehicle-based; many are attached to floating buoys that communicate with satellites to send the information to land- or marine-based research centers or command centers.

MOTION SENSING

Motion sensing can be accomplished in several ways, but two of the most common are optical passive infrared (PIR) devices and vibration sensors. Infrared sensing is discussed in the *Optical Surveillance* chapter.

Vibration Sensing

A vibration sensor is useful in situations where people might try to avoid triggering optical motion sensors. Optical motion sensors need a clear line of sight to detect movement, so it is harder to hide them than a vibration sensor.

Wall-mounted motion sensors can create alerts or signal other devices to engage, but they only work if the motion is within the sensor range. If someone is aware of the sensors, it is sometimes possible to get past them. Vibration sensors can be used to detect movement in flooring, walls, and sometimes windows. If someone tries to break in, the alarm can detect them even when they avoid cameras or other visible sensors. Vibration sensors are sometimes used as backup security on safes and vaults, since thieves sometimes disable nearby cameras. They can also be used on boxes that shield museum exhibits or in retail showcases with higher-cost items. Weather-resistant vibration sensors are incorporated into perimeter fences.

Vibration sensors are helpful in situations where it is difficult to mount a line-of-sight optical sensor, as on a car, motorcycle, or moped.

One of the capabilities now built into phones is a "tilt" sensor, the same technology used to measure tilting on the handle of a beer keg to monitor how much beer is dispensed. The tilt sensor can also be used to point at things to play games or to provide location information.

Tilt sensing is accomplished with an *accelerometer*, a device to measure movement. It is also sensitive to vibrations, such as the vibrations people create when they type. Thus, a mobile phone can be used to spy on a person's computer usage, if the phone is sitting near the keyboard. In newer phones, like the iPhone 4, that include gyroscopes to clean up the signal from the accelerometer, accuracy improves. Vibration diminishes as it travels, in many ways it is like an acoustical wave, so the phone needs to be near the keyboard.

A team of researchers at the School of Computer Science, Georgia Tech, has created a program that enlists accelerometers inside smartphones to detect and decode vibrations from nearby keystrokes with accuracy up to about 80%. Patrick Traynor explains how this enables a phone set casually on a desk or coffee table to monitor computer activity. [News photo courtesy of Georgia Tech, Oct. 2011, released.]

Vibration sensors are also used in warfare to detect intruders and in industrial applications to monitor machines to make sure they are correctly calibrated and functioning so they don't damage products or harm people operating them.

Seismic Surveillance

Seismic sensors are a form of vibration sensor developed for monitoring earthquakes, tsunamis, explosions, or volcanic eruptions—an ongoing form of surveillance that occurs around the world using *seismographs*—instruments that show patterns of movement in the Earth's crust. Historic seismographs typically recorded a row of continuous lines on a strip of paper or negative. Each line or set of lines represented a different detector.

 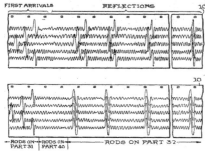

A seismograph senses motion and enables movement from earthquakes, tsunamis, volcanoes, and bomb explosions to be sensed and recorded. On the left, a historic seismograph from around 1912 is examined by three men. Right: A 1943 patent seeks to improve upon the prior art recordings shown here. Seismic events are shown as larger movements of the inking arm, or *perterbations* in the lines that are constantly drawn along a paper strip

Early seismographs translated physical movements directly into marks on the seismograph. Later, optical seismographs that measured interruptions of an optical beam were also introduced. By the 1930s, electronic seismographs based on vacuum-tube technology were being used.

Now, seismic events can be recorded digitally and viewed on computer monitors.

The biggest breakthrough in seismic surveillance didn't come in the form of physical and electronic improvements to the instrument itself, however, but to the ability to network seismic sensor data together worldwide, to reveal patterns of movements within the Earth's crust related to each other and the cyclic nature of some of the seismic events.

Global surveillance may not provide a crystal ball in terms of predicting major events, like earthquakes, but it does provide a fuller understanding of the dynamics of Earth movements and better information on which to judge the probability of something happening. It also provides a better communication network in terms of warning people when something has happened, like a major tsunami following an eruption.

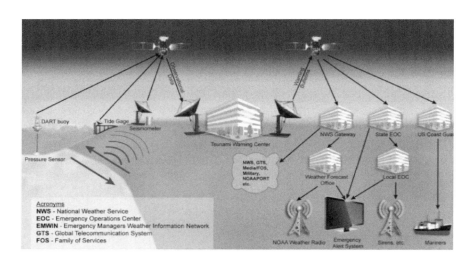

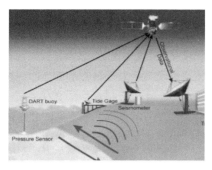

Seismic monitoring and warning systems are now much more sophisticated than historic seismographs, incorporating a satellite communications network to link up various weather , Coast Guard, and emergency alert agencies. [Photo courtesy of the Chicago Daily News, Inc., copyright expired. Illustration and photo courtesy of the National Oceanic and Atmospheric Administration/NWS, released.]

Seismographs are often used in combination with pressure sensors and other sensors, to give a clearer picture of what is happening in unsettled geological zones. Since seismic sensors are located in diverse environments around the world, including marine platforms, the data can be sent by satellite to control and data-storage center with banks of video monitors to visualize the geography and incoming sensor data.

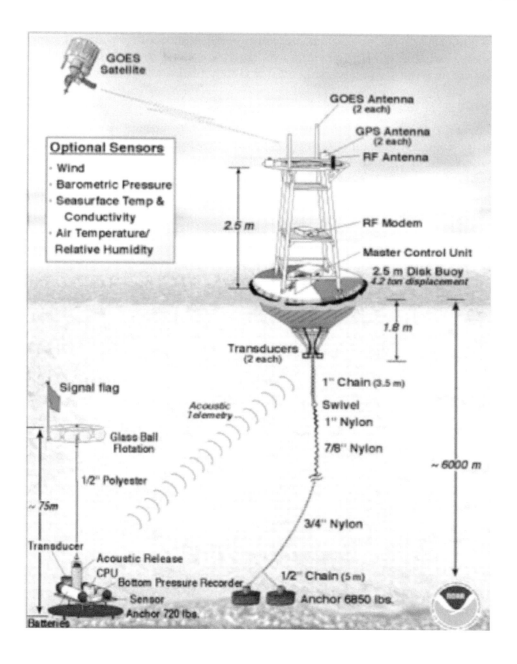

The Dart Mooring System is a good example of the wide variety of sensors that are used to monitor the environment and how the data can be transmitted via acoustic telemetry (through the water) and wireless satellite transmissions (through the air) to ground-based stations to send alerts about impending tsunamis or other unusual cataclysmic events. Many of these same technologies are used in building security applications and archaeological research. [Illustration courtesy of the National Oceanic and Atmospheric Administration/NWS, released.

Seismic monitors are sometimes installed on museum exhibits to protect artifacts, valuable jewelry, or works of art. If someone touches the case or box enclosing an exhibit, it triggers an alert and may turn on a security camera.

Pressure Sensors

There are two general kinds of pressure sensors: those that sense atmospheric pressure (barometers) and those that sense pressure from physical contact by substances heavier than air/gases. Barometers are commonly used in weather surveillance, pressure sensors are often used to sense pressure in liquids.

Pressure sensors can be combined with transmitters to send the pressure data to another location. Pressure sensors are extensively used in nuclear power plants and are useful in other industrial plants where pressure is an important factor in maintaining machines and making sure nothing is about to expand or explode. They are also helpful for calibrating machines in a number of industries.

Pressure sensors are built into tires in new cars, so the information can be wirelessly transmitted to the vehicle's computer system to warn the driver if the pressure changes or is set incorrectly for the tires. Unfortunately, any time an object is transmitting radio-wave signals, it is betraying its presence and components that are controlled by radio signals can potentially be hacked and controlled by someone or something other than the vehicle computer associated with the tires. In other words, wireless subsystems in vehicles can be subject to outside surveillance or tampering. Since tire sensors include unique IDs so that the vehicle can tell the driver which tire might be losing pressure, the information can also be used to track a vehicle.

Sometimes pressure sensors are activated when the pressure is reduced, rather than increased. This makes them useful for alarm systems. If a pressure sensor is placed under a keyboard or window that opens upward, and someone picks up the keyboard or opens the window, the pressure is decreased and an alert can be sounded. Changes in air pressure can also be used to monitor environments and trigger alarms. Opening a door can change the air pressure and indicate that someone may have gained unauthorized entry.

In museums, various sensors are used not only for security, but for protecting exhibits from damage or deterioration. Museums use temperature and humidity sensors, as well as sometimes using pressure sensors under exhibits to sound an alarm if they are handled or stolen.

Motion and tilt sensors are common in car alarms, but air pressure sensors are sometimes also included.

Location Sensing

The satellite-based GPS and the Global Navigation Satellite Systems (GNSS) are the best-known location-sensing systems. GPS is now built into many mobile devices and is especially prevalent in mapping applications and cell phones.

Most location sensing relies on satellites and cell towers to determine the location of a wireless device. Thus, at least three sensors are usually coordinated to get a rough location (hence the term "triangulation"), and more can increase accuracy. By calculating the distance between three or more points surrounding a device, a close estimate can be made as to where it is and also, with repeated polling, whether it is moving.

But satellites depend on radio signals and radio signals are blocked by certain physical structures and scattered by others. It is difficult to carry out location transmissions in caves, dense urban streets with few cell towers, and inside buildings. Sometimes satellite location is used in combination with indoor wireless systems to get a rough estimate that is then refined by the local system, but this has limitations too in that the local system must be installed and maintained by the property owners or those leasing the space and only extends as far as the FCC permits the wireless signals to travel.

More precise indoor location has some interesting applications. It would be possible for a shopper in a mall to find a product on a website and then trigger a "take me there" event that displays a map on a mobile device, like a phone, that walks the user directly to the store and shelf where the product is stored.

Visitors to museums, zoos, or amusement parks would never have to worry about getting lost and could easily find washrooms, medical aid rooms, and exits. They could also have information about an exhibit displayed interactively on their device and perhaps even spoken to them with audio files or text-to-speech applications.

COMPUTERS AND INTERNET SURVEILLANCE

KEY CONCEPTS

- *data repositories*
- *capabilities of data storage-related surveillance*
- *online computer surveillance*

Introduction

Computers are an intrinsic aspect of recent breakthroughs in surveillance technologies. From desktops to laptops, to multifunction phones and "smart" surveillance cameras, computer circuitry is fast being incorporated into even the tiniest devices. Smartphones are now as powerful as computers were a few years ago and are used in many aspects of surveillance.

Computers can serve as standalone data storage archives, or as "operations centers" for devices that are communicating from other locations. Laptops are often used as base stations in the field.

Computers act as relay stations to facilitate the transfer of information, or as firewalls for selectively deciding which pieces of information may pass through or which people may access an application or archive.

One of the most common uses of computers for surveillance is the creation and maintenance of data repositories that include personal or business information about an individual or firm. For example, consulting the *Better Business Bureau* database might indicate whether a business has had complaints. A search of the *National Instant Criminal Background Check System* can yield information on prior criminal activities if a person applies for bonded certification or purchase of a firearm.

A computer can be used to check a log of local or network activities or to access stored archives of mug shots, fingerprints, voiceprints, and DNA profiles.

COMPUTERS AND SPYING

Computers are spied upon and used to spy on others. With the Internet interconnecting millions of computers, opportunities for spying have greatly increased.

In the 1970s, *dumb terminals* (terminals with limited processing capabilities) were linked to mainframes that handled the central processing and ran shared applications. It was rare for desktop computers to be connected peer-to-peer. Gradually, system administrators and hobbyists created small local-area networks (LANs), and purchased circuit boards so they could dial community bulletin-board systems with a "suction-cup modem" that transmitted telephone tones, to text-chat with other computer users. Most computers, however, functioned alone.

Businesses were slow to interconnect desktop computers, partly because they had millions invested in mainframes and partly because of concerns about data security.

The most common way to spy on a computer at that time was to sit down and snoop through the files. If the computer were password-protected, the spy would look in nearby file cabinets or waste baskets to see if anything related to a password had been written down. Some dove in dumpsters to locate personal information that might provide clues—the password was often the user's name spelled backward, the name of a spouse, or a birthdate.

Computers networked by phone lines talked to each other by sending tones. The same concept could be used to make unauthorized telephone calls by bypassing the operator and thus the long-distance toll charges. Devices to generate tones to control the telephone connections were called *blue boxes.*

It was assumed that college students were the ones using blue boxes to steal long-distance services, but a study by a telephone company discovered the primary users of blue boxes were businessmen and well-paid professionals.

People in business still make heavy use of surveillance devices. Sales of spy devices on the Net are booming, and reports of businesses trying to offer bribes or steal stock tips, trade secrets, and private business communications are common news stories. Businesses also contract other firms to find business intelligence for them, often on the Web.

Governments and military organizations routinely use computer networks for finding and relaying intelligence information, as well as data that may be relevant to surveillance at some future date. Many national and international data store-houses are currently in use and more are created all the time.

The general population also engages in regular surveillance. Residences use motion detectors, IP cameras, and other devices that can send alarms or video through the Net. They watch children in nurseries and the people caring for them. They clandestinely spy on friends or relatives with cameras, cellphones, social networks, and stolen passwords to private accounts.

With the proliferation of handheld devices that are essentially small computers, there are many new ways to gain access to cyberspace.

COMMON WAYS TO BREACH COMPUTER SECURITY

Wireless Interception

Web 'cookies'
Forums
Social Networks

eMail Links

Keystroke Monitors

Smartphone Sensors

Spyware

There are a multitude of ways to breach computer security. This illustration shows some of the most common methods to gain access to a person's communications, files, Web activities, or system configuration information through Web-tracking software, wireless interception, packet sniffers, keystroke monitoring, spyware, and the various sensors built into powerful handheld devices like smartphones. [Illustration © 2012 Classic Concepts, used with permission.]

NETWORK SURVEILLANCE

By the 1980s, computer networks were catching on. Macintosh computers were networked so teachers could communicate with students and monitor their activities, and businesses were interconnecting smart desktop computers through miniframes, rather than relying on mainframes and dumb terminals.

Clumsy suction-cup modems were replaced by digital modems that were faster and easier to use. Once people realized the practicality of networks, they spread quickly. Most early networks relied on wired connections but, by the 1990s, there were many combinations of wired and wireless systems.

Even businesses concerned about data security began using wireless connections when data encryption became available.

Encryption was an important milestone in the balance between computer surveillance and personal privacy. There was a government plan to require security chips in people's computers to enable law enforcement or intelligence agencies to "unlock" encrypted files and look at them. The plan withered partly because of public opposition, but mostly because of the development of freely usable encryption schemes and the pleas of businesses to be allowed to create their own strong encryption to compete with those developed by other countries.

Laptops were once too expensive for the average person to buy, but prices dropped significantly in the 2000s. Soon laptops and netbooks costing under $500 were more powerful than $5000 laptops from 1999. At that point, they became popular consumer items.

Businesses eager to attract millions of new mobile computer users began offering free wireless hot spots in Internet cafés, coffee shops, and bookstores. Many of these free links to the Internet were unencrypted and spies started using *packet sniffers* and other data-snooping tools to monitor wireless laptop communications from a sidewalk bench, a quiet corner of a restaurant, or roving street vans.

Communication between computers requires a physical connection. Even a wireless connection is "physical" in the sense that transceiving antennas designed to resonate at specific frequency ranges require stimulation by electromagnetic waves.

Communications also require software protocols that both devices can understand. The most common protocol for LANs or devices that communicate with the Internet through routers and modems is called *Ethernet*. One of the ways companies can increase data security on LANs is by using less common operating systems and protocols.

Wired Communications

Most computers are interconnected with fiber-optic cables, copper wire, or wireless links.

Of the wired links, copper wire is easier to tap. Breaching the sheath that protects the wires and making electrical contact enables signals to be picked up or rerouted. Some instruments can read the electromagnetic emanations from the wire by holding it close to the wire without piercing the sheath.

It is also possible to insert taps at junction points such as phones, wall jacks, building-mounted junction boxes, and the point where the phone line interconnects with a city telephone switching grid. Taps can also be set up at the phone company's switching station, with a proper court order.

Wired taps are used to intercept both telephone conversations and computer data transmissions.

Fiber-optic cables are harder to tap than copper wires, but it is not impossible. Fiber-optic cables transmit light rather than electricity. Light wants to spread out, like the beam from a flashlight, and fades rather quickly. Laser light, which is light projected in a longer, straighter *coherent* line, is more suitable for projecting and reflecting through fiber optic channels. The signal may need to be refreshed occasionally—amplified by electrical junctions—but then it can be routed through more optical cables. At the business or residence, the light-signal is translated into electrical signals by a DSL or cable modem so the computer can understand them. Some computers are equipped with direct-connect internal fiber-optic interface cards. Fiber optic transmissions are vulnerable in places where the optical signal is converted to electricity or where cables are directly connected to computers, cables that could be temporarily rerouted. It's not possible to read optical signals from outside the sheath as with some electrical emanations but, with a little more technical expertise, the optical signals in an exterior cable could be diverted and read.

Wireless Communications

Wireless signals are transmitted in a variety of ways. The most common is by radio waves, but infrared optical networks also exist.

Unencrypted wireless links are easy to intercept since radio waves travel in all directions. Encrypted wireless communications are not difficult to intercept, but they may be difficult to decode.

Wireless communications enable computers, printers, and PDAs to intercommunicate—to relay information from one device to others or from one site to another. Since wireless signals become weaker over distance, they are sometimes relayed through wireless bridges, cell towers, and satellites, depending on the location and content of the communications.

In residential systems, a simple transceiver inside a laptop or desktop computer can link to a router in another room which, in turn, may link to a DSL or cable modem. Most handheld devices communicate through radio signals, although some, like remote controls, use infrared.

For field work, or military communications, portable wireless antennas in a variety of shapes and sizes can be temporarily installed.

Left: U.S. Air Force crew adjust a microwave communications antenna atop a communications tower to provide wireless network access. Right: This transceiving dish was installed in Kandahar, Afghanistan, to provide wireless in-the-field telephone and Internet communications via orbiting satellites. [News photo by Staff Sgt. Mike Meares and Sgt. James Harper courtesy of the U.S. Air Force, 2011 and 2009, released.]

Wireless communications are the target of many surveillance activities. If the communications are not encrypted, it is relatively easy to spy on them. Even if they are encrypted, it is sometimes possible to breach them. Even the strongest encryption schemes are broken by hackers who can't resist the challenge, or by security specialists who are testing the integrity of encryption systems.

In May 2010, Google, Inc., admitted in an official announcement that they had been collecting data from wireless networks from camera-equipped vehicles that travel the streets of more than 30 countries, collecting photos of everything within public view of the van, much of which is uploaded to the Google Street-view™ service.

Google isn't the only company that has monitored radio transmissions emanating from houses and businesses. Most wireless routers broadcast up to about 300 feet and Class 2 Bluetooth devices to about 30 feet, which is often far enough to reach the street. Thieves don't need to unencrypt signals to know something worth stealing might be sending the signals.

Encryption can help secure communications but it is not always sufficient. To make signals even safer, *spread spectrum* schemes have been developed to move the signal through a variety of wavelengths so it is difficult to capture the entire communication. Imagine watching traffic through a small hole in a fence (a specific wavelength on a wireless scanner). You would only catch momentary glimpses of the cars, even if the same ones are driving back and forth. This makes it difficult to determine information about the entire route traveled by a specific car. Someone monitoring a spread-spectrum transmission would have no way of knowing when to change the frequency or which specific frequency to scan.

Infrared is a more secure way to communicate wireless signals. Infrared can be sent in pulses to represent ones and zeroes. Because infrared is a form of light that doesn't pass readily through solid surfaces, it works through *line-of-sight* to make connections. A person trying to intercept the data generally has to be in the same room as the signals. Since infrared is less convenient than radio-frequency (RF), most people use RF.

SPYWARE

Spyware is software installed on a computing device to snoop on data. Whether computers are connected by wired or wireless communications, spyware can be inserted into their systems to log activities such as keystrokes, mouse actions, Web surfing, and communications via chats and videoconferencing. Some of these actions may reveal usernames and passwords that permit private areas to be unlocked locally or on the Web. Spyware is often configured to send the information to a remote location when the person logs onto the Internet.

Spyware can be loaded onto computers through Web links or links embedded into email, without users realizing they've been "infected." It is also possible to plant spyware directly on a computer or smartphone, if the person installing it has access and enough time to download the software into the device or to click on a Web link to initiate a download. Imagine someone leaving their phone on a table in a bar while they order a drink, dance, or visit the washroom.

Sometimes spy software is made known to the person using the system. Employers may keep keystroke logs of activity on workspace computers and are required to inform employees if they are logging their actions or monitoring them in other ways.

Spyware can be installed on someone's netbook or smartphone if they've left it unattended for a minute or two, especially now that smartphones and some computers are equipped with Quick Response (QR) code readers.

There have been allegations that a British-made spyware application was used by Egyptian state security to monitor and record personal Skype communications. There are also reports that the German police held meetings with the FBI, the French secret service, and other international law enforcement agencies about the use of surveillance software on computers. Court-approved taps for installing spyware for investigations had apparently been granted in Germany.

Digital Communications

A significant portion of computer surveillance is the monitoring of online communications, such as VoIP[16] calls, public chat-room conversations, posts on forums, blogs, and Internet-connected community videocams. Some of these are open conversations, but the people involved may not know the data is put together with information about them from other sources on the Web.

Much of the information on the Internet is private, but there are constant revelations about private areas not being as private as people were led to believe. Hackers breach security, and disgruntled associates steal passwords and open private accounts. There are also many business mergers that bring together separate databases that people assumed would stay separate.

Computers connected to the Web are not as safe as people think either.

Some operating systems have only one-way firewalls to protect Internet users from intrusions. Those with two-way firewalls are sometimes configured to be open (insecure) by default. The new user has to search for the settings and change them. Since many people don't know what a firewall is, they don't know their computer is vulnerable. This is one of the reasons for the dramatic increase in malware on people's systems. If security settings are open, computers get infected, especially those with operating systems that are inherently insecure.

It is also possible to turn on someone's videocam from another location without the user realizing the cam has started broadcasting over the Net. The applications are installed in the same way as spyware and may be difficult to detect.

Users can improve data security by setting up two-way firewalls and disallowing anything other than essential services. It is also advisable to accept cookies only from sites visited (not from third-party sites that link through the host site), and to block popups whenever possible. Turning off JavaScript can often help also. If a favorite site uses Java or JavaScript for games or other purposes, it is usually possible to *whitelist* the site by entering an *exception* URL to the preferences list so that other sites are still excluded.

---

[16] Wiretapping and VoIP are discussed in more detail in the Audio Surveillance chapter.

QUICK-RESPONSE CODES

Quick-Response (QR) codes® are starting to appear on business cards, doors, books, and many other places. Many people don't pay attention to them, and some have called them an "eyesore" but QR codes are changing the way people input information into PDAs.

A QR code is similar to a barcode—it is a high-contrast online or printed image that can be interpreted with an optical scanner and software. A QR code makes it easy to enter information without typing—just download the image or visually scan the code to download it to your computer or PDA and the software interprets it for you.

QR codes are license-free, which is why they are spreading quickly, and hold far more information than traditional barcodes, up to about 4000 characters of mixed letters and numbers. QR codes are commonly used in the manufacturing industry and for transferring business card information beyond what will fit on a normal business card.

QR codes were originally developed by Denso Wave, Inc., and are now an ISO standard. Tiny scanners can read Micro QR codes.

QR encoding of the letter A (left) and the number one (right). The coding system holds up to several thousand characters and can be read from a distance if the code is large enough and reasonably clear. QR codes can be used to instantly download computer instructions, URLs, text, small images, and other information to a computing device. By building in higher error-correction rates during the encoding process, design elements can be added to QR codes without making them unreadable. [QR code images courtesy of Kaywa™, open source.]

QR codes make it easy to input Web addresses (URLs), names and addresses, quotations, stories, coupons, information coded into tickets or menus, labels on products that include information about the company or product, a location on a map, cash register receipts, or almost anything that can be stored in a reasonable amount of space.

You could also mark belongings with a QR code, to include information about an email address or phone number and how big a reward you'll give if someone finds your lost item and returns it. QR codes could be attached to family photos to describe an event and the people in the picture.

What do QR codes have to do with spyware?

They make it easier to plant surveillance software on someone's computer or phone. A person with a QR code in a pocket can pick up a smartphone at a party, office, or home, scan the QR code into the device and insert a spyware network link, program, or instruction in seconds, rather than the minutes it takes to type or download manually.

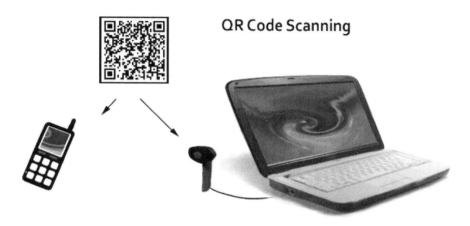

QR codes provide a quick and convenient way to enter pre-coded information into a computerized device, which makes it possible to instantly upload spyware that can lurk in the background and gather information about what's on the storage device, or log information about user activities and potentially send it to a remote location through the Internet. On a laptop, the code could be scanned with a USB scanner or, if the laptop is equipped with a videocam, by the camera built into the computer. [Illustration © 2012 Classic Concepts, used with permission.]

You don't necessarily need a hand scanner to decode QR codes. QR codes can now be read directly by security cameras.

In November 2011, Philex Enterprises, Inc., announced they had incorporated QR code-reading capabilities directly into their surveillance systems. If the code was in view of the camera, whether it was an analog or digital video stream, the software would scan it, log it and (if desired) trigger an event, depending upon user settings.

In the future, the capability will be built directly into smart cameras and would extend the distance at which codes can be read. Soon, every security-camera system and PDA may have this capability. Walking by a QR code would be enough to bring up the coded information on a PDA. It would serve, in a sense, as a visual RFID without the need for radio-wave transmissions, and conveys more information than RFID.

Smart RFIDs still have the advantage of being unseen and transmitting through physical objects, but RFIDs plus QR codes would turn the environment into a beehive of constant information transfer. The mall scene in *Minority Report,* with targeted ads popping up and addressing Chief John Anderton by name as he walks through the building, is no longer science fiction.

The most obvious application for QR codes is tags on store doors or windows to broadcast a sale or other information to nearby PDAs, but many other possibilities exist, including the transmission of safety instructions, directions to a nearby event or attraction, movie showtimes, or a few bars of music.

THE INTERNET AS AN INFORMATION REPOSITORY

Many companies and individuals scour the Net for information regarding others. Some of the information is privately archived in law enforcement or commercial databases, some is offered free on the Net to entice viewers and increase traffic counts to a site, and some is sold to subscribers on an individual-purchase, monthly, or yearly basis.

Rarely is the person whose private information is being archived or sold informed of the use of the information, asked for consent, or given an opportunity to correct errors or opt out. Even if they opt out, the information is usually already "in the wild" (being downloaded or replicated in other databases) and people are often magically opted in again when the host of the database does maintenance updates or implements policy changes.

The person whose information is sold, even if he or she doesn't mind being listed in commercial directories, is almost never compensated with a percentage of the profit.

How can companies take all this sensitive information and store it, display it, and sell it? Public information used to be available only to a very limited local audience, or to people who expended travel money and time to visit a courthouse or other repository to see it. Often they would have to show up in person at a government office and present ID to access paper or computer files. This was a more laborious (and secure) process than accessing a database online.

Thus, laws to protect a person's privacy didn't anticipate that an individual's public records, intended for a local readership, might someday be broadcast worldwide.

The information originally intended for a local public is now aggressively harvested by companies wanting to make money. They provide searches for a fee, or free access in the hopes of earning ad-click revenues.

Most employers now search the Web for information on job applicants, or contract the information from companies that amass large amounts of data on anyone who might be a potential job seeker.

Law enforcement agents regularly download information, seek out suspects or their associates on forums and in chat rooms, and subpoena Net-based companies for logs of Internet activity. A variety of databases, described later, are managed by the police to help them do their jobs.

Databanks and Databases

A significant part of surveillance involves gathering data and filtering through it to find something relevant and worthwhile. A detective may have to watch hundreds of people entering and exiting a mall or train station to locate a specific person or to sense behavior that might indicate something about to "go down."

Since it is difficult to be in two places at once, a security camera connected to a recording device sometimes backs up human observations. Sometimes a video feed goes directly to a local police department so a number of people can apply their expertise to assessing a situation. Images, conversations, and other data are increasingly captured in digital format, which makes it straightforward to save them on a memory stick or hard drive so they can be consulted later.

Data banks and databases are storehouses of digital information. They may consist of pictures, text, video, or a combination of these. A data bank doesn't necessarily have an organizational structure. For example, data captured from a digital streaming video camera may be saved as a series of long files without organizing them in any way, since it is often not known if they will be needed later.

Databases, on the other hand, are generally organized to be searched and sorted so the information can be readily found when needed.

Even before computers, law enforcement depended heavily on data stored in file folders and file cabinets, but it was often difficult to search a roomful of documents for something specific.

Rows of file cabinets with mug shots, fingerprints, wiretap logs, and crime scene photographs have always been an important resource for police agencies and intelligence services, but the files were typically on paper, which has many physical limitations such as vulnerability to water, fire, loss, and tampering. Paper also tends to become brittle with age.

Space and Speed

Data repositories no longer take up whole rooms or warehouses—millions of files can fit on a single hard drive. This greatly reduces the space needed for storage, and may significantly reduce the time to find a file, especially if seek-and-match software facilitates a search.

In the days of file cabinets, it could take weeks or months to match a criminal profile out of tens of thousands of pictures or prints. Now it is often accomplished in hours and sometimes minutes, as is described in fingerprint-matching examples in the Biometrics chapter.

Computers also permit multiple searches, something that wasn't possible in rooms-full of physical file cabinets. In the past, if a private investigator or police officer needed a file, a place-marker would be put in the file cabinet and the file used at the desk (or occasionally in a patrol car, in the days before photocopy machines). Other agents couldn't access the information until the file was returned, and the possibility of a file going missing (unintentionally or deliberately) was much greater. Fire or water damage could easily ruin it, as well.

With computer archives, an electronic dossier is available to anyone in the country with clearance to access it and several people may be looking at the information at the same time. Data redundancy and backup files stored in a different location can prevent data loss if something happens to the main archive.

Improved Witness Reporting and Victim–Suspect Matching

In addition to fast searches, one of the most important advantages of software archives over file cabinets is that *expert systems* (smart algorithms based upon professional knowledge) can be built into the software to improve informational efficiency. Software can be configured with pre-programmed priorities.

In the past, asking a victim to identify a suspect by looking at hundreds of pages of mug shots not only was wearying, but faces on the mug sheets would blur together, especially if the victim was traumatized by recent events or saw the suspect briefly under less-than-ideal conditions.

With computers, it is possible for an agent to put together an image of the suspect using a verbal description interpreted by a specialized software program similar to a paint program that selects facial features from pre-stored faces. It is also possible for a portrait artist to create a sketch, based on the victim's description.

The composite image can be presented to the victim for refinement. At that point, the computer can search the files for the best matches and the victim can carry out a more targeted searches of the filtered mug shots without being barraged with pictures that are dissimilar to the person being sought. It is not guaranteed that the right suspect will be found (or any suspect if he or she has no prior criminal record), but some of the mental fatigue and confusion that accompanies shotgun-style visual identification can be reduced.

Another method of matching sketches from victim descriptions has been developed by a team at Michigan State University. After a forensic sketch has been created, the picture is digitized and the software matches the photo to facial features in an existing database based on evaluating distinctive features. Accuracy, based on data from real crimes, was almost 50%, which is a good result compared to many other methods.

One of the problems of solving crimes is identifying victims of kidnappings who have been missing for some time. If they are young children, they grow and change quickly. Expert systems programmed with information on which parts of the head and face change, round out, or elongate as we grow older can produce an image of what a missing child might look like a few years later.

Search and Rescue

Software to construct faces is used in other ways, as well. Sometimes search-and-rescue workers and crime scene investigators discover an unconscious victim or body they need to identify. Software that helps construct a face that has been badly injured may aid law enforcement in identifying a victim who is in a criminal database, or may help rescue workers or medical staff to create a picture of what the person looked like before the injury in order to locate his or her relatives. Dental records are also commonly used for victim ID and can be stored in data-bases in the same way as other information.

In the military, identification is more often made on the basis of dog tags and DNA profiles that are matched with mandatory samples submitted by personnel.

Data Sharing

A further advantage of computer storage is shared resources. In the past, it wasn't uncommon for criminals to cross state lines to run from the police. Now, with data sharing and national databases, it is difficult to hide. If someone commits a crime in Florida and flees to California, a search is just as likely to bring up a match.

They say that justice is blind, meaning a good justice system is intended to give fair judgments regardless of socioeconomic level, ethnicity, physical features, or religious beliefs. Computers are also blind. This can be an asset and a liability. They may be less biased in some respects, but often they don't distinguish between law-breakers and the rest of the population unless social policy is developed to govern software and hardware design so that law-abiding individuals aren't represented and treated the same as criminals.

Criminals should not be divested of all rights. There are certain rights extended to all human beings, but the danger in too much overlap in electronic policing procedures, between those charged and those who are not, is that the privacy (and dignity) of those who have done nothing wrong begins to erode.

With electronic databases, computers have revolutionized storage and access to intelligence and personal information on suspects and convicted criminals. Digital storage media can degrade, just as paper degrades (CDs and DVDs have a limited life and hard drives can fail), but it is possible to make multiple copies of digital data that are identical to the original, and transfer them from one medium to another as storage technologies improve.

The same technologies that underpin criminal databases are used by private firms, as well. In fact, the distinction between government databases and commercial databases is blurring. Many license-plate recognition systems, for example, automatically access both government and commercial repositories when looking for a match. Even if the commercial database is carefully amassed and fact-checked, and stored in a secure facility, it is still not the same as a government database because commercial ventures are regulated differently. There are few restrictions on how long a private firm can store personal information or whether they can sell it to others.

EXAMPLES OF DATABASES USED FOR U.S. LAW ENFORCEMENT

Uniform Crime Reporting (UCR)

The UCR was established in the late 1920s by the International Association of Chiefs of Police and the FBI. It became a storehouse for crime statistics in 1930, with an emphasis on property crimes and violent crime. Data is now submitted by almost 17,000 American law enforcement agencies to create demographic and other statistical data. Some of the publications generated from this data include:

- *Crime in the United States*—an annual report since 1958 describing local and national crime rates and offenses.

- *Hate Crime Statistics*—an annual report on offenses and incidences that appear to be motivated in part or in whole by bias against a victim's perceived race, religion, ethnicity, sexual orientation, or disability.

- *Law Enforcement Officers Killed and Assaulted*—an annual report on accidental and felonious deaths and assaults of law enforcement officers in the line of duty.

Computers also facilitate searching, sorting, and building tables. The UCR site includes a table-building tool to customize information display.

National Incident-Based Reporting System

The *National Incident-Based Reporting System* (NIBRS) was formulated by the UCR Program as a comprehensive source of detailed crime information. The data is available for researchers, planners, students, the general public, and law enforcement personnel. The concept was developed during the 1970s, with the first pilot program conducted in 1987.

Data on individual crimes, within certain specified offense categories, is submitted by participating local and national agencies. Examples include arson, stolen property, narcotics offenses, criminal kidnapping, white-collar crime, domestic violence and abuse, pornography, organized crime, and alcohol-related offenses.

Group A offenses include greater detail than Group B offenses, which include only arrest data. Peeping Tom violations, which have direct relevance to surveillance, fall within Group B.

In 2007, more than 6000 law enforcement agencies contributed data (about 25% of overall crime statistics). A variety of reports and statistical compilations are available at *www.fbi.gov/stats-services/crimestats*. Development of the resource is ongoing.

National Instant Criminal Background Check System

The *National Instant Criminal Background Check System* (NICS) is a national database developed for background checks to determine if someone qualifies for possession of a firearm. It was established as a joint effort by the FBI, the ATF, and law enforcement agencies to comply with the *Brady Handgun Violence Prevention Act (Brady Act) of 1993*.

NICS was initiated in 1998 and, by 2012, included more than 61 million criminal history records, as well as persons named in restraining orders. Response time on queries is very fast, sometimes less than a minute.

Criminal history databases are also maintained by individual states. The laws about who may access them, and how much information they may see, varies from state to state.

As an example, the Pennsylvania Access to Criminal History (PATCH) system is open to the public to search criminal history records in accordance with the *Criminal History Information Act*. It also sets policy for dissemination of criminal history data to justice agencies and outlines restrictions on what may be sent to individuals and noncriminal justice agencies.

Many state courts provide online access to databases that display cases and court schedules.

Minnesota has acknowledged an aspect of privacy that many institutions overlook. Providing local access to a limited community is different from broadcasting to the world. The Minnesota Judicial Branch rules of public access enable certain information to be searched online, but street addresses and other sensitive information is only available by personal appearance at the courthouse terminals. This not only safeguards privacy to some extent, but also protects vulnerable people from stalking and other forms of harassment.

If someone is determined to find personal information, they can, but there's no need for law enforcement and other agencies to make it easier for a person with criminal intent to target victims just because the Internet exists.

International Criminal Databases

Leaving the country to escape criminal prosecution may no longer be possible. International databases have been established to help national police agencies identify travel documents and locate people.

INTERPOL, a global police agency with 190 member countries, maintains databases for solving international crimes, including information on lost and stolen travel documents, known international criminals, missing persons, and stolen art, vehicles and firearms. The databases include biometric information such as mugshots, fingerprints, and DNA profiles.

Access to the data, by qualified personnel, is through the National Central Bureaus (NCBs), which make the service accessible in realtime 24 hours a day.

There are also companies that claim to have extensive criminal records that can be searched on the Web. How these are collected is difficult to know—there are laws prohibiting distribution of most criminal records to individuals or businesses. There's no way to verify how accurate the information is, and many of the companies building these databases charge for access, even if the search comes up empty. They may be capitalizing on people's desire to find information rather than concerning themselves with accuracy.

License-Plate Databases

Most state and local law enforcement agencies maintain files of license-plate numbers that are considered suspicious in some way, either because the plates or the vehicle for which they were issued have been stolen, or because the vehicle was seen at the scene of a crime. Some private firms maintain license-plate data, as well.

Both government and private license-plate databases are accessible by qualified police officers and intelligence agents and those maintained by private firms may also be accessible by other entities, such as repossession companies and tow truck firms. The accessibility of license-plate information to anyone outside of law enforcement is controversial.

Some people consider license plate numbers to be private, as long as the individual isn't breaking the law. Others disagree and claim they are public because anyone can see them as a vehicle passes. One commentator pointed out that if you took the license plates of certain dignitaries or celebrities and posted them publicly on the Web, the perception of whether they are truly "public" might change.

The fact is, publicly distributing license-plate information beyond the local community can endanger a person's safety or sense of security. It may be publicly viewable, but only to those in the immediate vicinity, not to the whole world. As mentioned earlier, there is a difference between local and global visibility.

License-plate surveillance is discussed in more detail in the *Optical Surveillance* chapter.

Intelligence Community Databases

The Intelligence Community (IC) has had its share of controversy over the decades and has been reorganized a number of times, with the mandate being variously stated. Much of its current form was laid out in the *Intelligence Authorization Act of 1995*.

Computers have always been an important aspect of intelligence gathering. Not only are they used to store data and intercommunicate, but intelligence is harvested from people's computer files and network communications, as well.

The information gathered is stored. One of the characteristics of intelligence is that you never know how much you know or how much additional information may change a picture. It is an ongoing, endless pursuit.

In 2007, it was announced that the National Security Agency (NSA) would be building a huge data center in San Antonio, TX. In 2009, news spread that a massive data center would be built in Utah, on a 200-acre property south of Salt Lake City. The purpose of the facility would be to provide technical support for

Homeland Security objectives. The extensive facility is slated to begin operations around September 2013.

Given that most data these days is encrypted, it doesn't take much deductive thinking to assume that part of the operations may include data decryption research and activities.

Glenn Gaffney, Deputy Director, alluded to privacy concerns when he stated, "We will accomplish this in full compliance with the U.S. Constitution and federal law and while observing strict guidelines that protect the privacy and civil liberties of the American people."[17]

EXTRA EYES AND EARS—CITIZEN SURVEILLANCE

Public Assistance to Law Enforcement

Citizen-assisted surveillance is ages-old. *Wanted Dead or Alive* posters have been tacked on trees and community halls going back hundreds of years. The police often ask the post office and news media to display pictures of suspected criminals or missing persons so the public can help solve crimes.

Many government agencies have enlisted the Web to make it easier for people to file taxes, find forms, and apply for documents such as passports and licenses. There are similar mechanisms for enlisting the help of the public for policing business trade and other violations. In other words, the government has extended the idea of Wanted Posters to surveillance reporting through the Internet.

As a few examples, agencies like CBP, the FTC, the FBI, and the FCC provide ways to report violations. The *E-Allegations Online Trade Violation Reporting System* enlists the public's help in reporting import/export infractions by providing an online submission form.

Most police departments provide a way to submit anonymous crime tips via phone, cellphone texting, or Web links.

There are also tip lines people can call to report possible terrorism threats. Some communities, like the Bay Area, have posted transit posters encouraging people to report suspicious items or behavior.

> *"Report any item or behavior that is Hidden, Obviously suspicious or not Typical to the situation.*
>
> *Tell a BART employee, use the train intercom or call BART Police..."*
>
> "Bomb detectors" poster posted in BART vehicles since 2005

By enlisting the public, surveillance extends beyond the eyes, ears, and budgets of individual government agencies. Spurious reporting is discouraged through prosecution under the *Computer Fraud and Abuse Act of 1986.*

---

[17] J.N. Hoover, "NSA to build $1.5 billion cybersecurity data center," *InformationWeek Government*, October 29, 2009.

There are also ways to report crimes via anonymous tips to commercial crime-tracking sites. The crime incidents are displayed on map-based information servers. CrimeReports is a commercial firm that works with law enforcement agencies and charts U.S., Canada, and U.K. crime incidents and registered sex offenders on a Google map in near-realtime. The company provides an iPhone application so traveling viewers can monitor the information as it is added to maps.

One of the dangers of systems built on anonymous tips is that people might submit information as a way to mislead law enforcement agents and individuals away from a crime, or use them to defame or frame an innocent victim.

Submitting false police reports is a crime, but submitting a false report to a commercial agency may not be. Caution is advised when relying on citizen tips unless the information is verified by investigators or police agencies. There may also be legal considerations for the companies uploading information from crime tips. Any information specifically naming or implicating a person could be construed as libel if the allegations are false.

Environmental Monitoring

The Environmental Protection Agency (EPA) engages in many kinds of surveillance and keeps computer records of changes in the environment, environmental cataclysms and disasters, as well as coastal fisheries, marine life, and other environmental resources. This includes aerial surveillance of areas of environmental importance and observational counts of individual animal species.

The Clinton Global Initiative has generated programs for a variety of environmental concerns and described economic benefits related to addressing those concerns. One program, created by the Ocean Recovery Alliance, is called *Global Alert*. Its aim is to help reduce the millions of tons of trash thrown into waterways every year. Anyone with a smartphone or similar device can upload images of trash hot spots and the data will be incorporated into mapping systems like Google Maps™, enabling communities to monitor their waterways.

Citizen Fact-Checking

FactCheck.org, a site maintained by the Annenberg Public Policy Center, provides an online platform for the public to submit information that they feel sets the record straight on a wide variety of issues. Content submitted to FactCheck may be redistributed by the site personnel, but those submitting the information retain ownership insofar as it belongs to them.

Citizens are also sometimes encouraged to engage in political surveillance by finding statements or media coverage that appear to be false or misleading and rebutting them. Many forums engage in this kind of critical debate.

Political Surveillance

Spying on political rivals is ages old. In most dictatorial regimes, the leader enlists informants to make sure no other person achieves enough popularity and power to unseat the leader. In Roman times, *delatores* (informants) were regularly used to spy on supposedly illegal acts or otherwise denounce those who opposed the reigning administration. They were sometimes rewarded with the right of citizenship or a percentage of the assets of the person who was brought down.

Informants have been plentiful in other places such as eastern Europe and the Soviet Union. There have been periods of Soviet history where people were afraid to talk to each other if anyone else was within earshot, including neighbors, relatives, and domestic servants.

Intelligence and law enforcement agencies make use of informants to gather information and the IC may not have to tell informants that they are gathering information for surveillance purposes.

One of the ways in which surveillance using civilian spies is kept covert is by breaking the task into small pieces and parting it out. As an analogy, if you hand individual electronic components to a dozen different people, they may not know what they are for, but when they are shown together to a person with technical expertise, the expert may immediately recognize it as parts of a microwave antenna, or bug sweeper, or other identifiable device. Information functions in much the same way. Breaking it down and compartmentalizing it can make the overall structure of the intelligence "invisible" to the persons gathering it, yet it is valuable to those who put the pieces together to see where it leads.

In the U.S., the public makes a general distinction between intelligence organizations and law enforcement and perceives policing operations to be more open than activities of the IC. This distinction, however subjective it may be, does not extend to all nations. The German Gestapo established in 1933 was a shortening of the German words for Secret State Police. There are other countries where the distinction between the police, military, and intelligence agencies is somewhat blurred.

In democracies, the deposition of political rivals is based more on media manipulation and public debates than on Roman tactics, but bringing down an opponent by finding out information about a shady background, or illegal dealings with prostitutes or drug rings, is not uncommon. The public prefers to elect officials with reasonably clean slates.

The Internet has created new opportunities for political rivals and their campaign workers to poke and peck at each other, and to amass and contradict each other's public statements.

One example of a site specifically devoted to collecting and rebutting political comments was set up in January 2012 by the President's campaign committee, *Obama for America (OofA)*, a Chicago-based organization. OofA installed a site called AttackWatch in which opposition statements are quoted (or the original video uploaded) and debaters use email to submit debate responses if they feel the information is erroneous or significantly slanted. The site is controversial, with some opposition supporters claiming it is a way to suppress free speech and turn in those who oppose the current administration.

Big-Little Brother

Can public surveillance go too far? When do little brothers become indistinguishable from big ones?

In the years leading up to the Holocaust, Adolf Hitler encouraged citizens to inform on anyone who opposed the National Socialist party and to provide information on the Jewish population that could be used to marginalize and exterminate them. Similar actions may have occurred in Turkey where Armenians were "flushed" from the population and sent on death marches during World War I.

Not everyone wants to provide surveillance information in times of trouble, but many will do so if asked.

A *brittle brother* is someone who snitches under duress, for fear of losing a job or even, in past régimes, his or her home or life. Brittle brothers are those who don't want to divulge surveillance information, but do so out of fear.

There are also people who eagerly engage in surveillance, either because they support a cause or anticipate a reward. A *bennie brother* is a stool pigeon or other informant who conducts surveillance and divulges information to reap some behind-the-scenes benefit, such as a plea bargain, a promotion, an "in" with a particular crowd, or financial compensation.

SEARCH ENGINES AND SOCIAL NETWORKS

As they have grown, search engines and social networks have become significant sources of business and political intelligence.

Surveillance is not limited to recognizing individuals or tracking their movements; it also includes probes into their relationships, hobbies, occupation, lifestyle, economic status, and former residences. Within a decade or two, there may not be any aspect of a person's life that has not been archived in numerous databases and the information may continue to be distributed by third parties for commercial gain, as has recently become commonplace.

Search Engine Surveillance

Search engines amass a tremendous amount of information on the people who use them. When a person searches universities (looking for a good school), employment sites (looking for a job), or dating sites (looking for a mate), he or she is revealing a great deal about interests, ambitions, hobbies, and personal tastes. The administrators of search engines can configure the system to track cookies and IP addresses so that searches by specific individuals can be organized and analyzed in many different ways.

The search experience itself can be personalized. IP-based location information on recent searches can be tracked in order to make guesses as to what the person is seeking on the next search. If the search term is pizza, a personalized search engine can list the local pizzarias first. Some of these features can be very convenient. The Internet is vast and no one enjoys excessive wading to find what they're seeking. Sometimes the features can step over the line and intrude on a person's privacy.

Search engine companies that offer a variety of products and services may use search information to personalize a user's experience of other services, as well, by sharing information among them.

Google's search engine is a very clean, fast, popular Web utility, enjoyed by millions of users, but it is not the only service offered by Google, and certain of its corporate actions and policies have been criticized and scrutinized for a number of years.

In March 2011, the FTC alleged that Google, Inc., the company with the most popular Internet search engine "used deceptive tactics and violated its own privacy promises to consumers when it launched its social network, Google Buzz..." The FTC asserted that Buzz was launched through the Google Gmail system, "... leading users to believe that they could choose whether or not they wanted to join the network, the options for declining or leaving the social network were ineffective..."[18]

In January 2012, the FTC expanded its probe of Google to include the company's social networking service. Businesses and privacy advocates were concerned that Google might give its various companies preferential treatment in search-listing results, including information culled from its social networking site Google+ in general searches.

Concerns about search engine privacy resulted in a new search engine being launched called DuckDuckGo. The privacy page states, "DuckDuckGo does not collect or share personal infiormation. That is our privacy policy in a nutshell."

The DuckDuckGo site then explains some of the privacy concerns that may be associated with other search engines, such as "search leakage" (search information that is sent to the searched site that is next visited) and the broadcast of user-agent and IP-address information that may lead to identification of the user. In contrast to other search engines, DuckDuckGo has a stated policy of not storing a person's search history.

Some users have opted to go through proxy servers while surfing the Net. A proxy server is a relay station. Rather than going directly to a site, the user links into a proxy at some other location. Then the proxy interprets the location instructions and visits the site, sending back the information to the user. It is like asking a friend to ask a friend about something you don't want to know was originally asked by you—it shields the user from direct tracking. Not all sites will allow a user to log in through a proxy, there are limits to what can be done through a proxy server, but it can significantly cut down user tracking.

---

[18]Federal Trade Commission press release, March 30, 2011.

Social Media Monitoring

Social media sites have become a significant source of information for companies seeking product opinions, demographics, and background information. They are also used by unscrupulous people to dig up dirt on others, and by law enforcement to locate suspects or possible witnesses or associates in their social networks. Social networks are also hot spots for locating dates and mates and places where people sit back to chat and play games.

Harvesting Consumer Opinions

Marketing companies and firms seeking to promote themselves often search for opinions about their products in personal reviews and comments on chats and forums. Many people are utterly candid on the Web, and the give-and-take of opinions provides a range of consumer reactions. It is not an unbiased sample—it is a sample of people who have the resources and desire to be online and to express an opinion—but it is still consumer feedback that cannot easily be gathered in other ways. Those who speak out, with good arguments to back up their opinions, may win the respect of others and influence their buying habits in the future.

To monetize this feedback, there are companies that monitor the Web 24/7 and charge businesses for statistical summaries and lists of where the information can be accessed, based on the social buzz on blogs, social networks, video- and picture-sharing sites, and forums.

Popular social network sites like Facebook and MySpace provide a way for people to connect, to link to one another to share news, images, and play games, and make it easier to find individuals with similar interests.

Personal Data Sharing

"A Wall Street Journal examination of 100 of the most popular Facebook apps found that some seek the email addresses, current location and sexual preference, among other details, not only of app users but also of their Facebook friends. One Yahoo service powered by Facebook requests access to a person's religious and political leanings as a condition for using it. The popular Skype service for making online phone calls seeks the Facebook photos and birthdays of its users and their friends."

Julia Angwin and Jeremy Singer-Vine, Wall Street Journal Online, Technology section, updated April 8, 2012

There is concern about privacy on social networks because private areas sometimes become open during system maintenance, are sometimes viewed by staff, and are sometimes conveyed, in the form of statistical analyses, to other companies.

Private information on social networking sites is often shared with other sites. A box will pop up asking if the user wants to log in with their social networking profile and warns that if the user continues, personal information will be shared. How much personal information is involved is rarely explained.

In November 2011, the FTC announced a proposed settlement with Facebook, one of the most popular social networking sites. Allegations against Facebook included deceptive privacy assurances to users (areas that users thought were private were sometimes accessible to their linked-friends' applications) and retroactive changes in privacy policies that violated the *FTC Act*.

As cyberspace evolves, so do the laws that govern its use.

Data Aggregation Sites

Data aggregation sites are those that gather information from many sources and put them together in one convenient location for free or commercial access. Commercial crime sites, background check sites, and online phone directories are all examples of data aggregators.

Before the Internet, telephone operators did not divulge people's addresses to those calling for information. The address was available in phone books to the local community, but not to the world at large. Distribution of reverse directories, where the name could be determined from a phone number, was also carefully controlled and available only to certain businesses. These safeguards are known as "practical obscurity." Even though the information is public, it is only easily accessed by a small local audience, making it virtually obscure to the rest of the world. This confers a small measure of security and aids in public safety and the prevention of targeted thefts and stalking.

Entrepreneurs establishing data aggregation sites often obtain public records and use them in their services, in addition to anything they can find on the Web. Since laws to protect public safety and public documents did not anticipate worldwide dissemination in such a quick and easy manner, sites offer the information with little or no concern about how they might impact the person whose personal data is in their files. Even those that offer opt-out clauses do so in a cursory way, since the data often ends up back in the database and many opt-out requests are ignored.

Spokeo.com, a data aggregation and brokerage site that provides comprehensive information about individuals taken from dozens of sources, came under scrutiny in 2011 when the FTC was petitioned to investigate its business practices. Not only was the site providing personal information about age, relations, friends, personal habits, religion, ethnic background, and economic status, it further included Google Streetview™ images of the person's home.

Spokeo's president and co-founder informed a news interviewer that a visitor to the site could enter a person's email and Spokeo would search through 43 social networks to grab profiles, blogs, photos, videos and "anything that you can see..."

When these pieces of information are held in separate local files, opportunities to exploit the user are minimal. When they are combined and statistically presented on the World Wide Web, they take on a more invasive character.

There few checks and balances on the accuracy of the information held in commercial aggregation sites and many individuals do not know the site exists and can neither opt out nor correct erroneous, libelous or misleading data. Those who utilize the site, whether free or by subscription, may use the deeply private information to evaluate individuals for jobs or insurance eligibility. In fact, Spokeo promoted its site as a "great research tool to learn more about prospective employers and employees..."[19]

Unfortunately, such detailed information about people's lives can be used for darker purposes such as stalking, vandalism, or home invasions.

Most private information in public archives, up until the 1980s, was only available to locals who had to put in a personal appearance at a local government office to see them. Some offices required a signature and show of ID before releasing documents (e.g., court documents), and documents had to be returned within a specified time.

If people from out-of-town wanted to peruse the data, they would have to pay considerable travel and accommodation costs to each town of interest and laboriously hand-copy or photocopy each document (in come cases photocopying was not permitted). Once they had the information, there was no easy way to search and sort it, or to store it electronically. Even when personal computers began to catch on, in the 1970s, there were no mass storage devices. Data had to fit on a floppy diskette and it was difficult to transmit to other computers. BBSs, forerunners of the Internet, had slow connection speeds and limited storage.

Thus, the cost and inconvenience, and the necessity for person-to-person interaction with government clerks to see the data, prevented wide-scale exploitation of the information and the individuals concerned.

All that has changed, which is why the justification "it is already public" is being challenged.

Many data aggregation sites are free. Even if they require subscriptions, certain online subscription payment services will protect the payer's identity. There are few checks or balances on who can see it or how it is used. The aggregators themselves make those decisions and their motive, for the most part, is profit.

SURVEILLANCE THROUGH WEB BROWSERS

Web browsers have inherent characteristics that make it possible to spy on a person's browsing habits. They also have some security holes that spies are quick to exploit. As a result, some browsers have built-in security settings that can help

---

[19]*Ten Great Uses for Spokeo Free People Search*, Spokeo, June 22, 2010.

reduce the amount of information the browser broadcasts as users surf the Net. Browsers need to be continually updated to fix or improve security.

Everyone who uses the Web needs some kind of browsing client to get around. The client interprets the user interface instructions on websites. In the past, sites were mostly or entirely constructed with HTML tags and sometimes a few CGI scripts. These are software instructions that tell the browser how to display a page and respond to simple forms.

Site developers, or those who hired them, wanted more. Showing a few pictures with text and offering a few forms yielded a surprisingly rich browsing experience, given its simplicity, but entrepreneurs and Webmasters wanted the user to interact with the site in the same way as applications on their home or business computers. Increased versatility makes it possible to offer multimedia entertainment, games, and much more. So HTML was upgraded several times to make it more powerful, and languages like Java and JavaScript have been added to create a more interactive environment.

There are many popular browsers and choosing from a variety of browsers has many advantages:

- users can select those with features and interfaces that suit their needs,

- competition among browser developers helps advance the technology,

- a choice in browsers prevents companies from merging browsers into their operating systems, thus providing some consumer protection against spyware, and monopolies that might stifle product improvement and new entrepreneurial opportunities,

- a choice in browsers gives users more control over the security of their communications. Some offer more features that protect a user's Web-related activities from unauthorized snooping.

Browsers are the cars we drive on Internet highways and businesses would prefer users to drive in their cars and no one else's. Then they could add toll gates to the browsers and hide Web-surveillance software deep in the operating system, all of which would provide a competitive advantage—so companies will continue to try to win the browser wars.

If a single firm won a monopoly on browsers, it would be harder for users to choose their security settings and keep their Web surfing private. It would also make it easier for law enforcement agencies to subpoena the browser manufacturer to give up any logs of user travel through the Internet highways. Privacy is easier to maintain when there is a choice.

Quiz and Clickstream Surveillance

Data aggregation sites are not always obvious. Some lay out the data in an organized fashion and invite people to study it and pay subscription fees to obtain more. Others masquerade as quizzes or information sites, gathering data as people play games or answer questions to learn something about themselves.

Data Gathering Through Quizzes

Sometimes online entertainment is a way to gather information on people's interests, personal history, or medical concerns.

One example of online quiz surveillance is an age survey called RealAge® that asks health- and body-related questions and supplies a "biological age" result at the end. It is, in fact, a commercial information-gathering tool. Survey-takers are encouraged to join to learn more about staying younger through "nonmedical solutions" and RealAge promoters sell the results to pharmaceutical companies. Drug companies allegedly study the data to target individuals with certain disease-prone profiles and send them messages about a disease they might be afflicted with before they've had a proper evaluation and diagnosis by a doctor.

Vice President of Marketing Andy Mikulak admitted that their newsletter focused "on the undiagnosed at-risk patient."[20] At the time, RealAge's privacy policy stated that information was shared with third parties but did not explicitly mention drug companies.

Data Gathering Through Clickstreams

A clickstream is a series of user interactions with a computer. Actions such as clicking a mouse or moving a stylus to select a button, network link, or icon can be logged for later study and analysis. A log is helpful for improving system performance, site organization, and interface design, but may also glean information specific to an individual, often with marketing research and ad presentation as the goal.

Clickstream logging is a transparent process (one that occurs without user awareness). In some cases, network service providers use the information to improve their service; in others, they sell the information to third parties.

Charting a person's search engine keywords and link-selection habits is one example of a clickstream. Another is tracking a sequence of visits to specific Webpages or sites.

In 2006, AOL uploaded a body of search results to a public site without the users' knowledge. Usernames had been coded with an ID number to protect their privacy. As it turned out, this wasn't enough.

The action was not intended to harm anyone. The data was for academic analysis, but once something is "in the wild" on the Internet, it is downloaded and reuploaded on hundreds (sometimes millions) of sites, and cannot be retracted. As curious parties studied the AOL data, it became clear that recoding a name

---

[20]S. Clifford, Online age quiz is a window for drug makers, *The New York Times*, March 25, 2009.

into a number doesn't obscure the personality or necesssarily protect the identity of the person who performed the search.

The author viewed the information before it was removed from the original site by AOL and noticed profiles of high school students studying for exams and searching for prospective post-secondary institutions, people searching for information on same-sex partners or certain ailments, and others doing genealogical research—all which might be sensitive in one way or another in a work or social context.

Investigative journalists and amateur detectives gleaned sufficient information from the search patterns to track down some of the users and verify that they had done the searches.

Since it is difficult to know when someone is tracking or targeting individuals unless they openly discuss the process, privacy advocates have looked for ways to technologically exempt computer users from intrusion.

One example is empowering browsers to prevent sites (particularly third-party sites) from setting a software *cookie* (a unique identifier) to track a user's Internet movements. Another example is the *Targeted Advertising Cookie Opt-Out* (TACO), by Chris Soghioan. This is an *opt-out cookie* software plug-in that sends messages to more than two dozen major ad networks that the user doesn't want to be tracked.

Job Applicant Screening

The Internet has become a significant source of information for companies screening job applicants and many are paying online companies for personal information. Companies offering background check services aggressively amass large amounts of data on anyone who might potentially hunt for jobs.

Searching is not totally unrestricted. There are laws that require that job applicants be informed about certain forms of searches. For example, according to the terms of the *Fair Credit Reporting Act,* prospective employers must notify job applicants if they plan to do a criminal background check. Whether all employers follow the law is difficult to know. If they can run a check with a low probability of an employee finding out, some might do it.

The information gathered by commercial firms doing background checks is not guaranteed to be accurate. Since many people hide behind personas on the Net, or represent themselves as playful characters on forums and social networks, the image they convey may have nothing to do with their work qualifications or work ethic. On the Internet, people are fairies, gnomes, goddesses, giants, bullies, clowns, and dopes, and some sites even encourage imaginative characterizations, since the medium of cyberspace makes it possible.

The *Fair Credit Reporting Act* does not prohibit seeking information on job applicants on forums, social networking sites, blogs, or other public sources of information. It also does not prevent employers from rewording questions to skirt the laws. They may be restricted from asking if the applicant has ever had a bankruptcy, but they may ask, "Have you ever *filed* for bankruptcy?"

This should be of considerable concern to job hunters because even if an applicant's record is clean and blameless, even if he or she doesn't don a wizard hat or clown suit on forums, if the individual has any kind of presence on the Net, he may be damned by his associations. Web users have no control over what other people say about them or who else their online buddies may have linked to after linking to the person looking for a job. A person could appear to a prospective employer to have undesirable associations and not know it.

Employment agencies have cautioned job hunters to be careful about what they say publicly on the Web. They recommend that anything objectionable or racy be discussed only in private areas (or be moved to private areas).

Although well-intentioned, there are several problems with these suggestions:

- *Private information isn't private if the hosting company's staff can see it.* Nor is it private if the hosting company is selling the information to third parties.

- *Moving racy comments to a private area prior to job hunting doesn't remove them from the Web or from public scrutiny.* Thousands of organizations are harvesting information on a continual basis and storing it in private databases for the express purpose of using the data later if a name comes up with a match. There are companies building huge vaults of personal information to sell to other companies seeking to hire.

- *Comments on many sites cannot be retracted.* Many forums disallow edits, or may only allow them for a few minutes after a message is posted. After that, comments may be part of the permanent content of a site. Sometimes moderators will relent and take down contents if the user makes a reasonable request, but just as often the answer is no. Forum administrators view content as an asset, since it attracts more viewers. If a high school student posts comments that are normal for high schoolers, he or she may regret it four years later when hunting for a job.

- *In time, software vaults will contain years and decades worth of data—the entire history of a person from birth to the time of application may be accessible.* Anything that happens in the teen years, the brief time in a person's life where he or she gets to experiment, try things, speak out, or dress unconventionally, is part of those records and any advice to keep quiet in case it might hurt job prospects 10 years down the road amounts to significant suppression of freedom of speech and opportunities to be a kid.

- *Personal areas are not invulnerable. Hackers* brag about gaining inside access to popular sites, especially those offering free services like Google Mail.

- *Personal information is sometimes published on the Net without the users' knowledge or consent* for a variety of reasons. As mentioned, AOL published search results intended for academic research, but identifiable patterns enabled journalists to locate some of the people who performed the searches—something AOL didn't antici-pate. In 2012, Comcast included unpublished numbers on their EcoListing™ site without informing users that this would happen. In fact, Comcast reassured their customers, after the fact, that no changes would occur in the directory information—which is untrue if unpublished numbers were inadvertently included.

- *Bankruptcies, juvenile offenses, and other information that traditionally was struck from the records after a certain period of time are no longer temporary.* Once captured and cached, they are part of a permanent software legacy that can come back to haunt an otherwise law-abiding, respon-sible citizen.

- Throughout human history, there have been cases of mistaken identity and, with the whole world coming online, information pertaining to a specific name may be errone-ously assumed to be one person. *Those seeking to make money selling background checks based on harvesting Web information don't necessarily fact-check the information* to make sure it is the right person or that all the information is the same person rather than an amalgamation of information from two or more people with the same name. Since back-ground check results are rarely revealed to job applicants, the applicant doesn't have the opportunity to rebut incorrect information.

- *Humans make mistakes and companies change settings.* There have been sites where changes in privacy policy or where site maintenance has temporarily opened up private sections. Sometimes when policies change, users have to manually re-engage their privacy settings. If they are on vacation or busy with work or family, they may not know for weeks that their account is wide open.

About half of companies now use the Web as a research resource for screening job applicants and this number is expected to rise.

Network Carrier Tracking and Monitoring

LAN administrators and Internet service providers use a wide variety of software monitors to operate and tune the efficiency of their networks and to integrate them with communications from other networks, such as mobile phone services. The monitors provide an intimate look at all activity on the network, including ingoing and outgoing communications, bandwidth usage, email communications, Web surfing patterns, game-playing activity, and more.

Unfortunately, not all computer users are responsible Netizens (*Net*work cit*izens*), so administrators have software to manage users, some of whom attempt to use network services to defraud or exploit people through *phishing* scams, identity theft, Trojan horse information-stealing software, malware (including viruses, worms, and spyware), or stalking.

Monitoring network activity is necessary to maintain efficient operations and prevent criminal activities, so the law gives system administrators more leeway, where privacy is concerned, to do their jobs.

In general, system administrators are smart and computer savvy. They can quickly assess and apply technical information to connect, maintain, and optimize network systems, and many are involved in the continual development of network protocols and applications.

There are also specialized companies providing statistical programs to ISPs to gather information on network traffic, optimal configurations, bandwidth allocation, bandwidth conservation, and data flow.

Managing a large network, especially one that interconnects with the Internet, is an ongoing challenge.

What this means, in practical terms, is that system administrators engage in constant surveillance as part of their job of keeping a network running in an efficient and orderly way. Collectively, they have access to every aspect of a system, sometimes even private passwords on systems that don't automatically encrypt the data, and can read most private messages or inspect most data streams (such as text messaging or audio/video chats).

The only thing preventing broadscale abuse of this knowledge is personal integrity. In the early days of bulletin board systems, the system administrator was usually the owner and operator of the system, and many of them, despite snooping on user communications themselves, are privacy advocates who don't share the information beyond local systems.

This situation is changing, however, as smaller companies are merged into larger ones, and company executives make decisions that focus on profitability rather than personal ideals or civil rights. Resale of statistics and content is not uncommon and business mergers make it possible to combine content from one system to another, creating a fuller and sometimes significantly different portrait of users than when databases were kept separate.

MILLIMETER COMPUTERS

If you thought smartphones and netbooks were small, get ready for "smart dust"—computers the size of the head of a pin.

In early computing days, computers were huge systems that required warehouse-sized rooms and thousands of expensive, fragile vacuum tubes. They cost millions of dollars and yet were less powerful than handheld calculators from the 1990s.

A fundamental change happened when transistors were developed. Transistors enabled electronic functions to be scaled down onto breadboard-sized circuit boards that were more powerful and flexible than the early computer behemoths. Further improvements made it possible to develop personal computers that were affordable to consumers and small businesses, and paved the way for even smaller devices like netbooks and PDAs.

Now computing is poised to enter the millimeter format. Researchers at the University of Michigan are working on systems that include a processor, memory, and wireless transmitter in a package so small it is dwarfed by a penny. This shows great promise as an implantable sensor for medical conditions.

Researchers are also working on transceiver systems that would enable millimeter computers to intercommunicate. When that is possible, tiny systems could be arrayed or set up in wheel-and-hub configurations to create sensing or processing "nets" for thousands of new applications. Incorporated into tiny aerial robots, millimeter computers could make instant assessment of local situations and radio their observations and analyses to a central databank. Thus, a team of surveillance robots the size of bees could collaborate in search-and-rescue operations or fly out to survey an industrial accident or natural disaster.

At the University of Michigan, D. Sylvester, D. Blaauw, and D. Wentzloff have developed a pinhead-sized solar-powered millimeter-scale computer as a prototype for implantable sensors that could potentially monitor glaucoma patients who suffer from swelling in their eyes that can cause vision impairment. Combined with radio transceiving capabilities, other potential applications include ecological monitoring, structural monitoring, or as "smart tracking" devices that could challenge RFID chips. [News photo by Gyouho Kim, 2011, released.]

There will doubtless be personal surveillance applications for millimeter computers. Bee-sized aerial vehicles could follow people everywhere. Celebrities worrying about harrassment from paparazzi journalists may be followed by paparazzi-puters.

A child walking to school may be accompanied by a not-so-imaginary friend the size of a moth that could call home if anything unusual happens.

Milli-computers are so small, they could be built into shoes, sunglasses, or other apparel to serve as tracking devices for parolees or inmates on field trips.

George Orwell's *1984* was prophetic in many ways, but he didn't anticipate computers the size of pinheads.

DATA CARDS

- *basics of data cards*
- *how data cards are breached*
- *data-card surveillance*

Introduction

A data card is a portable card equipped with magnetic, radio-wave, microcircuits, or other data storage or transmission capabilities. Common data cards include debit/credit cards, retail store membership cards, building access cards, and ID cards.

The term *smart card* is used rather broadly to mean cards that can be read without directly contacting a surface or which contain more sophisticated tracking capabilities or communication protocols.

While many cards are not issued specifically for surveillance, those that are swiped or radio-transmitted through networked storage systems can indirectly provide information on a person's whereabouts and activities. Others are specifically designed to track customers, employees, or other people and objects.

The Smart Card Alliance is a multi-industry, not-for-profit organization that studies, discusses, and promotes smart-card technology through publications, seminars, and certification exams.

Eurosmart, an international not-for-profit association promoting smart security services and standards, reported that more than 6 billion smart cards were shipped worldwide in 2011.

How do smart cards apply to surveillance? Smart cards can reveal a person's location and activities. From a trip to the ATM or one to the washroom in a secured building, networked card-access points can log entries, exits, and checkpoints, providing a picture of a person's daily habits. An unusual pattern of behavior can trigger alarms or automatically record notes in a computer personnel file.

BASICS OF DATA CARDS

Data cards are wallet-sized cards that carry information in coded format in addition to what is visually apparent. The information may be coded magnetically or electronically or both. With the right equipment, the information on the cards may be read by someone other than the person who owns or carries the card.

Data cards can be categorized in a number of ways, based on what forms of data storage they use, or whether they are passive or active technologies. The most common distinction is whether they are passive or active.

- Passive magnetic cards do not require a power source and data is only transmitted when it is passed through a reader that recognizes the medium and file format. Passive radio-frequency (RF) cards transmit only when queried by a nearby RF reader.

- Active cards are capable of transmitting whether a reader is present or not and generally require a power source.

RF cards are a common form of access card and RF has recently been incorporated into credit cards, as well. The transceiving antenna is built into the card. SAW ID cards are similar to RFID cards except that the signal is translated to an acoustical wave and the transmission range is longer than typical RFID cards.

The electronics to create RF cards can be printed on a machine resembling a laser printer that sells for about $3,000. This is inexpensive enough that larger companies can fabricate their own cards.

Radio waves can pass through wallets, clothing, walls, furniture, and other structures, so some people are storing cards in RF-blocking sleeves to prevent transmissions from being triggered or intercepted.

Recent innovations have substantially increased battery life for active transmitters to several years and transmission ranges to several thousand feet. Transmission distances are regulated by the FCC. RF tags, which are similar to cards but often attached to goods or keychains, are discussed further in the *Radio-Wave Surveillance* chapter.

Access Control

Data cards are sometimes promoted as an alternative to password-entry systems and they are commonly used to make financial transactions.

Passwords are easy to forget, sometimes easy to steal, and difficult to associate with a specific user. A card may also be forgotten or stolen, but offers more data security than a password, and may be associated with a specific user through biometrics so that a stolen card might be unusable by a third party. It can also be designed as a multi-access device, offering access to doors, computers, vending machines, and photocopy machines on the same card.

Left: Magnetic or RFID cards and keys are used as *access control* to selectively open doors to buildings, offices, hotel rooms, and other restricted areas. They may also be used like phone cards to purchase photocopy or vending machine services, such as are available on college campuses. Right: ATMs and point-of-purchase machines commonly accept debit and credit cards with magnetic stripes for transactions at numerous locations including banks, gas stations, supermarkets, and department stores. The data is often logged and stored and combined with other information to track movements or to chart behavioral or demographic profiles. [Photos © 2000 Classic Concepts, used with permission.]

Data cards can help track deliveries or protect access to dangerous, restricted, or "clean room" areas. They can facilitate financial transactions, insurance-covered medical care, subway rides, or open a gateway to online services. Cards can track purchases for customer rewards programs, print photocopies at a university, or enable library books to be processed more quickly. They can be used for metered services such as time-share computers or printers in commercial service outlets. Data cards provide access to long-distance phone services and may act as gift certificates.

Identity theft affects about five million Americans per year, so the National Strategy for Trusted Identities in Cyberspace (NSTIC) was launched to support the development and issuance of trusted credentials for data access. While this may have advantages for security and convenience, one must consider whether the data will be logged and monitored by issuing authorities. It is intended for the system to be voluntary, and for users to choose from a variety of government and private issuing authorities, but "voluntary" doesn't always mean a user can opt out and use another system such as passwords—voluntary sometimes means "no access" if the organization providing a service, such as a bank, provides access or special perks only to those with NSTIC credentials.

If a user chooses NSTIC as an access credential, what protections does he or she have for how data is logged, used, and archived? NSTIC service providers must abide by the Fair Information Practice Principles (FIPPs). These are policies of Homeland Security (arising from the *Privacy Act of 1974)* that regulate how information is collected, used, disseminated, and maintained and, further, that the user be informed of these activities.

How many card issuers abide by FIPPs is not known.

Internet Data Cards

Internet data cards are like phone cards, except they provide Internet access from any location where the issuing carrier offers wireless services. It is a form of Internet *roaming* service.

Authorization can be through a *dongle* (a small security device that looks like a memory stick and plugs into a USB or other data slot) or by swiping a data card. Data cards are sold by telecommunications companies with a variety of access levels, usually on two-year subscription plans. There are roaming services in more than 200 countries that hook wirelessly to the Internet, usually through local cell towers. Users don't have to rely on free wireless hotspots to connect to the Net and most companies don't restrict the card to one computer—the dongle can be moved from one laptop to another.

While USB data cards are not technically cards (they resemble memory sticks), they are often called data cards by retailers.

Retail Profiling Through Data Cards

Reward card membership is encouraged by retail stores and most will offer discounts for customers to use them. They provide information on the buying patterns of groups and individuals as well as making it possible to print targeted ads on the back of sales receipts. Data cards are also widely used in workplaces to control access to secured areas.

In 2008, The Netherlands began phasing in a smart card as the only means of payment for rides on public transportation systems. There were immediate concerns about privacy and theft, and worries that it might limit transit access for visitors. The security of the system was also questioned when a "hack" to get into the cards was posted on the Internet.

Smart cards log data in a way that is valuable to marketing firms and other third parties, especially if they can combine information from several organizations. Logged information is often sold in statistical or literal form, and when firms are bought out, smart card information from one venue (e.g., a supermarket), may be combined with those from another (e.g., a department store), thus turning individual puzzle pieces into a unified picture unanticipated by users.

Smart-card logs can be subpoenaed by law enforcement or used against a plaintiff or defendant in a lawsuit without secondary people who are represented in the logs being individually notified.

Even if FIPPs standards are applied and employees or other card users are notified of logs, this does not guarantee all forms of abuse can be prevented.

The workplace is highly competitive, with people vying for bonuses, promotions, raises, and movement within departments or across branches. Like other forms of surveillance data, smart-card logs in the hands of a system administrator, manager, or colleague could be used for political gain.

Employee Monitoring

Data cards are commonly issued for entry/exit access control on campuses and in businesses. Hotels and some apartment buildings rely on them as well.

Data-card access is not limited to doors. Data cards can also control or log access to plotters, photocopiers, washers and driers, phones, computers, and vending machines. This can be very handy for determining how often and when certain machines are utilized and convenient if they are prepaid.

Corporations have shown considerable interest in using data cards to monitor and track employees and sometimes use RF cards rather than magnetic cards for this purpose. RF cards can provide continuous monitoring without a person having to swipe them through a machine. Instead, an antenna transmits identity and location information to receiving or transceiving units and sends them to a central processor. A recent example is an RFID tag that transmits through an antenna attached to an employee's chair.[21]

A basic RFID card transmits only identity, usually through a serial number.

They are not especially secure—it is relatively easy to intercept the radio signals because many of them work on the same frequencies, and the small bit of information that is transmitted is not difficult to "hack."

RFID smart cards may include stronger encryption and may also store logs and transmit additional information about the user or his or her location.

Magnetic-Stripe-Card Surveillance

Magnetic stripes are familiar on credit cards, but they are now also included on some driver's licenses and may be swiped through card-reading machines at liquor stores, supermarkets, and department stores for proof of age or in conjunction with credit cards for proof of ID.

Those presenting the cards as ID for purchases are typically not informed as to why the card is swiped, what information is on the card, or how the information is used or stored, but it is possible for the retailers reading the cards to combine the card information with cash register information to build customer profiles.

Magnetic-stripe cards are also commonly used for financial transactions.

Security around ATMs has increased due to theft of PINs and other information about people's financial transactions. Most ATMs and card-readers in supermarkets and department stores now have a hood over the device so the keyboard and screen are less visible to people nearby (or to security cameras overhead).

---

[21] S. Srinivasan, H. Ranganathan, and R. Srivel, "Employee monitoring & HR management using RFID, electronics, communication and computing technologies (ICECCT)," *2011 International Conference.*

Most credit cards and many access and ID cards include a magnetic stripe on the back encoded with information about the user and company issuing the card. ID cards for small companies, community organizations, and colleges may have simpler encoding that is easier to breach than those intended for financial transactions. Cards can typically encode up to three tracks but often only one or two are used. Some cards have both magnetic stripes and RF tags. Information from the cards can be stored in databases or *spoofed* to gain unauthorized access or to commit identity fraud. [Photo © 2012 Classic Concepts, used with permission.]

One way thieves gain access to personal information on magnetic-stripe cards is to purchase a magnetic-stripe reader that can be used to read, write, and reprogram the cards. Data readers are portable and can be powered by plugging them into the USB slot on a laptop. It's the same technology common to bank machines.

So imagine this—a clerk works at a convenience store on a quiet shift and uses a card-verification system. A customer makes a purchase, the clerk swipes the card under the counter and hands it back. Did the card get swiped through the store's machine or a covert machine hidden next to it? Some systems display the person's age on a small screen to show that the card went through the system, but not all systems are the same and there could be two card readers under the counter.

Some universities issue magnetic-stripe cards to selectively permit access to services and buildings to registered students and authorized staff. The encoding on these is often quite simple. With a data-card reader, a thief could potentially find or guess a student name and number, code it on a card, and gain access to buildings or computing services.

Knowing what's encoded on a card makes it easier to create fictitious cards for use in less secure systems. For example, an access card for a small company might only include the name of an employee. Programming a new card with an employee name would give immediate access.

Reading information on a card to use the information for some other purpose or to create a new card with similar or identical information, is called *skimming*. Thieves have gone so far as to fit unauthorized card readers over legitimate ones to capture data as people try to use the machine. A user walks away thinking the machine is out of service without knowing the information on the card was read and could be duplicated. The PIN may even have been recorded by a hidden video camera aimed at the keypad.

A third-party credit-card processor announced in March 2012 that as many as 1.5 million North American credit cards may have been skimmed for their informational content.

The problem of stripe-card surveillance or theft will probably increase now that newer PDAs are equipped with built-in stripe readers so friends and business associates can make financial transfers through their handheld devices.

It is not always necessary to use a card to trigger a card-reading device. Rather than making the effort to determine card formats, numbers, or names, sometimes all that is needed to breach a card-based security system is an electronic device that manipulates the reader's magnetic field in the same way a card would stimulate the sensors when it is being swiped. This is called *card spoofing*. More secure systems use three data tracks along the length of the stripe, but many only use two. Spoofing is a technical trick, but there are always people who will sell these devices for a quick dollar and, apparently, many people buy them.

Bank machines have surveillance cameras aimed at the machine, sometimes from different angles, to prevent unauthorized access by card spoofing. Most will also have software that senses failed attempts at accessing the machines, so that alarms can be triggered if someone tries to hack in. Some include extra security in the form of sensors that detect the emanations from the second card device and trigger an alarm.

Despite all these precautions, there will likely be continued attempts to capture or modify card-related transactions.

BIOCHEMICAL SURVEILLANCE

Introduction

Biochemistry is a big part of surveillance but we're largely unaware of it because most of it happens behind-the-scenes in laboratories, some of which have carefully monitored temperature and humidity or which are designated as *clean rooms*—areas that are kept as sterile as possible to prevent contamination of samples or evidence.

Biochemistry is an essential aspect of surveillance and a very interesting one. Shining a black light on something to make it fluoresce, spraying Luminol™ to reveal bloodstains, or using a chemical or light to reveal hidden inks or signs of counterfeiting is like hunting for treasure. You're not sure if you're going to find anything, but when you do, it is often a rewarding surprise.

Recently, with smaller, more powerful components and better stabilization of optical imaging systems used to observe chemical processes, biochemical surveillance is coming out in the open. It is now possible to send mobile forensic labs or portable kits directly to a crime scene. Biochemical surveillance is particularly helpful for investigating cases of counterfeiting, arson, violent crime, theft in which traces may have been left behind, or products queued for import or export. Many of these situations can be better understood with help from high-powered microscopes or chemical tests.

Chemicals are sometimes used to mark items to prove ownership or to mark a thief who steals them. Dye packs and special powders are sometimes added to sacks that are handed to a bank robber. If the dye or powder gets on fingers or clothing and the thief is apprehended, it makes identification easier.

In the case of a murder investigation, where poison or drugs are suspected, tests can sometimes determine if there are any unusual chemicals in the person's bloodstream or organs, and may also reveal whether the chemicals were introduced before or after the person died.

Biochemical surveillance also involves the study of how bodies decay after people pass away. It is not a pleasant subject, but a body goes through distinct changes after death that can reveal how long it has been since the person died and what kind of environment he or she may have been in if the body has been moved.

If someone died by suffocation, but was later dragged to a chair, shot in the head, and the gun placed in a hand to make it look like suicide, specific biochemical clues help a forensic pathologist determine the real cause of death. Changes in temperature, stiffness, coagulation of blood, bruises, and other biological markers are all studied to determine if the person died before the gunshot occurred.

DNA profiling is another aspect of biochemical surveillance that has been highlighted on television and is now available in kit form to the general public for determining familial relationships. Consumer kits are most often purchased for paternity testing.

DNA profiling can identify familial relationships or provide evidence to help solve cases of rape or murder. Genetic evidence is sometimes able to solve crimes that are more than ten years old where someone was falsely accused, or where insufficient testing was available at the time of the original arrest to make a positive ID. DNA is discussed in more detail in the *Genetics* chapter.

BIOCHEMICAL WARFARE AND CHEMICAL-AGENT DETECTION

Biochemical Warfare

Chemical agents are constantly monitored in the armed services. Billions of dollars are spent on chemical agent suits, masks, detectors, and strategies to deal with biochemical events. Do we need chemical-agent detectors? Are biochemical attacks really that much of a threat?

Biochemical warfare has a long history. Rivals would poison the food or drinks of high-ranking officials to attain their status or wealth. Kings had *tasters* to test food before eating it themselves.

Fortresses, moats, and walls were built to defend inhabitants from attack. Attackers would throw smallpox-infected objects or disease-infested rats over the walls.

Sometimes armies would maintain hives of bees to unleash on their rivals. Scorpions and poisonous snakes were flung or slipped into holes. Sometimes combatants would "sting" with poison-tipped arrows. Catapults were used to launch dung or a mixture of toxins and human excrement into walled cities. Piranhas were stocked in moats to prevent people and horses from fording them.

Castle walls are difficult to scale and it took equipment to launch biochemical weapons over the parapets. Equipment couldn't easily be transported from one town to the next on foot, so attackers sometimes burrowed under the walls instead. A U.K. researcher examined evidence gathered over the years that suggests poisonous sulphur dioxide, created from burning bitumen and sulphurous compounds, was used to overcome Romans in tunnels when Persian soldiers attacked a city ca. 256 A.D.

Crude oil could be mixed with other chemicals to create a petroleum compound that would burn readily. This was ignited and thrown from castle parapets or spread sludge-like over water to wreak havoc on mariners who tried to get too close.

In ancient times, warships were equipped with storage bladders full of liquid concoctions that could be spread over water and ignited, or directed through a hose-like spout to create flame-throwers (siphones). The petrol mixture could be dispensed by mouth-blown pipe, as well, like a pea-shooter. [Tenth-century Byzantine manuscript annotated in Greek.]

In Asia, clouds of limestone were used to suppress peasant rebellions, and chemical bombs, which have a shared history with fireworks, could be lobbed at enemies like grenades. Since that time, additional lethal chemicals such as mustard gas, which was used in World War I (WWI), and Sarin gas, a deadly nerve agent hundreds of times more toxic than cyanide, have been developed.

Anthrax, a deadly natural bacterium, can be spread by direct contact with anthrax spores or by ingesting the meat of a diseased animal. Since it occurs naturally, it has long been used as a weapon of warfare. Since development of a vaccine in the 1880s, it has been eradicated in most developed nations but can still be grown in labs.

Napalm is a mixture of petrol and a chemical thickening agent. White phosphorus, which burns for quite a while, is used as a chemical "fuse" to ignite the napalm which, if it comes in contact with a person, creates a human torch, with the burns cutting deep through the skin into muscle and bone.

Prior to WWI, there were agreements to not use poisonous gases and prohibitions on manufacturing them. This didn't prevent the use of mustard gas during the war, so additional treaties were drawn up after the war.

In 1925, with input from several countries, the Geneva Protocol was established to prohibit the use of poisonous gases and bacteriological weapons. Prior to World War II (WWII), the protocol was ratified by a large number of countries but not by all of the great powers, and combatants still wanted leeway to retaliate with biological warfare if their enemies used it. The U.S. President, in 1943, stated they would not be used unless an enemy used it first.

In spite of treaties and prohibitions, chemical agents such as *Agent Orange* (so-named because of an identifying stripe on the barrels) were extensively used in the Vietnam war.

Agent Orange is a chemical defoliant to eradicate plants in forests and farms that might provide hiding places for ground militia. It is a highly toxic combat-grade chemical that causes genetic damage to humans. *Agent Blue* is similar, designed to destroy crops and disrupt food sources. Both chemicals were sprayed extensively by the U.S. in Vietnam until the early 1970s.

The U.S. Department of Veterans Affairs (VA) reports that more than 19 million gallons of "rainbow" combat-herbicide combinations were sprayed in the 1960s and that Agent Orange was most frequently used. The Vietnam Red Cross reports as many as 3 million Vietnamese have been affected by exposure.[22]

Left: This aerial photograph illustrates defoliated areas (left) sprayed with combat-grade herbicides. Right: Students at the chemical defense training facility in south-central Missouri wear protective suits while finishing up a course in chemical detection and defense training. Chemical detection paper and actual nerve agents manufactured at the site are used for final evaluation, after several training runs. For practice, nerve gas antidotes are administered to a uniformed dummy. Ceiling-mounted gas chromatographs inside the buildings constantly monitor the air to make sure there's no on-site contamination. The air test results are combined with entry/exit logs and video monitoring and kept in DoD files for a generation. [Photo courtesy of U.S. Veterans Affairs. News photo courtesy of American Forces Press Services, 2000.]

---

[22]"Agent Orange: Diseases associated with Agent Orange exposure," Department of Veterans Affairs Office of Public Health and Environmental Hazards, March 25, 2010.

It is estimated that approximately 650,000 people in Vietnam still suffer from chronic conditions as a result of the chemical warfare and half a million have reportedly died. The Red Cross estimates that about one million people in Vietnam and neighboring areas were disabled by Agent Orange.

Close to 400 barrels of Agent Orange were also provided to South Korea in 1968 for defoliating the demilitarized zone between North and South Korea.

Upon returning to the U.S., exposed U.S. Vietnam war veterans claimed their children had a high incidence of genetic defects such as spina bifida, and a possible increased incidence of acute myeloid leukemia. Other forms of cancer in the veterans themselves did not appear significantly higher but exposed veterans may have a higher incidence of diabetes. The *Veterans' Benefits Act* granted benefits for afflicted children in 1997. Vets settled a class-action suit out-of-court with the defoliate manufacturers in 1984. The VA offers health exams for veterans who served in Vietnam, the Korean Demilitarized Zone, or Thailand during the years the chemicals were deployed.

As a result, extreme safety measures are now taken when training military personnel in the detection and handling of deadly chemical agents.

Some countries manufacture chemical weapons as an alternative to nuclear bombs. Not every country has the capability to produce nuclear bombs, nor access to the materials, so they tout the existence of chemical weapons (even if they don't have them) as a "deterrent" against nuclear-capable countries. Much of the proliferation of weapons of warfare is ancient sabre-rattling carried out with bigger and more powerful chemical "sabres."

With the increase in biochemical agents development over the last half century, one might think piranha-stocked moats are long-forgotten aspects of Byzantine bio-warfare but there were appeals when the Geneva Protocol expressly prohibited piranhas in the updated 2011 version. Ukrainian delegates asked for an exception so crocodiles and alligators could continue to be used as moat security.

Playing with biochemical weapons is playing with fire. Poisons that are added to water sources or air currents not only kill civilians either directly or by decimating food crops, but may also kill or genetically damage the people who initiated the attacks so that they suffer health problems themselves, some of which lead to having stillborn children or children with birth defects. Natural diseases such as viruses and plague can be unleashed as biochemical weapons, according to nature's design rather than man's preferences, but they may also infect those who spread them and their allies.

Another result of brutal warfare is that the relatives of those injured or who lose their lives are often provoked to retaliate in kind and the original attackers have to live with the worry of reprisal—sometimes for generations.

As can be seen by historical examples, biochemical agents continue to resurface when war erupts, hence the development of biochemical surveillance systems and devices.

BIOLOGICAL AND CHEMICAL DETECTION

Chemical detectors fall into two general categories: those that involve a traditional chemical reaction[23] and those that are "sensed" electronically. Electronic sensors are becoming increasingly compact and powerful so they can be used in the field. Traditional chemical mixing, incubation, or centrifuging are still common procedures in labs and may also be used to confirm an electronic analysis or determine other information about a chemical or chemical compound that might be beyond the scope of portable electronic detectors.

Chemical Tests

Chemical reactions are often used to determine the composition of materials that are unknown, as sometimes happens with crime scene evidence, or which are known, but which may not be made in accepted ways (e.g., milk chocolate with suspected melanin contamination).

Taking a small sample and combining it with certain chemicals and watching for changes in temperature, color, or cloudiness, can help determine the character of a substance. Spectrometers are also used to optically inspect materials as a cross-reference to chemical tests or during or after a chemical reaction. Samples may be grown in incubators for a few hours or days to test for the presence of dangerous or prohibited bacteria or mold. This has to be done under carefully controlled conditions, since the incubation of dangerous substances creates larger amounts that might be hazardous to lab personnel.

M8 detection paper, which looks like a "ticket" and is hung on the outside of clothing like a badge, was developed by Edgewood Chemical and Biological Center (ECBC) to provide a convenient, portable means to detect nerve agents in the field. Ideally, the M8 is followed by tests with other equipment to confirm results, or check for agents that might not register on the M8 ticket. M8 is not a fast-response detector, but in situations where other devices aren't available, it can provide important information on whether an area has been contaminated.

Biochemical Research Facilities

The ECBC, headquartered in Maryland, was established in 1917 for research and development of chemical defense and, later, was expanded to include biological defense. The agency develops detection, protection, and decontamination systems, equipment, and technologies. It currently works out of three facilities.

The Detection and Engineering branch provides support for the development, acquisition, and maintenance of chemical defense systems used by the armed forces.

The Advanced Design and Manufacturing division provides assistance to military personnel working overseas in locations such as Afghanistan. In 2011, the ECBC established a field-assistance program at the Bagram airfield. Civilian

---

[23]As an example, chemicals can be combined in a test tube in measured amounts to note a change in color, texture, or consistency that may reveal their identity or properties.

engineers travel to outposts to provide technical assistance with chemical/biological detection-related equipment.

 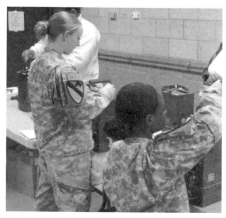

Left: The Edgewood Chemical and Biological Center (ECBC) develops technologies for detecting and identifying chemical and biological agents and works with other research facilities, such as the University of Delaware, to improve care for those contaminated and injured by biological and chemical exposure. Right: The EDBC also helps the army repair chemical/biological detection equipment such as Automatic Chemical Agent Detector Alarm (ACADA) units that have been returned from active duty. [News photos courtesy of the U.S. Army ECBC, released.]

Working with the ECBC Advanced Design and Manufacturing Division, Priv. 1st Class Zachary Root of the Mobile Public Affairs detachment positions a sensor that uploads information to an application program in a laptop. [News photo courtesy of the U.S. Army ECBC, released.]

Applications of Detection

Chemical detection is an important aspect of industrial safety and environmental monitoring for spills or other toxic releases. First responders to explosions or fires have to determine if any dangerous chemicals, such as chlorine or industrial pollutants, have been released into the air or local waterways.

Customs and Border Protection (CBP) has an ongoing job of monitoring items that cross the border to make sure they are genuine, safe, nontoxic, and made to certain standards. To make these determinations, they use a wide assortment of chemical tests to check the characteristics of samples of any goods suspected of being poached, or of not meeting import/export standards or regulations.

Chemical detection is an integral part of battle preparedness and response, and the Department of Defense (DoD) sponsors many programs related to biological and chemical agent detection, protection, and decontamination.

Between 2001 and 2004, DoD spending for chemical/biological (CB) defense increased from about $800 million to $1.1 billion. It is tempting to assume the increase was a response to the events of September 11, 2001, but looking farther back, spending had increased from about $400 million to $800 million between 1996 and 2001. Thus, a steady upward curve between 1997 and 2004 can be seen.

There are many biological- and chemical-detection systems under the auspices of the DoD. A few examples are listed here.

- The Joint Bio Standoff Detector System (JBSDS) is an early-warning system for detecting biological warfare agents in near realtime that can be mounted on the back of a vehicle.

- The Joint Chemical Agent Detector (JCAD) is a compact, lightweight detector intended to detect and identify low levels of chemical warfare agents.

- The Joint Services Lightweight Standoff Chemical Agent Detector (JSLSCAD) is a compact passive infrared detection system that senses the presence of chemical warfare agents that show activity in infrared wave bands. The system has the capability of distinguishing infrared signatures that stand out from what is commonly found in certain environments. A chemical release can be sensed up to about 3 miles away, depending on the wind and amount of chemical in the air. A vehicle-mounted Mobile Chemical Agent Detector (MCAD) is a similar device that has been tested as an alternative to JSLSCAD.

- The Joint Biological Point-Detection System (JBPDS) is a modular suite of instruments that can be mounted on service platforms for automated operation. It is designed to identify local biological agents within 15 minutes of exposure.

- The Joint Biological Tactical Detection System (JBTDS) is a lightweight system for detecting, warning, and providing preliminary identification of biological agents.

- The Joint Bio-Agent Identification and Diagnostic System (JBAIDS) is designed to detect and identify biological disease agents.

- The Joint Service Chemical/Biological/Radiological Agent Water Monitor (JCBRAWM) is a portable unit designed to detect chemical and radiological contamination in water. With immunossay "tickets" the JCBRAWM can also perform biological detection. The device is intended to supplement water testing kits.

The sensing of chemical vapors at a distance is usually accomplished with infrared absorption spectroscopy. Substances that don't absorb infrared, like chlorine gas, are detected using other technologies.

Chemical detectors are also incorporated into chemical protection gear like masks and suits or used separately to test the integrity of the suits and make sure those inside them are protected from leaks and haven't been contaminated. The Joint Service Mask Leakage Test (JSMLT) is used to check masks in the field.

Remote data relays (RDRs) that transmit over satellite links and computer networks are an integral aspect of most of the recent biological and chemical sensing technologies. The results can be logged and transmitted to other sites through radio links or cables to create event logs for statistical analysis to chart contamination levels over time or to identify patterns in their detection or use. The logs can also be analyzed to provide information for command decisions and emergency responses whether in military or civilian contexts.

The JBSDS is a transportable biological agent detector for detecting biological agents at a distance. [News photo courtesy of the DoD, released.]

In recent years, the number of governmental departments dealing with chemical detection and analysis has increased. The Department of Homeland Security includes a Chemical and Biological Division and Chemical Security Analysis Center (CSAC) tasked with chemical threat awareness, assessment, and analysis. It initiated an Autonomous Rapid Facility Chemical Agent Monitor (ARFCAM) system in 2004 intended for continuous facility monitoring.

Various government labs have been developing chemical detectors and contracting them out so they can be manufactured for military and industrial applications. The Department of Defense has a Joint Chemical Agent Detector (JCAD) program for acquiring and deploying chemical detectors in military ships and aircraft. The devices are handheld and suitable for installation in vehicles as well and can store logs on the chemical-detection data. Chemical-detection device simulators have also been contracted by departments like the FBI, for training purposes.

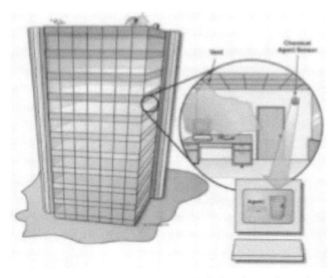

The Department of Homeland Security's ARFCAM program is designed to continuously and autonomously detect the presence of a broad range of toxic chemicals. It is aimed at facility protection with information relayed from detectors to centralized computing systems. [Informational diagram courtesy of the Department of Homeland Security, 2011.]

The Automatic Chemical Agent Detection Alarm (ACADA) is a point-source (local) system capable of detecting and identifying common blister and nerve agents. Once installed, it functions autonomously to sense chemicals and provide audio and visual alarms as area warnings for various branches of the military. ACADA simulators are used for training. The ACADA has been in use since the late 1970s.

The U.S. Naval Surface Warfare Center (NSWC) announced in early 2012 that their fleets would be equipped with automated chemical detectors.

One of the ways chemical or biological agents can be dispersed is through missiles or other airborne explosives that shower the surrounding area upon impact. This works better for some agents than others, since some, like vapors, may ignite and burn off quickly. This kind of chemical attack is more difficult to detect, since it arrives quickly from another location and disperses quickly. Once chemicals have been detected, however, protective measures can be taken and other incoming projectiles from the same source tracked.

Detector Trends

The U.S. Army uses a mobile Biological Integrated Detection System (BIDS), mounted on the backs of Humvees, that is designed to detect four different chemicals simultaneously.

Many chemical detectors rely on mobility spectrometry, with ion-mobility being the most common. Ions have a detectable velocity (speed and direction) that can be assessed when traveling through an electrical field. Examples of chemicals that are commonly sought with mobility spectrometry include chlorine gas, hydrochloric acid, phosgine, arsine, ammonia, and hydrogen cyanide. Those associated more directly with chemical warfare include mustard gas, sarin, VX, lewisite, and tabun.

The MicroChemLab is a handheld system that enables sample handling, chemical separation, and detection in one system. Gas chromatography is used to separate chemical compounds and SAW (surface acoustic waves) sensors then make a determination of a chemical. The MicroChemLab is suitable for detecting biochemical weapons, explosives, and for use in DNA sequencing in the field.

The portable bio-detection MicroChemLab developed by Sandia National Laboratories was found capable of detecting and analyzing toxic agents in laboratory tests, and was field-tested in the fall of 2003. The gas-phase device was initially developed for chemical warfare detection and emergency response, but also has applications for ecological monitoring and industrial process control. [News photos courtesy of Sandia National Laboratories, 2003 and 2011, released.]

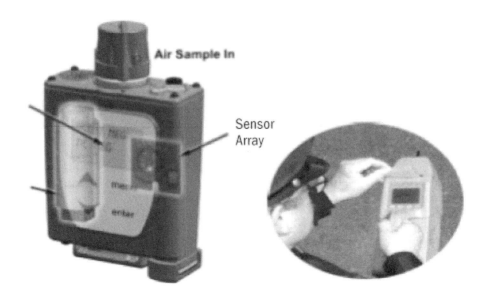

Air Sample In

Sensor
Array

The Lightweight Autonomous Chemical Identification System (LACIS) is a network-capable handheld chemical-detection unit designed to respond to a wide range of chemical hazards and identify contamination. [Illustration and photo courtesy of the Department of Homeland Security, 2011.]

The Lightweight Autonomous Chemical Identification System (LACIS), also developed at Sandia Labs, is an electronic handheld chemical-sensing device intended to detect and identify toxic industrial chemicals and chemical-warfare agents.

Chemical detectors use different approaches and some are specialized. Some form of spectrometry is used in most detectors, although some are based on gas chromatography. The most common method is *ion mobility spectrometry*, but *differential mobility spectrometry* may have some advantages.

Another project within the DHS is the development and testing of chemical detectors that can sense low levels of toxins with a device that doesn't have to touch the surface of a possibly contaminated object. Systems like this would help prevent the spread of further contamination and would be helpful in crime scene investigations so as not to disturb the evidence by direct contact.

DHS has also outlined plans for an Integrated Chemical, Biological, Radiological, Nuclear, & Explosive (CBRNE) system by which chemical surveillance information collected by responding agencies, such as police departments, fire departments, emergency operations, and transportation centers would flow through networked repositories in federal data-sharing agencies.

A mobile laboratory (a van equipped with chemical-detection systems) was also developed and transitioned to the Environmental Protection Agency (EPA) in 2007.

Left: Sandia National Laboratories maintains a state-of-the-art chemical and radiation-detection laboratory in California designed for researching chemicals, sensors, and detectors. Right: In 2008, prototype handheld chemical detectors were tested for emergency response suitability. Chemical suits protect first responders until it is known whether any toxins are in the area. [News photos courtesy of Sandia National Laboratories and the Department of Homeland Security, released 2008.]

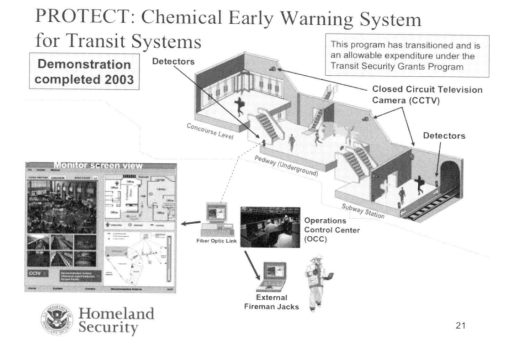

In 2003, the Department of Homeland Security demonstrated a chemical early-warning system for transit systems consisting of chemical detectors and surveillance cameras gathering data and providing alerts that are relayed to an Operations Control Center (OCC). [Informational diagrams courtesy of the Department of Homeland Security, 2011.]

Field targets as seen
through the PORTHOS™
viewfinder

In March 2012, Johns Hopkins University awarded Block Engineering, a Marlborough company, a contract to support development of handheld versions of the Portable Hazard Observation System (PORTHOS) on behalf of the Intelligence Community. The current model is transportable, but there is a continued desire for devices that are handheld. [News photos courtesy of Block Engineering, released.]

PORTHOS® is a multiple-chemical-agent detector and identification system able to quickly sense hazardous vapors up to about three miles. The infrared spectrometer is designed to detect nerve and blister agents as well as toxic industrial chemicals such as ammonia or nitric acid. The interferometric raw data, identification information, and time are logged. Block Engineering also developed the Mobile Chemical Agent Detector (MCAD) passive infrared detector, which has been tested as an alternative to the Joint Services Lightweight Standoff Chemical Agent Detector (JSLSCAD) mentioned earlier.

Chemicals often occur in combination with one another or with ambient environmental chemicals in air and water, and it helps identification if they can be separated or otherwise distinguished. Different molecules have different weights and boiling points, providing clues as to their identity. With gas chromatography they can be funneled through a narrow channel so that their individual differences cause them to emerge at different times. Then mass spectrometry ionizes the chemicals within a magnetic field to determine their mass-to-electrical-charge ratio. This information can be compared to a database of stored values for different chemicals to make the identification.

In 2009, the Marine Corps contracted for a portable gas chromatography/mass spectrometry (GC/MS) system for evaluation as a chemical-detection technology for use in the field. This resulted in improvements to the Torion Technologies/ Smiths Detection system that featured a smaller size and faster detection times.

Systems that used to be transportable in the 1990s, requiring a truck to relocate them, were becoming portable devices in the 2000s.

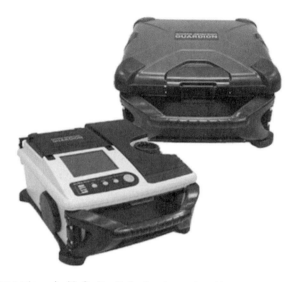

Torion Technologies partnered with Smiths Detection to create a high-speed, high-resolution GC-MS Guardion™ system for detecting and identifying chemical warfare agents and toxic industrial liquids and vapors. The unit is about the size of a carry-on case and can analyze up to 30 samples before the battery has to be swapped out or recharged. [News photos courtesy of Smiths Detection, released 2012.]

It is clear from the previous examples that faster, smaller, more powerful chemical detectors are desired for military, governmental, and industrial security and that significant improvements in the technology have occurred in the past decade.

Customs and Border Protection Chemical Detection

Customs and Border Protection (CBP) makes extensive use of laboratory chemical tests to determine the composition and quality of materials crossing the border. It also enlists portable chemical detectors to locate and identify chemicals smuggled in luggage, open boats, ship's hulls, and sometimes in the inner walls and linings of land and marine vehicles.

CBP uses a variety of other technologies to check cargo and movements across borders, such as ultraviolet light, infrared detectors, and X-rays for large containers. One of the problems of using X-rays is that high-energy radiation is needed to penetrate the walls of shipping containers and sometimes there are human refugees or illegal migrants hiding inside. The X-radiation machines are costly and require careful safety measures to operate, and may cause physical harm to people exposed to high levels of radiation.

Chemical detectors to detect "human cargo," people hiding in boat hulls, trains, and shipping containers as a prerequisite to other more invasive surveillance technologies, such as X-radiation, would make it possible to pre-screen large containers before applying X-rays and to detect humans hiding in a wider variety of locations, such as border tunnels and marine vessels.

CBP is on constant alert for smuggling. People have tried to smuggle drugs, jewelry, coins, parts of endangered species, and other contraband items inside luggage, hollowed-out books, dogs, snakes, and their own bodies. In December 2011, CBP X-ray inspection revealed $140,000-worth of methamphetamine hidden in containers of nacho cheese sauce and jalapeños.

Smugglers often hope that the smell of food can mask the smell of illegal drugs, but that doesn't stop X-rays from revealing the contents and dogs can often distinguish drugs from food. Sometimes fake foods are used to hide drugs, but they have to be very good fakes to pass visual inspection.

Chemical detection is an important aspect of forensics. Detectives and forensic lab personnel frequently test for blood stains, blood types, DNA traces (especially in hair and saliva), tire tracks, paint chips from cars (especially those involved in hit-and-run accidents), and identity markers on stolen objects.

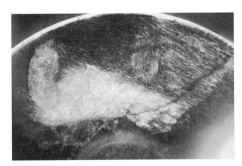 

Top: Customs and Border Protection is constantly challenged to monitor imports and exports for quality, counterfeiting, contamination, or other anomalies that might make them unsafe or unsuitable for cross-border transport. Chemical testing is an important aspect of this work. Here lab workers prepare samples for analysis of imported goods. Bottom left: Chemicals and lights can be used to reveal faults, fingerprints, fractures, or other traces that provide information about items that are faulty or counterfeit. Bottom right: Poaching is always a concern, as is drug smuggling inside animals and objects. This alligator head is a replica, but anything that appears to be contraband needs to be checked. [News photos by James Tourtellotte courtesy of Customs and Border Protection, 2006.]

ANIMAL CHEMICAL DETECTORS

Canines as Chemical Sniffers

Biological chemical sniffers (dogs) are still used regularly by a number of government departments, including CBP and the Bureau of Alcohol, Tobacco, Firearms and Explosives (ATF).

Sniffing dogs have long noses (compared to humans) and extra chambers in their noses to further distinguish individual smells. They are also eager to work and able to take instructions from a distance, which is not possible with handheld chemical detectors.

CBP enlists sniffing dogs to check luggage in airports and other border terminals for illegal agricultural products, drugs, and explosives. The ATF uses dogs to sniff out explosives and various chemicals that may be associated with arson investigations.

Dogs are valuable aids in search-and-rescue operations, robbery and missing persons investigations, as well as customs and border missions. But dogs tire after a couple of hours of constant sniffing and need breaks. They also have to be trained for a year or two to learn their jobs. As a result, microelectronics have been used to design sophisticated chemical sensing tools to locate drugs, explosives, and other articles of interest to law enforcement personnel and investigators in situations where a dog isn't available or isn't the best choice for a particular sensing task. They are also useful in environments that might be dangerous to a dog, such as the site of a toxic chemical release.

Left: An instructor at the CBP Canine Enforcement Training Center demonstrates chemical-detection abilities of one of the trainees as the dog sniffs packages on a conveyor belt. Right: The ATF often has to seek out chemicals in industrial accidents, arson investigations, and building security. Explosives-sniffing dogs in the ATF must be able to individually detect 20 different explosives scents, two of which were not included in their initial training. [News photo by Gerald L. Nino, courtesy of CBP, released 2003. ATF news photo, released.]

Public Terminals and Large Events

Airports and Sports Stadiums

Most people have heard of body scanners and X-ray machines in airports, but they may not know that there are also chemical sniffers in some locations. Starship-style "portals" with holes that poof jets of air can be used to "sniff" for chemicals that might indicate explosives. Wind is blown across the person's body so that it travels down a channel that separates the chemicals to make them easier to identify.

By making some adjustments to the sensing channel, it was possible to take this technology and make it smaller, and cabinet-sized units now dot strategic locations in stadiums that host large public events, with the data sent to a central location. Handheld versions of the sniffers exist, as well.

It is a challenge for scientists to create practical devices that can distinguish potentially dangerous chemicals in public venues from ambient smells such as food, cigarettes, and sweating crowds.

Sports Surveillance

Doping Detection

Athletes are very competitive and sometimes try to gain an advantage by using restricted or prohibited chemicals to build muscles or otherwise enhance performance. Drug testing is a common form of biochemical surveillance carried out in competitive sports and sometimes also in the workplace, especially in jobs that require the use of dangerous machinery. Urine and blood tests are most commonly used, since byproducts of chemicals may be detected in urine or blood.

The World Anti-Doping Agency (WADA) maintains a list of prohibited substances that may not be used before or during competition. The most common drugs tested for in sports are steroids and growth hormones or hormone modulators. Diuretics, which encourage the excretion of fluids, are also banned. In recent years, gene doping has also been added to lists of banned products and procedures. During competition itself, stimulants such as cocaine, ephedra, cannabinoids, and sometimes caffeine may also be prohibited.

Athletes have clever ways to substitute someone else's urine for testing, going as far as attaching prosthetic genitals with urine reservoirs that can be squeezed to make it look like the athlete is peeing naturally into a sample cup. The "clean" urine is purchased from donors with no known medical conditions or banned substances in their bodies. Many suppliers openly advertise on the Net. Given this much commerce in substitute urine and prosthetics, clearly doping still happens despite chemical surveillance.

For over a decade, Sandia National Laboratories has been developing and field-testing electronic chemical detectors, often called "sniffers," in various law enforcement labs, airports, and public venues.

Top left: This chemical-detecting portal is designed to sense various chemicals used in the production of explosives. The portal, a form of chemical "body scanner" was tested at the Albuquerque International Airport. A puff of air is passed through the holes to guide chemicals to the sensing units. Top right: Project leader, Kevin Linker, looks through the chemical-sniffing air-intake valve.

Bottom Left: Researchers found ways to reduce the size of the chemical pre-concentrator and thus the size of the system so that portable versions could be carried to crime scenes, industrial sites, or other locations where chemical detection was desired. Bottom right: Next to a sports stadium, this cabinet may look like an abandoned hot dog stand, but it contains a wide array of high-tech equipment, including a chemical sniffer. The unit is networked to other sniffing nodes around the stadium, with surveillance cameras for extra security. The nodes are designed to sense more than 40 different chemicals, including chemical warfare agents and industrial chemicals. [News photos by Randy Montoya, courtesy of Sandia National Laboratories, 1997, 1999, 2004, and 2006, released.]

AGRICULTURAL SURVEILLANCE

Monitoring Food, Fish, and Agricultural Practices

Chemical reactions are used to test a number of agricultural products. Constant surveillance of the agriculture industry is necessary because food, especially meat and poultry products, may contain excessive amounts of growth hormones or other chemicals, or may be overloaded with pesticides or industrial contaminants. Food also has to be examined for signs of infestation (e.g., worms) and for diseases that might be communicable to humans, such as avian flu or mad cow disease (bovine encephalitis).

As the population increases, demand for food increases, and agricultural businesses increasingly look for ways to produce more product in less space and time. Chemical contamination and unsanitary practices happen more often in situations where there are cutbacks to ensure or increase profitability.

Fish stocks also have to be carefully monitored to make sure endangered species aren't harvested along with permissible stocks and to ensure that existing fish stocks are not decimated by over-fishing.

Genetic Engineering

A new aspect of food production surveillance is monitoring the results of genetic engineering. Genetic engineering involves manipulating genes or inserting them into plants or animals to change their characteristics, capabilities, size, or manner of growing.

Mother Nature adjusts to changes in ways that can't always be predicted and agricultural businesses have ways to increase profitability that aren't always safe for humans or the environment. Genetic engineering has created many novel products, such as goats that can produce spider silk in their milk. The silk can be harvested as one of the strongest microscopic "ropes" in existence. Natural pesticides can be genetically inserted into certain plants to increase crop production, but then humans may be ingesting the pesticides along with the food.

It is also possible for the seeds of genetically modified plants to get loose and hybridize with other plants. The other downside to built-in pesticides is they may kill important insects that are food sources for birds and animals, or which are vital for pollinating other plants such as nearby food crops.

Sometimes "biological police," such as bees, are enlisted to find and track pollen that can be used for forensic purposes. Pollen can be tested to get a sense of which plants are in the vicinity of the bee hive (which may include drug-related plants or unlicensed genetically altered plants). Pollen-collecting bees might also be used by business competitors to "spy" on plants that are genetically altered or which are being grown by pharmaceutical companies or by companies seeking to make sure their engineered plants are not growing elsewhere—either because the seeds have spread or because they have been stolen by amateur growers or competitors for cultivation.

ENVIRONMENTAL SURVEILLANCE

Toxic Spills

In past millennia, human garbage was all biodegradable. Clothing, shoes, adornments, and implements were all made from stone, bone, wool, silk, leather, and other materials that are readily recycled into soil and plants by nature.

Industrialization and chemistry changed this. Trash is no longer nontoxic.

Because of the vast quantities of synthetic products that are manufactured and discarded by humans, surveillance is necessary to keep our environment healthy, to protect our food stocks, livestock, and wildlife. Environmental surveillance also enables us to detect natural disasters or the consequences of human error that might cause injury to people or their habitats.

Oil, gas, and toxic effluent leaks can poison drinking water and food resources such as crops and fish. Some spills and leaks result from tsunamis or earthquakes, others because of human negligence. Early warnings and response are vital to containing a spill and making sure there are no gases that might ignite or contaminate the air.

Some industrial systems have chemical detectors built into equipment or building structures. They may trigger local alarms, as well as network alarms to remote locations.

Often visual surveillance technologies, such as aerial photographs, are used to confirm and monitor environmental contamination.

CHEMICAL DETECTION IN HUMANS

Gulf War Syndrome

The question has often been asked whether Gulf War veterans were exposed to chemicals dispersed by the Iraqi military or by chemical agents released by the U.S. military. It has also been surmised that inoculations given to service personnel may be responsible.

The U.S. government reports that they had seventeen chemical detection systems in the Gulf region during the Gulf War. The Office of the Special Assistant for Gulf War Illnesses has stated

> "We have no evidence that biological weapons were used during the 1990–1991 Gulf War. In addition we have not found any reports of verified biological agent detections nor are we aware of anyone, soldier or civilian, who has reported experiencing symptoms consistent with exposure to a biological agent.
>
> To date, there is also no evidence that any warfareagents [sic] were released as a result of coalition bombings during the air war campaign, January 17, 1991 through February 23, 1991..."

One of the problems with detecting and clearing chemical spills is that sometimes equally toxic or more toxic chemicals are used to douse flames, contain leaks or, if it is a marine-based leak, disperse crude oil. First responders may also be exposed to chemicals due to inadequate protection, either because information about chemical dangers is unavailable in emergency situations, or because equipment to cope with a specific incidence may not be available.

This is a problem in industrial accidents or explosions involving nuclear fallout, where it is difficult to avoid both chemical contamination related to explosions and atomic damage from the nuclear fallout itself.

Like chemical spills, nuclear fallout may further contaminate local water and milk supplies, not only directly, but indirectly, when cows or goats eat contaminated grass. The explosion may also damage other facilities and release toxic chemicals.

Chemical contamination can become a problem in unanticipated incidents such as these:

- Materials that are fabricated to create large buildings include toxins that are normally in solid form or contained within protective surfaces. They are not considered harmful when incorporated into internal building structures. The collapse of the World Trade Center released materials that were never meant to be breathed such as glue, binder, asbestos, heavy metals, plaster, concrete, fire retardants, and plastic coatings—many of which were crushed to fine powder and released into the air in large quantities.

- A gas pipeline leak in the Pacific Northwest caused an explosion that roared through a park along a watercourse. Thick gray clouds billowed hundreds of feet in the air and blanketed many acres. Since the explosions were near a chlorine plant, it was unknown, at first, whether toxic chlorine was mixed with the clouds. First responders risked contamination in order to contain the fire to prevent it from spreading throughout the park and nearby houses before it was known how many chemicals the smoke might contain.

When chemical toxins are released into air and water, it is important to monitor people near the site to see if they were affected by contaminants spread by leaks or explosions, or by the chemicals used to douse or contain the incident. Usually blood is drawn at intervals over time, but urine tests and other chemical tests may be necessary depending on the nature of the contamination. Since people with chemical exposure may have pre-existing conditions, like diabetes or heart disease, additional tests are often necessary to distinguish pre-existing problems from new ones directly related to chemical exposure.

A chemical dispersant was used to sink the crude oil in the 2010 Gulf Coast spill to prevent it from spreading farther through the gulf, and was later claimed to be detected in humans near the coast, some of whom had never been in the immediate vicinity of the spill.

In 2010 a massive oil spill, the largest in American history, occurred off the Gulf Coast and the DoJ filed a civil suit against the oil company. Further lawsuits followed by local residents claiming chemical damage to property and health. [News photo courtesy of the Department of Justice, released.]

Residents filed lawsuits in Alabama, Louisiana, and New Orleans federal courts. One of the claims was that planes flying at night dispersed a toxic chemical that caused property damage and illness. This too entails chemical detection since lawsuits are based on evidence, and forensic evidence of chemical exposure or evidence of symptoms specifically known to be a result of certain chemical exposures would be expected in court.

There has been controversy over whether chemical dispersants used in the Gulf were safe. The makers of the dispersant released a statement that the chemicals were a blend of well-established, safe ingredients that were biodegrade and which don't bio-accumulate. In congressional testimony, Lisa Jackson, Administrator of the U.S. Environmental Protection Agency (EPA) stated

> *EPA has also established an extensive network to rigorously monitor the air, water, and sediments for the presence of dispersants and crude oil components that could have an impact on health or the environment.*

Lisa P. Jackson, Administrator, EPA, July 15, 2010[24]

The EPA took samples from petroleum waste and conducted daily sampling from air, water, and sediment. Hourly readings were collected by the mobile Trace Atmospheric Gas Analyzer van. Information from sampling and other tests were forwarded to the Centers for Disease Control and Prevention (CDC).

---

[24]The EPA established a website to document its response to the BP spill. This includes what chemicals are being monitored and how *(www.epa.gov/bpspill/)*. Results are reported on another site that includes general information about the health of the Gulf region *(www.restorethegulf.gov/)*.

After a federal court decision, the Environmental Protection Agency released a report on dispersants used in the Gulf. When evaluated by an independent agency, the list was said to include more than four dozen chemicals, some of which are known to cause skin irritation or respiratory irritation, and some which are known or suspected carcinogens. This appears to contradict reports from the EPA and the CDC regarding chemical danger and cleanup progress.

The National Institute of Environmental Health Science (NIEHS) began a project to study the health of Gulf Coast residents to see what long-term effects there might have been from handling or exposure to the agents sprayed on the oil. Investigators were enlisting residents of Florida, Alabama, Mississippi, and Louisiana to participate.

At the request of BP, the National Institute for Occupational Safety and Health (NIOSH) is conducting a health hazard evaluation of Deepwater Horizon Response workers.

Potential risks to human health from contact with oil from the spill, listed by the name of each chemical and its toxicity level, have been posted by the EPA at *www.epa.gov/bpspill/health-benchmarks.html* along with cautionary warnings to avoid areas that appear to be contaminated or which smell of oil.

Incidents like the Gulf oil spill illustrate that chemical surveillance is not only important for public and environmental safety, but is part of the process of sorting out who is responsible for the original incident and who is liable for the aftermath and cleanup. It is also used for longer-term studies monitoring people and environmental resources that may have been impacted by chemical releases.

Armed Forces Chemical Detection

Chemical detectors are found throughout the armed forces and some of the handheld detectors discussed earlier are utilized by the U.S. Army in locations like Afghanistan and Iraq. Transportable, vehicle-mounted, and handheld devices are favored by ground forces and are particularly important in combat zones where chemicals may be included in flying projectiles or exploding bombs.

Toxic, Industrial, Chemical Protection and Detection Equipment (TICPDE) training and certification is conduced by the U.S. Army to provide technical expertise to chemical soldiers for reconnaissance and sampling operations.

The U.S. Navy uses chemical detectors in their marine vessels and, in 2012, initiated a six-year plan to upgrade the chemical-agent detectors on their cruisers, destroyers, aircraft carriers, and other warships. The Improved Point Detection System-Lifetime Replacement (IPDS-LR) system is based on spectrographic analysis of ions that have been separated as they move through a magnetic channel. Air samples are taken outside the ship and channeled through intake valves on either side of the hull. If suspicious chemicals are detected, an alarm is sounded.

Chemical sensors are often used in mine-clearing missions, as well, since the explosives inside the mines, such as C4, can be detected. Metal detectors are sometimes used in combination with chemical detectors.

Chemical detectors are less controversial than many other aspects of armed services. Since they are intended to protect rather than kill, few people oppose them. Other aspects of the detectors have been criticized, however. Since chemical detectors are funded with tax money, taxpayers want to be sure they are effective under combat conditions, which may include unusually hot or cold situations that could affect results. Thus, continued support of research and development and upgrades occurs.

MICRO-ELECTRONIC CHEMICAL DETECTORS

Cellphone Sensors

The quest for ever-smaller chemical detectors continues and some are now so small they can be plugged into a cellphone or incorporated into the housing of the phone.

An idea originating at NASA/Ames was demonstrated in Sept. 2011 by the DHS Science & Technology Directorate and the Los Angeles Fire Department, who used a commercialized version of the Cell-All developed at NASA to detect a mock carbon-monoxide release. This tiny detector can fit in a pocket and provide preliminary information at a suspected chemical incident.

Even smaller detectors may be possible. At UC San Diego, a porous flake of silicon changes color if exposed to certain chemicals. By engineering the structure of the pores, the component can be "tuned" for specific chemicals.

Jing Li, a scientists at NASA/Ames has developed a compact electronic chemical sensor (left) that can be interfaced with a cellphone docking station (right) to detect small amounts of chlorine gas, methane, and ammonia. The chip includes 16 nanosensors and can relay the information via the cellphone to another phone or computer. A newer version of the chip includes 16 nanosensors per side, for a total of 64 nanosensors. [NASA/Ames news photos by Dominic Hart, 2009, released.]

The concept, like the Cell-All, is to create a subsystem small enough to incorporate into a cellphone. The prototype demonstrated the ability to distinguish between certain chemical warfare agents, like mustard gas, and industrial chemicals like toluene. The color changes can be imaged with a tiny camera based on a fluid lens rather than a glass lens. A fluid lens makes it possible to focus dynamically so that each pattern of sensing pores doesn't have to be individually wired to report results. This greatly reduces the size of the component.

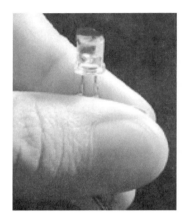 

At UC San Diego, researchers have worked with startup company Rhevision, Inc., to develop a tiny silicon chip that can be fabricated to respond to airborne chemicals by selectively changing color, depending on the pattern of the pores in the chip. [News photos courtesy of Sailor Lab/UCSD, 2010, released.]

Detectors for First Responders and the Public

Products like the Cell-All represent a fundamental shift in chemical detection. Not long ago, highly trained response teams, with mobile labs and access to larger labs, were the only ones with the expertise and equipment to assess local chemical situations or possible contamination. First responders and the public would have to wait to see if an area was safe or whether they may have been exposed to toxic pollutants.

Now, with increasingly portable devices, first responders can make an initial assessment, to help decide on how to go about their jobs, and the public can be warned of possible dangers prior to more extensive lab tests and evaluations. Tiny devices are still limited in what they can detect and need to be supplemented with more comprehensive testing, but they are likely to improve in both power and versatility in the next few years.

Tiny chemical detectors represent another important change, as well. Since a cellphone is a communications device, results from the environmental sensing components can be relayed to authorities, an idea discussed by the Department of Homeland Security (DHS). Whether DHS means to continuously monitor environments where the public goes, or receive specific incident reports, is not yet clear.

Commercial Applications of Networked Public Sniffers

In a few years, when these tiny chemical sensors are able to detect a wider range of chemicals, commercial applications of the technology are likely to follow. A network of public testers, whether volunteers or those hired by an agency, could sense much more than chemical leaks, they could theoretically sample the perfumes, shampoos, and other scents that are associated with personal hygiene and the purchase of commercial products. This, once again, is consumer demographic data that advertisers are hungry to collect so they can capitalize on the information.

Consumer chemical sniffers may also be used by store personnel to pre-assess customers before asking if they "need any help?" at the personal hygiene or makeup counter or, in a more covert way, to sample chemicals emanating from customers when they pass through checkouts. Since the checkout process records who the person is, either via payment method or reward-card profiles, it would be easy to add the chemical assessment to a growing log of information on that specific customer.

One can envision this kind of technology finding its way into movies. Imagine a confrontation in which one spouse accuses the other of cheating and "proves" it by matching the scent of perfume, lipstick, shaving cream, or shampoo to another person.

ANIMAL SURVEILLANCE

KEY CONCEPTS

- *how animals are used*
- *animals of particular interest*
- *bio-robotic hybrid surveillance systems*

Introduction

Animals are frequently used as surveillance aids because they can fit in places humans can't, fly to places we can't easily fly, and hear, see, and smell things we're unable to sense without technological aids.

Animals like dolphins and bats can "see" with sound, which enables them to visualize things behind cloth or sand or other materials that enable sound waves to penetrate.

Dogs and dolphins are eager to work. Their body language conveys a sense of pride and accomplishment when they figure things out, much the same as humans.

Sea lions are willing to work, but they are also eager to play and while they are helpful in marine surveillance, especially in seeking out mines, they take frequent breaks to engage in marine antics.

Animals are also helpful as surveillance "alarms." Dogs bark or growl when they sense intruders. Certain breeds of cats will growl also—Siamese cats were used as guards in past Asian dynasties. Birds regularly sound alarms, with each species having a different "danger" call. Parrots and jays are quick to squawk at intruders.

For hundreds, perhaps thousands of years, dogs have helped chase thieves or search for missing persons. Blood-hounds have been specifically bred for this task, but beagles, certain spaniels, golden retrievers, and a number of other species are good at following or finding things by scent. Dogs are also trained as seeing eye dogs for the visually impaired or to scent blood-sugar-levels for diabetics.

Dogs are particularly helpful in search-and-rescue operations. They can locate a skier buried in an avalanche or crawl through the rubble of a collapsed building to find a person pinned under debris.

Even bees can be enlisted as surveillance agents as they travel out to flowers and bring back pollen whose genetic and chemical characteristics tell something about what is happening beyond the hive.

Some creatures are very sensitive to environmental chemicals. Canaries were traditionally used in coal mines to test whether the air was safe to breathe. Frogs and salamanders are very sensitive to industrial pollutants, contaminated water, and other environmental anomalies and will show birth defects or die-offs if toxins are present—a form of biological chemical sensor.

Sometimes animals themselves are the subject of surveillance. Over-fishing, poaching, industrial spills, and smuggling are all problems that involve animals. Wildlife protection requires careful animal surveillance to make sure there is sufficient habitat and that nature's balance isn't upset by intrusive development, over-harvesting, or illegal hunting.

Dogs are one of the most common animals used for surveillance. They protect homes and ranches by patrolling fences and doorways. They are enlisted by search-and-rescue teams, CBP, police departments, the ATF, and even hospitals and nursing homes.

Dogs

Dogs have been mentioned in other chapters because they perform surveillance tasks for many different organizations and governmental departments. They help detect human cargo, smuggled goods, victims of disasters and accidents, missing persons, and victims of crimes. They also assist in civilian search-and-rescue operations and medical diagnostics. Their keen noses can sniff out scents that humans can't smell, with special chambers to separate the smells so they can be more easily distinguished. Dogs have made some remarkable "hits" in terms of finding smuggled objects on behalf of the CBP, even locating drugs that were hidden inside food or dog biscuits to try to throw the dog off the scent.

Counter-drug enforcement programs in several government departments utilize dogs to detect narcotics. Sometimes the same dogs are trained to sniff for explosives. They also help locate humans who may be trying to cross borders or hide from police investigations. In addition to surveillance, if a dog finds a refugee, combatant, or other person of interest, it will sometimes "down" the suspect until personnel arrive or "hold" the suspect until personnel can return.

Canines are still the best way to locate many things and often they can go where a person can't, slithering through small openings in tunnels or under collapsed buildings or mine shafts, to locate injured people or bodies.

For the 2009 Super Bowl, CBP dogs were used to check the seats for explosives prior to the game and the insides of vehicles carrying packages, since the staging of a major event means stocking large quantities of food and souvenirs and all the dispensing equipment and stands that go with them. There were also tall, outdoor electronic scanners monitoring vehicle traffic. A mobile van was stationed outside the stadium with computers and equipment used for forensic analysis and a mobile command center handled communications as helicopters and boats swept the region around the stadium. Dogs weren't the only aspect of the security team, but they fulfilled a security role not covered by other technologies.

The military utilizes dogs in much the same way as CBP and the ATF. The canines check for explosives or sound an alert if they sense intruders. They can also hunt for missing persons or locate victims of disasters or explosions. Dogs accompany military personnel on patrol to sniff out possible snipers or other forms of ambush hidden behind grass or rugged terrain.

CBP was brought in to help with security at the 2009 Super Bowl at Raymond James Stadium in Tampa, Florida. In 2012, many of the same technologies were used, plus full body scanners were brought in for game security. [CBP news photo by James Tourtellotte, 2009, released.]

Left: A dog learns how to negotiate a conveyor belt similar to those used in mail facilities and sniff packages at the same time. Right: An explosives-detecting dog methodically checks packages in a warehouse. CBP has to constantly monitor goods leaving and entering the U.S. [CBP news photos by James R. Tourtellotte, 2012, released.]

Dogs are intelligent and loyal and often do more than sound alarms or locate objects or hidden people. Sometimes if their handler is injured, they run for help or lie down next to him, keeping him warm until a rescue team arrives.

Dogs have also been known to lead people out of burning buildings, or to pull an unconscious or injured child to safety. With steady improvements in technology, one wonders if chemical-sniffing boxes will replace dogs. Handheld devices are convenient, reliable, and suitable for many applications, but they don't provide moral support to a wounded soldier or avalanche victim that might keep him or her alive long enough to be rescued and they can't run for help if their handler loses consciousness. There will always be situations where a dog may be the best choice.

"Dan," a German shepherd, assists F. Rapp in checking vehicles for drugs during a Canine Narcotic Operations training course. [U.S. Navy news photo by MCS 2nd Class Maebel Tinoko, released.]

Left: "Nero," a German shepherd trained to sniff out explosives goes on a training search in the Whidbey Island supply warehouse. In addition to daily exercise, dogs receive specific training five times a month. Right: "Renzo," a military working Dutch shepherd, accompanies Staff Sgt. B. Miller in the fields along Byaa during Operations Aqrab, in search of weapons caches and individuals of interest, in 2006. [U.S. Navy photos by C.R. Hutchens and J.L. Wood, released.]

Dolphins

Dolphins have very large complex brains, and special sonar-sensing organs that humans don't have. Bottlenose dolphins in particular are curious, intelligent, can learn from watching videos, and appear to have an individual sense of identity. They are also creative problem solvers, able to come up with new ideas and to communicate those ideas to their companions, as demonstrated by training exercises in which they make up a new behavior among themselves.

Dolphins are like humans in many ways and often enjoy the company of humans. They are eager to learn and willing to work and, together with sea lions, have been enlisted by military services to seek out marine mines, unusual objects, and hostile divers.

Dolphins have been trained to locate objects that appear to be mines and signal their presence. They also watch for suspicious behavior on the part of swimmers or divers. They were used in Iraq in 2003 to help servicemembers locate and clear mines. Since the mines were intended to disable large objects like warships, they are not triggered by light contact by dolphins or sea lions. Dolphins can also be equipped with imaging devices and tracking beacons for surveillance and research purposes. [News photos courtesy of U.S. Navy NewsStand, released.]

Dolphins have been assisting the armed services since 1970. In the Vietnam war they were brought in to help protect piers.

The U.S. Navy Marine Mammal Program evolved from studies of dolphins in the late 1950s at independent research programs at China Lake and Point Mugu. The initial studies were not to enlist the dolphins, but to study their hydrodynamics to see if the knowledge could help create vessels and ordnance that would move more effectively through water. Then it was realized that dolphins could go places divers couldn't and emphasis shifted to training and researching how they might be capable of assisting in other tasks. Dolphins weren't used for surveillance at the start; they were taught to take tools to divers, but it became apparent that dolphins could be used to locate and tag objects.

Dolphins have remarkable biological sonar abilities. They use acoustics to locate food and objects and, since sound waves can pass through certain objects, they have a form of acoustic "X-ray vision" for seeing into or behind certain materials. As an example, if a trainer puts a cross-shaped toy inside a box, the dolphin can "image" the cross using sonar, similar to how we image a fetus with ultrasound, and tell the trainer which shape is in the box out of a set of options drawn on cards. Dolphins can also find objects hidden under sand by taking *soundings* around the area to locate the buried shape. They can also image objects through thick murky water.

Dolphins have been trained to note suspicious behavior in the water, such as divers planting an unusual object. If they note something unusual, the dolphins will move a buoy to the vicinity of the divers.

Dolphins and sea lions can also be equipped with other sensors, like small cameras and radio-frequency or sonar beacons to provide additional information about their whereabouts or about the environment surrounding them as they do their jobs.

In 2009, four dolphins were flown aboard a C-17 Globaster III to Noumea, New Caledonia to participate in an international humanitarian project to clear mines that had been seeded into twelve minefields during World War II.

Research on dolphin sonar is conducted by the Marine Mammal Program to discover whether the new knowledge can lead to improved electronic sonar.

Sea Lions

Sea lions are not as smart as bottlenose dolphins, but they are nevertheless intelligent and energetic and have been trained in many surveillance tasks, including mine detection. They have excellent eyesight and can see much farther underwater than humans. Sea lions have external "directional" ear flaps and keen hearing on land and in water.

Sea lions are agile and fast, swimming in bursts up to 25 mph. They are also surprisingly fast on land, considering they are designed for speed in water. Their cousins the seals are rounder and don't have external ear-flaps and are almost helpless on land, but sea lions can move across a sandy beach at speeds up to

7 mph. For over 100 years, they have entertained crowds in circuses and marine shows due to their ability to learn a variety of behaviors.

Sea lions have been trained to note unusual behavior, especially in shallow water and in and around piers. As part of the Navy's Shallow-Water Intrusion Detection System (SWIDS), they can be on the lookout for suspicious divers and attach ankle cuffs to detain them until personnel arrive to assess the situation. Since the sea lions have good eyesight under water and are quick, the cuffs are often in place before the diver realizes what has happened.

Sea lions in the Navy Marine Mammal program have been trained to search for mines and, if they find one, come back and give a signal, such as hitting a target with their noses. They are then equipped with a tethered marker to attach to the device. They can also seek out other unexploded ordnances sitting on the seabed and attach a cable so it can be raised to the surface or towed to a safer place where it might be disabled or detonated. [News photos courtesy of U.S. Navy NewsStand, released.]

Rats as Chemical Detectors

We may not think of land mines as chemicals, since many have metal or plastic housings, but they are filled with chemical explosives and animals like dogs and rats can smell them. Since dogs are intelligent much-loved companions, most people object to dogs being used to sniff out land mines—there's a possibility they'll be blown up if they step on one.

Rats have good noses too, and are also intelligent and trainable. It is possible to breed many of them in a short time to enlist as sniffers and they are lightweight, so they are unlikely to trigger a mine that might kill a dog.

The world is full of mines. They are spread around military bases and major transportation routes by the millions. Every year tens of thousands of people, many of them children, are injured or killed by mines, not just in combat zones but in places where war happened long ago. The oceans are full of them too and dolphins and sea lions are sometimes enlisted to find and tag them. It is estimated there may be as many as 80 million mines buried around the world.

Government labs and commercial firms have developed many kinds of mine detectors, ranging from metal detectors and ground-penetrating radar to radio-frequency C4-explosive detectors. Not all countries have sufficient financial resources or political access to these technologies however, and they suffer substantial human injuries and financial loss from thousands or millions of uncleared mines.

In Vietnam alone, it is estimated that more than five million hectares of land need to be demined or cleared of other forms of unexploded ordnance. Vietnam is the second-largest exporter of rice in the world, but cannot reach its full potential as a nation or as a supplier of food due to millions of mines hidden on residential and agricultural land.

Colombia, a site of decades of internal strife, has a mine problem as well. Deaths and injuries affect more than 500 people per year from mine explosions, so a government project to breed rats as mine sniffers was initiated in 2006, in hopes of clearing the mines. Trainers have even enlisted the mother rats to help teach baby rats to scurry toward metallic parts similar to those found on mines— for which they earn a small treat. The rats are selectively bred for the ones that perform best at the task, just as dogs are bred for specialized abilities.

In Tanzania, a larger breed of rat, African giant pouched rats, have been trained from the time of weaning to sniff for mines and also to seek out injured people or those who may be infected with tuberculosis.

Members of the U.S. military traveled to Tanzania and Mozambique to watch mine and bomb-sniffing rats in action, to see if the U.S. military could benefit from their use. Perhaps mine-sniffing rats could be helpful in Vietnam.

Bees

Bees are natural data collectors. They make daily trips to flowers to collect nectar and pollen to nourish themselves and feed their larvae.

Pollen is a natural substance used by plants for reproductive purposes. As bees gather pollen on their legs, they move traces of pollen from plant to plant, which helps the plants reproduce. Pollen typically adheres to the stamens of plants (although there are many variations on what constitutes a stamen in the plant world) and is released by pressure, temperature changes, or wind. The pollen holds traces of other substances from the surrounding environment that can be used to construct information about the environments where the bees are traveling. Bees gather it and bring it back to the hive.

Pollen can be used in another way—instead of analyzing what has been collected from bees, it can be obtained from crime suspects or victims of crime. This is called *trace evidence*. The study of pollen in relation to crimes is *forensic palynology*.

Chemical analysis of pollen found in clothing or hair can reveal if a person was in a certain area, or downwind from the area, which might link him or her to a recent crime scene. When combined with other trace evidence, such as mud in the cracks of shoes or dirt under fingernails, accurate assessments of locale can sometimes be made. Pollen may also indicate whether a body was relocated after death.

Traces of pollen can sometimes be found in illicit drugs that are derived from plants, giving clues as to the origin of the drugs.

The bees themselves will provide information on where they collected the pollen. When they arrive at a hive, they convey interesting information to the other bees by doing a *waggle dance*. This tells other bees which direction and how far to go to find what the returning bee discovered, like a new field full of flowers. This dance can be recorded with video and decoded.

Dogs and rats are natural "sniffing" animals so it's not surprising that they've been trained to sniff out mines, but bees can detect unexploded ordnance as well. It was noted by researchers at the University of Montana and Sandia Labs that they would swarm where mines were leaking chemicals into the air and it didn't take more than a few hours to train them by rewarding this detection behavior. At that point, bees passed the information onto others.

Bionic Birds and Bees

Imagine a flying bug you could control from a distance like a radio-controlled plane.

Remarkable advancements in micro-robotics have occurred over the last few years that enable the creation of tiny flying robots and tiny parts that can be attached to insects. Tiny robots that fly, navigate, and intercooperate now exist. But real bugs still have some advantages over robots, and scientists have been combining microelectronics with living insects and animals.

Sometimes surveillance using insects is nonintrusive. Taking a sample of nectar or honey to see where a hive of bees has been foraging minimally disturbs the bees. Attaching a micro-miniature camera to certain species of birds doesn't hinder them either. Tracking herds of larger animals with tiny radio-frequency chips can yield important information on their migration patterns, numbers, and welfare, that can help ensure their survival.

These are passive uses of surveillance that provide information with minimum or even positive impact on the animals.

There are also active modifications, like implants or attached electronics, that are more intrusive and sure to provoke controversy. Bionic control mechanisms—components that can be attached to animals or implanted—have been developed to "steer" them or control them in other ways.

What advantages do bionic birds and bugs have over robots? They are easy to breed, more versatile than most robots and, in the case of birds, they can live a long time, sometimes up to 20 or even 80 years. Birds are easy to tame and many, like intelligent species of parrots, bond strongly to humans.

Harvesting them for surveillance would raise issues of species depletion and concerns about their individual welfare, but research continues, nonetheless, partly because the research is open-ended. It isn't always initiated to turn the tiny bugs into spies. Sometimes it begins as a way for wildlife conservationists to study them in order to protect them.

Other times, the research occurs in military labs, or university labs funded with defense grants, with the express purpose of studying how animals and insects could be used for reconnaissance or other missions.

The Hybrid Insect Micro-Electro-Mechanical Systems (HI-MEMS) project consists of miniaturizing electronics so they can be fitted on insects without hindering their movement. The tiny wires are inserted early in life, so the body grows around them. They enable the insects to be controlled remotely, with information relayed back to a computer. [News photos courtesy of DARPA, released.]

University labs have already demonstrated insects that can be selectively controlled to move one wing or the other and to flap them at controlled speeds. As part of the HI-MEMS program at a group of cooperating universities, a Carolina sphinx moth larva was implanted with electronics in the thorax to integrate with neural systems that control the wings. In a moth, there is an abdominal nerve that is similar to the spinal cord in humans except that it is not wrapped inside a bony spine. A GPS system and air-sampler for chemical detection could then be added when the larva pupates into a moth. Since beetles can also fly, similar experiments have been done to install bionic flight control in beetles.

Bionic insects could be used to test air quality in hazardous areas or to gather imagery of a hostage situation. They could fly through openings into collapsed mine-shafts or buildings to see if there are survivors or bodies inside. They could swarm an intruder or swarm around a toxic leak to indicate the location of the leak.

Bio-Robots

Animals can be implanted with micro-electronics, but electronic devices can also be implanted with biological nerve cells.

It has been demonstrated by a lab in Reading, U.K., that rat brain cells can be used as biological controllers for robot mechanisms to prevent a robot from running into walls. The rat brains are grown in culture and incorporated with the electronics so that electrical impulses interact with the neurons and receive signals back from the cells. Since biological brain cells have an innate capacity to "reconfigure" themselves, to make new neural connections, the robot appears to have some ability to learn. This has remarkable implications for medical research and for bio-robotics. Bio-robots that use living nerve cells to navigate and learn might be more flexible and adaptable than wholly electronic robots.

Implications of Bionics

Surveillance is often instituted to protect people, to ensure their physical safety, and to protect property. But every aspect of surveillance seems to have a dark side.

Scientists in the U.S., Israel, and China have successfully created implants to steer birds, fish, and rats. By taking the concept one step farther and equipping them with tiny cameras, they become mini biological spies.

Implanting radio controllers into birds, bees, rats, or moths to control them like a remotely controlled toy plane also raises issues of animal welfare. Not everyone thinks of insects as nuisances or as expendable. Since they have nervous systems and most creatures respond in similar ways to pain (by trying to avoid it), it is assumed they feel pain even if they don't process that knowledge or experience it in exactly the same way as humans or other animals.

The idea of lab-created bionic bugs will probably be disturbing in other ways as well, since this kind of research increases the possibility of robot spies in every corner of the globe and every residence on the planet. Tiny creatures with implants are often indistinguishable from those without, especially if the implants are installed early in life and the tissues grow around them. Individually, insects may not be too difficult to deal with, but anyone who's encountered a swarm of ants, bees, or mosquitoes knows the situation is different when there are many of them.

With the success of bionic insect programs there may be additional research grants to develop control electronics for implantation into humans. Bionic parts have already shown promise in helping people regain some or all of their hearing or sight, but it may further be used to enhance people in ways that give over control to someone else.

Imagine police officers or bionic soldiers with implants that suppress natural fear responses or which program them to pull a trigger when a commander in a remote location watching video from a surveillance drone makes the decision and gives the signal.

There was a time when science fiction was seen as a fanciful exercise of imagination. With genetic modifications, micro-electronics, and nano-electronics, all those fanciful ideas may be possible within 10 or 15 years.

BIOMETRICS

KEY CONCEPTS

- *passive and active biometrics*
- *varieties of biometrics and changes in biometrics*
- *applications and accountability*

Introduction

Biometrics is short for *bio*logical *metrics*—measures of biological identifiers that are traits of individuals. For surveillance purposes, it is preferable to record traits that remain unique throughout a person's life.

Biometric systems often use chemical tests or optical scanners to identify a person. Common examples include a fingerprint or iris scan offered at an access point, like a customs kiosk, or a fingerprint or saliva sample left on a glass or cigarette stub at a crime scene. Fingerprint dusts, chemicals that reveal blood stains, and scanners that image a person's facial proportions and contours are common ways to recognize or reveal biometric characteristics.

Mugshots have long been used in law enforcement to identify suspected or convicted criminals. More recently, photographs of faces have been combined with thermal imaging to provide a composite portrait that maps not only the shapes and colors of a face, but individual "heat zones" to increase the probability of a correct ID.

We usually think of biometrics as physical traits such as fingerprints, voice patterns, eye colors, or genetic makeup, but biometrics also encompasses unique patterns of movement, such as the way we run, the way we write, or the way we type.

One of the biggest motivations for using biometric identification is to deter fraud. Identity theft, multiple aliases for collecting social security or insurance payments, use of unauthorized credit cards, human smuggling, counterfeit green cards, or unauthorized access to buildings or equipment are some of the ways people attempt to defraud companies and government institutions.

Biometric characteristics can be categorized as *physical* or *active*.

- Physical biometrics include traits such as facial structure, DNA, fingerprints, iris color and patterns, retinal structure, height, build, bone density, tooth shapes, vocal patterns, and other intrinsic characteristics.

- Active biometrics include gait, facial expressions, breathing rate, resting pulse, handwriting, speech habits, brain patterns, and typing patterns.

Some biometrics change over time. Aging or a change in diet can alter resting pulse or cholesterol levels. Glaucoma tends to cloud the iris, bone density decreases with age, and handwriting may be affected by disorders such as Parkinson's disease. Sometimes it is possible to guess what the biometric was in the past or might be in the future. Age progression is a form of biometric prediction that uses growth patterns to estimate what a missing child might look like in a few years.

Most biometric identification systems rely on physical characteristics. Fingerprints are favored for their stability, but there is increasing interest in using speech and other active identifiers to pick someone out of a crowd with surveillance cameras or to recognize them for access control by their gestures, typing patterns, or a spoken command.

Biometric identifiers coupled with optical scanners, RFID cards, and computer databases are increasingly prevalent access-authorization tools. Customs and Border Patrol uses many of these technologies for monitoring travel security. Border checkpoints also include license-plate scanners, as well as X-ray scanners for large cargo containers. Shown here is a biometric fingerprint device emitting a red light to illuminate the fingertip. [Customs and Border Patrol news photo by James Tourtelotte, released.]

FINGER, PALM, AND FOOTPRINTS

Fingerprints

Fingerprinting is one of the oldest and most common forms of biometric identification. Finger and toe prints stay the same throughout life and are unique to each individual. Even identical twins have unique prints—they're often very similar, but rarely identical.

In the past, fingerprints were captured with ink. The ink was a nuisance and hard to clean off the fingers. If the finger wasn't rolled evenly, the ink could smudge or come out too light. If it were too dark, it would fill between the ridges. Traditionally, prints were stored on index cards. Since ink is inconvenient and paper files are difficult to search, the trend is to scan inked prints and organize them as searchable computer files. More recently, non-ink scanners have been developed to image the fingerprints directly into digital files.

Left: The fingerprint division of the Soldiers' Bonus Bureau, 1924, stored thousands of cards in long racks of file cabinets. Right: A youth is fingerprinted using traditional ink. Military recruits, immigrants, criminals, and certain government personnel have routinely been fingerprinted over the last century. [Photos courtesy of the Library of Congress, public domain.]

Digital fingerprint matching has significantly improved in recent years.

In November 2011, a German biometrics manufacturer announced that its digital fingerprint identification system set a world record in correctly identifying a set of prints from one individual out of a database representing 10 million people. If the same task were attempted with paper index cards, and the individual's prints were in one of the last file cabinets to be searched, it could conceivably take more than 200,000 hours to ID the prints.

Until recently, most identification cards, such as passports and driver's licenses, required only two biometric samples: a photo and a signature. Now fingerprint scanners are increasingly used for ID cards and electronic passports, and have been tested for use as identification authenticators for cash machines. They also help expedite travel and entry to events by season's pass holders. The justification for additional biometrics is to reduce the possibility of theft, and to deter sharing of cards by family members or friends who are not authorized by the issuing authority to use them.

In April 2012, the Internal Revenue Service (IRS) instituted a new requirement that tax preparers submit fingerprints before being issued a preparer tax ID number (CPAs and certain law agents are exempt for the time being). This illustrates an increasing trend toward collecting biometric information on individuals seeking certification.

Increasing Accuracy and Improved Identification

Fingerprint identification goes beyond matching one set of prints with another. Software developers have devised ways to fill in information on partial prints, using fractal, directional, and other algorithms that make informed guesses on the shape of the complete print so it might better be compared with existing records. This is a boon to forensic investigations, since crime scenes rarely yield perfect prints. Many materials do not hold finger oil well and even when they do, "lifting" the prints can be subject to error. If too much or too little dust is applied, the prints can be damaged or hard to discern.

Fingerprint readers used for access points, like doors to secured buildings, or gates into industrial yards, have become more sophisticated. A system that reads not only the part of the print we see, but the capillaries that echo the pattern under the surface of the skin, has been developed by a Connecticut company.[25] The firm announced development of a multispectral optical reader that could read patterns that extend beneath the skin's surface to match the surface print. This form of *deep-scan identification* is not hampered by dirt or dampness on the surface of the skin and makes it difficult to fool the machine with fake prints.

Tongue Prints

Tongue prints may seem unlikely as a biometric identifier, but tongue shapes are unique and Chinese researchers have figured out ways to use a laser to image a tongue in three dimensions. It may not be the first choice in biometrics (most people would be uncomfortable having a laser light stuck in their mouths to take a 3D picture of their tongue) but it is one more way in which human bodies are being studied and cataloged.

Footprints

For decades, hospitals have taken footprints of newborn babies to prevent deliberate baby-switching or inadvertent mistakes. Babies are also footprinted to deter kidnapping or to make it easier to identify a child that has been found. About 10 newborns a year are abducted in the U.S.

In 1983, the American Academy of Pediatrics and the American College of Obstetricians and Gynecologists warned hospitals that studies showed that hospital footprints were technically inadequate for identification. This was partly due to the difficulty of getting clear ink prints from wiggly babies, but also because the

---

[25]Accu-Time Systems, Inc., news release, 2011.

ridges on an infant's foot are closer together and more difficult to capture than on an adult.

Hospitals have gradually added DNA sampling to their arsenal of identification tools, but even this is not universally adopted, since it requires invasive piercing of the heel to obtain a blood sample. Some already pierce the heel to test for genetic anomalies that might harm the child if not treated early, and use the blood spots as samples for development of new testing and screening technologies.

Research now indicates that umbilical cord blood can provide a noninvasive source of DNA identical to heel blood. Optical palm print scans have also been suggested as a way to ID newborns more accurately.[26]

Bearder vs. State of Minnesota

In a controversial lawsuit in Minnesota, the Department of Health made arrangements to destroy newborn blood spots as a result of a Minnesota Supreme Court decision on genetic privacy.

For decades, the Department of Health has kept the blood spots on file but citizen concern about indefinite retention of private information, such as blood spots that contain DNA, led to a lawsuit in which the *Genetic Information Act of 2006* was cited. This act mandates that informed written consent must be obtained from parents or guardians to retain samples for anything other than the original stated purpose.

The health department made arrangements to begin destroying the spots but, ironically, since the court cases were not all resolved, and it is not permissible to destroy legal evidence, the health department had to seek a further protective order for permission to destroy the spots.

In the future, Minnesota will seek written consent from guardians to retain specimens that test positive for medical anomalies, as they may be needed for further diagnostic reference and medical procedures.

Palm Vein Scanners

A palm vein scanner is a "deep-scan" technology that looks beneath the skin to the patterns of vessels that circulate blood through the hands. Palm vein scanners have been installed as an alternative to iris and fingerprint scanners and can now be found in dining halls on college campuses to prevent meal card sharing.

---

[26]R.P. Lemes, et al., "Biometric recognition of newborns: Identification using palmprints," *Joint Conference on Biometrics (IJCB)*, 2011, IEEE..

This brings up an issue of where the line between humanity and technology lies, one that occurs often in ethical discussions about surveillance. Giving up a meal that you've paid for to a fellow student who is short on grocery money is not an uncommon practice on university campuses, but those issuing the cards are counting every penny and often the contract stipulates that the card may only be used by the original issuer. Technology makes it easy to track and prevent unauthorized use of services, but what is the cost to society if people go hungry when technology imposes inflexible rules on basic needs on a widespread scale? Computers are impartial. In this example, the catering company is entitled to the terms of the contract, but if all of society becomes automated to the point where there is no leeway, then disadvantaged segments of the population have nowhere to turn in short-term emergencies that they might otherwise weather and overcome if there is "wiggle room" in the system.

Iris and Retina Scanners

The iris is the colored portion at the front of the eye. The retina is the imaging surface at the back of the eye. Both are unique biometric features that can be scanned and recognized. Iris patterns are more complex than fingerprints and could potentially be used to identify every person on the planet.

In the past, retinal scans required the user to put his or her eye close to a cup so that the scanner could "see" the back of the eye. However, newer systems enable scans from farther away.

The National Institute for Standards and Technology (NIST) has developed recognition software designed for iris identification. Called the Video-based Automatic System for Iris Recognition (VASIR), it can analyze video captured as a person walks through a portal at a distance, determine which part of the image is the eye region, and extract and encode the relevant data.

In 2010, the Inmate Recognition and Identification System (IRIS), financed by the Department of Justice, was demonstrated in Nevada. The system will link prisons through a national database of inmate biometrics.

It was an escaped inmate that motivated installation of iris scanners. He impersonated a fellow inmate. But it wasn't a failure of technology that facilitated the escape—they had fingerprints on file—it was a failure in procedure, like cashiers who don't check the signature on the back of a credit card. If electronic devices replace people because they don't do their jobs correctly, what happens if powerful technology is in place and those responsible for managing all this sensitive information don't do their jobs correctly?

Iris scanning is powerful technology.

In 2010, the city of Léon, Mexico, began installing real-time iris scanners capable of scanning people as they walk by. Like license-plate scanners that are surprisingly fast and accurate, iris scanners can identify up to 50 people per minute. The Léon scanners use infrared and are integrated with a tracking database that follows the movements of everyone in the city who is imaged as they pass a scanner. Scanners can be placed in locations like ATMs, storefronts, light poles, and major transportation terminals.

Iris-tracking is somewhat analogous to strapping a radio-wave-transmitting cuff on a person's ankle to send information back to a control center each time a person passes a checkpoint—except that there's no way to take off the "cuff." Even sunglasses don't deter iris scanners. As mentioned in the optical surveillance chapter, infrared wavelengths can pass through sunglasses as though they were clear. Masks can't hide a person's identity either, if there are eye openings in the mask. People have fooled scanners by putting images of someone else's iris across their eyes, but computer algorithms to detect this kind of subterfuge would not be difficult to program and could be designed to sound an alarm.

The city of Léon has begun scanning the eyes of convicted criminals, but is also asking citizens with no criminal record to submit voluntary scans.

The CEO of the company that supplies the scanners is eager to spread the technology to every aspect of our lives, such as opening doors, accessing medical records, getting a prescription, or entering a place of work. With scanning distances up to 30 feet, it is possible to scan everyone walking through a hallway, doorway, or passing on the street.

The justification for the Léon system is to try to stop fraud and theft. The problem is that tracking information housed in a central computer can only be seen by police and city officials. The populace doesn't have equal access, and tracking information of this nature can be used for many other purposes besides deterring crime, such as business and political surveillance.

In November 2010, the New York City Police began scanning the irises of prisoners to guarantee that a person appearing before a judge is who he or she says she is. Would existing fingerprint databases be sufficient for this purpose considering that the new non-ink deep-scan technologies are very reliable? Fingerprint ID provides good identification without impinging on the privacy of everyone nearby, as happens with an iris scanner. Is it necessary to pile one technology on top of another, and can taxpayers afford multiple-system IDs?

The U.S. military uses iris scanners in the Middle East and prisons have purchased them, as well. The Department of Homeland Security tested them in 2010 to see if they would be useful at border crossings. A privacy impact assessment released by DHS said that the images would be destroyed following the research period, but private companies are not subject to the same requirements as the government when installing security devices. They need only inform workers that they are being monitored and to keep the scanners out of toilet stalls. The government is also at liberty to purchase surveillance information from private companies, just as they purchase many aerial reconnaissance photos from private firms.

Iris scanning isn't used only for tracking criminals and military personnel.

In 2009, the New South Wales Department of Corrective Services announced that its prison system would require iris scanning in addition to fingerprinting for visitors entering the prison. Thus, being associated with someone with a criminal record, regardless of whether a person's own history is clean, means people will be subjected to scans.

Iris scanners are used at airport checkpoints in the U.S. and European Union. Frequent flyers voluntarily provide iris "prints" and pass through the scanners to expedite travel times.

In the U.S., tracking is illegal, in the general sense, but tracking of prisoners on parole and people in the workplace is usually permitted and, once a technology takes hold, it is not uncommon for it to bleed into other areas of society such as hospitals, nursing homes, nurseries, and large-scale events such as concerts and sports meets. Banks in the U.S. are already beginning to use iris scanners and convenience stores are probably considering them as a security measure. Once iris information is in one private database, it can be sold to other companies. Even those who prefer to opt out of a system have little recourse unless there are laws to maintain the balance between individuals and institutions.

Anyone who has applied for a mortgage in recent years, where the bank requires automatic debits from the loan recipient's account, knows that big institutions can "voluntarily" require many concessions from borrowers because they have something individuals need. If they want an iris print as a condition of a loan, most people would surrender it rather than lose the opportunity to buy a car or home. While many vendors refer to this as "consumer choice," with regard to surveillance technologies, it is actually lack of consumer options. If every bank requires an iris scan, and cash is no longer accepted at many retail outlets, there are no opportunities to use alternate payment methods or to shop around.

Iris scanning technology has become so powerful, it could be used to clandestinely capture people's iris IDs. Many companies on the Web sell personal information, such as credit reports, economic status, names and ages of relatives, and information on where people live now and have lived in the past. Since data aggregators already know where everyone lives and where most people work, setting up iris scanners on a sidewalk or near a doorway to add iris IDs to those data repositories is not unthinkable. A street-viewing van could be supplemented with a scanner to image pedestrians without them knowing their irises had been scanned. Since iris scanning is an optical technology, there are no radio-wave signals to give away clandestine scans.

Keystroke Surveillance

Keystroke surveillance can mean 1) identifying a person by their keystroke patterns or 2) capturing the informational content of a keyboarding session.

People have different typing speeds, different pattern combinations that they have learned better than others, different cadences, and differing abilities to type numbers. Mimicking another person's typing pattern may not be impossible, but it is difficult, just as mimicking another person's handwriting is difficult.

Like speech recognition systems, a keystroke identification system has to be trained to recognize an individual, but once the pattern is logged, it could conceivably be used for a number of years, or as long as a person's typing patterns don't significantly change.

Since distance education over the Net is increasing, and the concept of keystroke identification is affordable compared to many other forms of biometric identification, it may catch on.

The Defense Advanced Research Projects Agency (DARPA) initiated calls for active biometric identification systems, called *active authentication,* to replace online passwords and the Common Access Card authentication used on DoD IT systems. Passwords are seen as unwieldy when it is necessary to remember so many, especially when some systems require that they be changed periodically. DARPA suggested the monitoring of keyboard and mouse input as one way to approach the problem of authentication. The agency aims to have a pilot system ready to test by 2015.

Depending on the software design, active monitoring and recognition of keystrokes carries with it a payload of information on everything a user does on a computer, so it is more than a simple authentication technology. Commercial developers aiming at a broad market would probably build both identification and keystroke monitoring into the system. In the workplace, where keystrokes are sometimes already logged, a combination authentication/keystroke logger might be used if it became available.

Researchers at Pace University[27] have developed a keystroke system that can identify and authenticate a keyboard user through the Internet by his or her typing patterns. This was conceived as a way to authenticate online users taking a test or to help prevent e-commerce-related identity theft, but clearly this could also be used by law enforcement to identify and monitor keystrokes of a suspected stalker in a chat room.

It is not a fool-proof system, however. Creating keystroke software to impersonate someone (e.g., a person taking an online test) is not out of the question. If a person's keystroke patterns are learned and stored by a computer application, a software application could alter the keystrokes to fit the stored pattern.

The concept of keystroke monitoring is not just another identification system— it represents a fundamental shift in control. At the present time, the user decides when to log on, what password to use, and when to log out. The DARPA strategy would change that equation. Simply sitting at the computer and typing would identify a person without any choice or active involvement.

Banks have made a concerted effort to switch their customers to paperless statements, which means online access could eventually become the only way to monitor a bank or credit-card account. If vendors such as banking systems begin

---

[27] V. Monaco, et al., "A keystroke biometric system test-taker setup and data collection," *Proceedings of Student-Faculty Research Day, CSIS*, Pace University, May 2011.

requiring keyboard authentication to access their online services, users have little choice but to allow them to capture a "keyboarding print."

If a keyboard print captured by a vendor makes its way to other companies through mergers or sale of biometric info to third parties or the Intelligence Community, then chat hosts on the Web could theoretically link up pseudonyms on forums and chatrooms with their associated user identities.

It is also likely that people seeking to mask their identities in chat or gaming rooms will develop keystroke-cloaking software that changes keystrokes on the local machine before they are transmitted. Some may even put up keystroke-masking servers, just as there are proxy servers to hide people's email or Web-surfing identities. For every system that seeks to pinpoint an individual on the Internet, there are several that attempt to circumvent it.

Can a keyboarding pattern be impersonated? Capturing a person's typing patterns with spyware is not difficult. If the patterns can be captured for authentication, they can be captured and reproduced via a keyboarding Trojan, so it is not an unbeatable authentication system unless the system is secure against spyware (few are) or the identification process can be backed up by other identification.

Facial Recognition

Facial recognition software used to be crude and easy to fool, but it has reached the stage where even people wearing disguises may, to some extent, be recognized in a crowd, even without iris-scanning technology. Facial recognition software analyzes proportions, size, and individual features such as ears, eyes, mouths, and noses. Facial recognition through streaming video may also be facilitated by active biometrics—expressions and patterns of movement that are unique to individuals.

Facial recognition can be used for access control at doorways and gates and may, in the future, be used for financial transactions. Since many ATMs are already equipped with microphones and cameras, it wouldn't be difficult to add facial recognition to the system.

Facial analysis has been built into marketing billboards. The system scans a person passing by, analyzes biometric features and makes some decisions about what a person who looks like that might want to buy, and displays a targeted ad based on gender, age, and possibly also style of dress or colors. Once again, it is similar to the concept illustrated in the movie *Minority Report*.

It doesn't stop there. Facial recognition software for smartphones can now assess the people in a room. You hold up the smartphone to scan the room with the built-in camera and it tells you approximately how many people are present, the ratio of men to women, and the average age of the attendees. This kind of software could be used at business meetings, trade shows, parties, and bar scenes.

Facial recognition isn't limited to cameras taking pictures of people entering a bus terminal or standing in a police line-up. It can be used to record and identify faces from photos, as well. Mugshots from different sources could be scanned to see if there's a match.

In 2010, Facebook began using facial-recognition technology on the information uploaded by certain users in the U.S. The pilot program was apparently started before users had a chance to opt out or delete any pictures they didn't want included in the scan and those who don't check their accounts or communicate with their linked friends on a daily basis might not know about the program until their photos are tagged by others.

The software is designed to make it easier to tag friends. Imagine uploading a picture from a party with you and three friends. The software scans the faces, looks for probable identities elsewhere on the site, and suggests tag names. It is convenient but it is also powerful technology that could theoretically look for every other instance of a similar face anywhere on the Net, without the person whose image is being scanned knowing they've been searched and stored.

At a security conference in 2011, Alessandro Acquisti, a Carnegie Mellon CyLab researcher, described a study using off-the-shelf facial recognition software and information publicly available on social networking sites to identify individuals on a popular dating site where the posters use pseudonyms. They were also able to identify a number of students walking on campus and to predict some Social Security numbers. The researchers pointed out that this information is available not only to online sites, but to anyone with the software and a camera-equipped smartphone. A smartphone can't compile the massive data repositories available to large social networking sites or police departments with access to national databases. However, in the future, it may be possible for a staff member from a social networking site (or a police officer) to walk into a room, hold up a phone, scan the people, and have their names pop up on the screen along with their biographies and vital statistics.

Facial recognition may soon be added to the marketing arsenal already in place in retail outlets. Currently, many stores log purchases from rewards cards and credit cards, and have overhead cameras to chart customer buying patterns at the cash register. In the future, they may add face recognition to log the purchases of "uncooperative" customers who don't carry rewards cards and who pay cash. At that point, everyone's buying patterns would be recordable, whether they opt in or out of store card programs. In fact, the cards (and accompanying discounts) may no longer be necessary.

This kind of system, whether for banking or retail shopping, would probably be presented by the marketing department as a customer service to prevent identity theft, but clearly also takes away the choice of opting in or out of retail profiling databases.

Voice Recognition

Voice recognition is matching unique sound patterns with a stored *voice print* in order to identify an individual.

Sound patterns may include various combinations of cadence, tone, loudness, pitch, and sometimes the smoothness or raspiness of a person's voice. Speech recognition is related to voice recognition, and sometimes takes into consideration some aspects of voice recognition, but focuses mainly on the content of the speech rather than voice characteristics. Voice recognition software can also be programmed to recognize some aspects of high emotion, such as quickened pace, changed pitch, or changed volume (shouting or whispering).

Voice recognition was once an obscure area of biometrics because it requires more complex computer processing than fingerprint scans or mug shots but, with improved voice software, may now be used by law enforcement agencies to filter wiretapped communications and, with increasing use of mobile services for e-commerce applications, is being investigated for its authentication potential for access to online financial services.

To use a voice-print service for authentication involves more than talking into a microphone. Just as fingerprints must be compared to a database in order to determine a match, a voice print must be created and put on file for comparison. This means showing up in person, presenting identification, and giving an audio sample for matching purposes before using online authentication systems or ATMs. The problem with systems like this, besides the potential for the voice-prints to be used for other purposes, is that a poor microphone, a bad connection, or even a case of laryngitis might lock a person out of his or her account. Nevertheless, despite technical obstacles, more than 10 million individuals have voice-prints on file and the number is expected to grow.[28]

Speech Acquisition

Speech acquisition is obtaining the content of communications. With new technology, tapping a phone is easier than listening in on a conversation in a public place or private golf course.

One of the more intriguing technologies for speech acquisition is *vibrometer* speech-surveillance systems. These are laser detectors that can sense vibrations from a distance, such as vibrations on glass if two people are having a conversation near a window. It is both a sophisticated and imperfect technology in that distance, environmental vibrations, and the positioning of the instruments must all be calibrated to get useful data.

While imperfect, vibrometer surveillance has applications as a triggering device. It is costly, in terms of bandwidth and storage, to have cameras running all the time. An auditory detection system could be used to alert operators to activate floodlights or could automatically trigger cameras to begin sensing or recording, and may even provide information for aiming the cameras at the source of the sound. By creating software that can recognize the difference between ambient noise and target noise, such as a whispered conversation in a parking lot at night

---

[28]"Voice biometrics authentication best practices: overcoming obstacles to adoption," ValidSoft and Opus Research, January 2012.

or a vehicle breaching a security zone, cameras could fill in the gap by collecting visual information to accompany the audio.

Speech Recognition

Speech recognition is the interpretation of human verbal communications so that the content may be recognized or analyzed. Often the speech is interpreted by speech-to-text software and recorded as written text.

Speech-to-text programs are still in development and the 90% recognition rate claimed by some advertisers is typically over-optimistic (50% to 60% is more realistic after optimization by speaker training), but it is often good enough for surveillance purposes.

Speech recognition is an active biometric, if the patterns of speaking are analyzed for characteristics such as choice of words, accents, and grammatical patterns that might be unique to an individual. It is of particular interest to law enforcement agents, especially if voice communications between a number of parties have been tapped and recorded and attempts are being made to identify the speakers. Correctly identifying a person who made certain statements generally goes beyond voice recognition and involves an analysis of what was said and, biometrically, how it was said.

When voice and speech recognition software became available off-the-shelf, it provoked an increased demand for wiretapping authorization. Three significant changes have occurred that raise questions about whether historic wiretapping legislation is adequate or suitable to regulate rapidly changing wiretapping capabilities.

1   **Accountability**. In the past, physical access to a structure or a communications line leading to a target was necessary. This ensured a level of accountability regarding agents installing and uninstalling legal taps. Their physical presence required a commitment of time, and gave some assurance that taps had been correctly placed and removed according to the terms of a court order. Now most taps are electronic and initiated remotely through service providers' switches, which means agents can tap from a distance and, depending on the technology, there may be no assurance, other than the organization's word, that the tap has been discontinued when authorization expires.

2. **Logs**. Before tape recorders, agents had to create handwritten or typewritten transcripts of the informational content of taps if they wanted to refer to them later. This physical limitation significantly restricted tapping to those that were most urgent or more likely to be fruitful in terms of law enforcement priorities. Now taps are automated, can be recorded digitally, and stored in massive digital databanks, perhaps indefinitely, making it easy to record and archive communications far beyond the purview of a specific tap.

3. **Mandate to Obtain Warrants**. Historically, the process of obtaining warrants (and the staffing necessary to carry out taps) limited how many could be installed, and somewhat put the brakes on tapping abuse. Now, with technologies such as speech recognition and remote access to communications switches, large-scale tapping can be facilitated with software, with voice and speech recognition programs filtering mass communications and triggering a recording if a specific voice, an unusual emotional outburst, or certain set of terms is detected. Thus, automation opens the door to tapping at unprecedented levels. Broad-scale tapping is more likely to include the communications of innocent parties, and is far more easily abused. Broad-scale tapping far outstrips the ability of the courts to evaluate and approve tap requests and issue congressionally mandated court orders. Public and legislative constraints on tapping are rapidly diminishing.

Polygraph Tests

A polygraph or lie detector test is a form of biometric surveillance based on the fact that people cannot control unconscious physiological reactions. Under stress, a person's heart rate, breathing, blood pressure, and galvanic skin response change in ways that can be measured with biometric sensing technology. A polygraph is a machine that interprets some of these unconscious signals into a graph with peaks and valleys. Stress reactions might provoke a spike in blood pressure, for example, and suggest that a person is lying or familiar with some event or person he or she denied knowing.

Lie detection by written tests or polygraph tests is familiar in law enforcement contexts, but some corporations also test job applicants. A written multiple-choice test has been given to employees of certain tech retail companies, for example. Requirements like this are hard to avoid, since job hunters are rarely in a position to turn down job offers.

BIOMETRIC REPOSITORIES

Biometric surveillance is not restricted to local databanks. Global data repositories consisting of shared biometric information by countries allied with the U.S. have been proposed by the FBI. Thus, data held by nations such as the U.K., Canada, New Zealand, and Australia would yield fingerprints, palm prints, and other identifiers, with a satellite-based *Server in the Sky* to relay the information around the world. Not all countries had agreed to the arrangement in January 2008, when the idea was made public, but it shows the trend in data-sharing that law enforcement agencies are seeking to achieve.

The FBI has stated that the data would be limited to "the worst of the worst" in criminal terms, but wiretapping and DNA sampling have shown that as soon as technology makes it easier to throw a wider net, agencies tend to extend their original mandates beyond what they promised and sometimes beyond what legislation allows.

GENETICS

KEY CONCEPTS

- *the nature of DNA*
- *legislation protecting genetic information*
- *DNA databanks*

Introduction

Within our cells are genetic storehouses that utilize information passed on from our ancestors through the union of egg and sperm and, in combination with environmental factors, determine the shape and functioning of our bodies and some aspects of our abilities and personalities. DNA is an important component of this genetic material and it is now possible to study it and to map out patterns and characteristics of our chromosomes that can be identified from tissue samples from bone, hair, and saliva.

The invention of the microscope, in the 17th century, was the most important tool unlocking our understanding of cells. Yet, even after we could see into these important building blocks of life, it took time to unravel how they function and replicate.

The research of Regnier de Graaf, Gregory Mendel, and other scientists who studied human reproduction (and experimented with breeding plants and animals in a systematic way) greatly increased our understanding of genetics. Animal breeders intuitively understood that certain traits might be passed to offspring by pairing certain animals with desirable characteristics, but a more specific understanding of how and why this occurred came later.

The decades-long observations of Charles Darwin regarding speciation led to publication of the *Origin of Species,* in which Darwin offered theories about natural selection and biological inheritance.

These discoveries and theories were not immediately accepted by the public. The general idea, in those times, was that familial traits were inherited through blood (thus the term *bloodlines*). The understanding that genetic material stored in reproductive cells passed on genetic traits took time to catch on.

What is DNA?

DNA stands for *deoxyribonucleic acid*. It is contained mostly within the cell nucleus of living organisms, although a small amount can also be found in mito-chondria, the cell's "energy factories." The hereditary information in DNA is stored within chemical bases that combine in pairs to form a *double helix*.

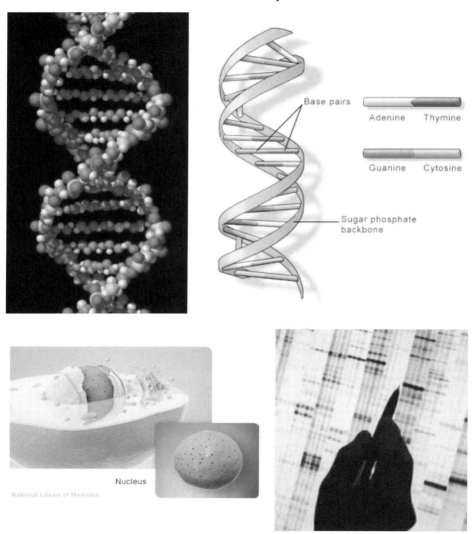

Top: Conceptual rendering of a ladder-like DNA double helix that comprises a gene. Numerous genes are contained within each chromosome. Four bases, often abbreviated as A & T, C & G, combine as pairs and are supported by nucleotide strands. Bottom left: Most genetic material is contained within the centralized cell nucleus, but a small amount is also present in lozenge-like structures called mitochondria. Bottom right: A chart, originally called a "fingerprint'" but now known as a profile, showing genetic patterns resulting from a process called electrophoresis in which genetic patterns are transferred to an autoradiograph membrane for analysis and storage. [Top left rendering and genetic chart courtesy of *www.all-about-forensic-science.com*, public domain. Double helix and cell illustrations courtesy of the U.S. National Library of Medicine, public domain.]

DNA sample collection, analysis, and identification have become intrinsic aspects of law enforcement, military service, and forensics investigations. The remains of an unidentified war victim known as the Unknown Soldier were entombed for 26 years until DNA testing made it possible to identify him as Air Force fighter pilot Michael Blassie who had been shot down in Vietnam.

Like other biometric characteristics mentioned earlier, DNA patterns are retained throughout life and are strong indicators not only of a person's identity, but of his or her familial relationships. Siblings (especially twins), parents and their offspring, cousins, and grandparents all have similarities in DNA patterns that suggest a probability of their being related.

One of the first consumer applications for human DNA profiling, in addition to seeking medical anomalies or tendencies, was to verify paternity. It is not hard to identify a child's mother as she goes through pregnancy and childbirth, but the identity of the father is sometimes in question, especially if the mother has had multiple sex partners. Paternity testing is so prevalent, it has become a common theme on television talk shows and test kits can now be purchased off-the-shelf, like a pregnancy test, with the sample sent to a lab for analysis for under $200.

When DNA testing was first developed, it wasn't admissible as evidence in court. It took time before the legal system was confident that it was an accurate means of identification. Even now, it should be remembered that humans are all genetically very similar. While an exact match probably indicates genetic material from the same person (or an identical twin), identification of familial relationships is based on probabilities.

The infamous O.J. Simpson murder trial in the mid-1990s illustrates some of the difficulties of submitting DNA evidence from a crime scene. It is possible to argue that DNA evidence was poorly collected, mishandled, or contaminated and the Simpson trial defense did so, with considerable vigor.

DNA Data and Sample Banks

DNA data banks are maintained by law enforcement agencies such as the FBI, the Royal Canadian Mounted Police, and the Forensic Science Service (U.K.), and by research societies such as the Human Genetics Society of Australia. By 1998, each state in the U.S. had enacted legislation to create DNA banks.

The FBI labs began using DNA testing to investigate cases in 1998 and found that they could pinpoint people more precisely than with serological tests, so DNA forensics were added to their crime investigation tools.

The FBI Laboratory Services provide a website with DNA archive statistics, including breakdowns for each state. As of March 2012, the National DNA Index System (NDIS) contained over 10 million offender profiles. These include what are defined as convicted offenders, arrestees, detainees, and legal profiles at NDIS. Over 400,000 forensic profiles were included, as well. It was reported that approximately 170,000 investigations had been aided by information in the Combined DNA Index System (CODIS) program.

CODIS and NDIS

In the U.S., the CODIS and NDIS within it were established as a result of the *DNA Identification Act of 1994* and are managed by the CODIS Unit of the FBI.

- CODIS supports the exchange and comparison of DNA data for violent crime investigations. Crime scene DNA profiles from the Forensic Index are also used in comparisons. CODIS does not store personal information—DNA profiles are identified by number and the name of the submitting agency, and accessible only by FBI-approved personnel with security clearance.

- NDIS has a broader mandate that includes DNA data for convicted offenders, arrestees, forensic cases, missing persons and their relatives, and unidentified human remains. NDIS data is required to be removed if a conviction is overturned or if a final court order documents that a charge has been dismissed. Only accredited agencies may submit DNA data.

Expansion of DNA Sampling

Following Operation Desert Storm, the federal government set up a deep-freeze warehouse for housing genetic tissue samples. By 1998, DNA from more than 1 million servicemembers, almost two-thirds of those on active duty or reserve, had been collected, with the rest expected to be sampled within a couple of years.[29]

The program was set up by the Armed Forces Institute of Pathology to improve upon fingerprints and dental records held in FBI files as a means of identification. Servicemembers killed in action or in accidents such as air crashes don't always yield good fingerprints and sometimes the fingerprint cards aren't readable due to smudges or other inking problems. Even if only small traces of the body remain, there's a good chance of making an identification through DNA sampling.

Each sample card in the warehouse is populated with two round drops of blood the size of fifty-cent pieces. The cards are then placed in envelopes and vacuum sealed.

Despite collection and handling issues that made the process difficult in the early days of DNA sampling, law enforcement agencies have strongly supported the collection and storage of DNA samples, or scanned or textual data associated with the samples, in order to solve or prosecute crimes.

At first DNA collection was aimed only at convicted criminals of specific violent crimes. Then it was broadened to include most violent crimes. With

---

[29]Douglas J. Gillert, "Who are you? DNA registry knows," *U.S. Department of Defense News*, July 13, 1998.

Title 42, it broadened further to include those arrested or facing charges, even though they might later be found to be innocent.

> *"The Attorney General may... collect DNA samples from individuals who are arrested, facing charges, or convicted or from non-United States persons who are detained under the authority of the United States."*
>
> Title 42 U.S. Code § 14135a—Collection and Use of DNA Identification Information from Certain Federal Offenders

It wasn't the federal government alone that broadened its scope. By 2008, 38 states were collecting DNA samples for a broader range of crimes, including certain misdemeanors.[30] In many cases, individual precincts forwarded the information to state bureaus for upload to national databases.

In New York State, legislation to expand DNA collection to those charged with misdemeanors was opposed. Nevertheless, in 2012, district attorneys were encouraged to take samples to add to the state data bank.

In some places, DNA is now banked for people who have been arrested and released without further legal procedures. Most law enforcement agencies promised this wouldn't happen when the storehouses were first initiated and yet the mandate continually broadens.

Since anomalies or ethnic patterns in a person's genetic makeup might reveal sensitive family histories, or affect eligibility for employment, life insurance, or medical insurance despite laws forbidding this, the populace is understandably concerned when DNA samples of law-abiding citizens remain in institutional files.

Using DNA to Solve Crimes

DNA has tremendous utility in identifying victims and perpetrators of violent crimes, and of exonerating those falsely accused of a crime. The Innocence Project is a nonprofit organization dedicated to exonerating wrongfully convicted criminals using the results of DNA tests. By 2012, almost 200 people who had served an average of 13 years in prison were exonerated and released, including some who had been on Death Row.

As mentioned earlier, DNA evidence that has been collected under imperfect conditions or which may be contaminated or mishandled, may not be admissible in court, but as the science improves and the courts become more open to DNA evidence, it is increasingly used as a forensic tool.

In cases of terrorist bombings, bomb fragments can sometimes yield DNA connected with suspected bombers, but something that has been detonated is generally not a good source of biological samples. More recently, investigators have turned to containers that held the bombs, such as boxes or bags, and have had better success.

---

[30]Harry G. Levine, et al., "Drug Arrests and DNA: Building Jim Crow's Database," *Council for Responsible Genetics Forum on Racial Justice Impacts of Forensic DNA Databanks*, New York, June 2008.

In the past, violent offenders often escaped to other states or provinces, and sometimes to other countries. With distributed computer networks and remotely-accessible databases (or personnel who can forward local information to another precinct), there's nowhere to run, especially since these resources are increasingly exchanged between foreign nations, as well.

Collecting, storing, managing, and retrieving DNA will help solve a certain number of crimes, but it also adds to the public tax burden.

DNA testing extends beyond domestic law enforcement. In 2011, the Department of Homeland Security announced a pilot program to use quick-test portable DNA devices to verify kinship among immigrants seeking asylum or refugee status. The rationale is to prevent immigration fraud and human trafficking but may also result in a few surprises for the families themselves since children are sometimes adopted or raised by other branches of a family without being told.

DNA is also an invaluable tool in anthropological research, making it possible to chart human evolution and global migration patterns. It can even reveal what kinds of goods were crafted and sold in ancient times.

DNA Technology

In the 1990s, DNA analysis was bulky, expensive, and time consuming, sometimes taking weeks to complete the full evaluation. Analysis in the field was impractical and sending samples to a lab always created the possibility of damage, loss, or tampering while in transit. This dramatically changed near the end of the 2000s.

DNA extraction has steadily improved. While large samples were once necessary, it is now possible to extract DNA information from as little as $0.025 \mu l$ of blood[31]—an amount equal to about 1/1000th of a drop of blood.

The Department of Homeland Security takes an active interest in DNA technology. In 2011 they announced testing of faster, more portable DNA devices developed by a Boston company and, in 2012, they expressed interest in new, faster amplification technology from a Pittsford, NY, company.

DNA is often available only in tiny quantities, especially if it is trace evidence from a crime scene or anthropological find. Duplicating the sequences so there's enough material for use in analysis was one of the bottlenecks that used to slow down the process. The new technology can accomplish the task in about five minutes rather than hours.

NEC has developed a portable DNA analyzer that can analyze DNA in less than half an hour and, in February 2012, Oxford Nanopore announced a gene sequencing tool the size of a memory stick that could be interfaced with a laptop computer.

---

[31] Jason Y Liu, Chang Zhong, et al., "AutoMate Express™ Forensic DNA extraction system for the extraction of genomic dna from biological samples," *Journal of Forensic Sciences,* March 5, 2012.

The accuracy of some of the tiniest devices may not be sufficient for certain applications, but it shows the direction things are going and accuracy no doubt will improve. In addition to portability, the price of tiny devices is only a fraction of full-sized sequencing machines and, even if less accurate, could have applications for war zones and wildlife management.

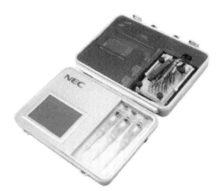 

Left: A portable DNA analyzer developed by NEC that can integrate DNA analysis from extraction to polymerase chain reaction (PCR) to electrophoresis and individual profiling intended primarily for law enforcement applications. Right: An ultra-portable single-molecule testing device that can be adapted for DNA sequencing and interfaced with a laptop for testing in the field. [Left: NEC news photo, 2007, released. Right: Oxford Nanopore Technologies news photo, 2012, released.]

DNA profiling has changed from an unwieldy and somewhat obscure lab process to a trusted and sought-after technology for a wide range of security and scientific applications. It has the capacity not only to unlock clues at a crime scene, but to reveal our relationship to past civilizations and extinct hominids. Concerns about its use in relation to privacy are discussed in later chapters.

PREVALENCE

EFFECTIVENESS

PREVALENCE OF SURVEILLANCE

- *prevalence of different kinds of surveillance*
- *use and abuse of spy devices*
- *public controversy and accountability*

Introduction

How prevalent are surveillance devices?

Many devices sold for other purposes are used for surveillance. Many devices sold for surveillance aren't regularly monitored. Some devices are fake—they only look like video cameras or motion sensors—there are no electronics installed.

One security and communications company alone has installed almost 50,000 cameras on trains and buses in the EU, and there are tens of thousands of companies selling similar products. Optical surveillance is only one of many forms of surveillance.

The truth is, no one knows how many surveillance systems are in active use but there are many millions, and billions of dollars in federal grants have been awarded to companies to develop more.

Science fiction books have long used surveillance societies as a backdrop for social commentary, but the situation is no longer fiction. We are living in a surveillance society and books and movies that depict people going to extremes to try to avoid being tracked may not be fiction either, a few years hence.

A surveillance society isn't just about security or a certain number of devices; one must consider gradual restructuring of the economy around the sale of surveillance systems. Installation of the technology becomes a vested interest—a form of job security—regardless of whether studies show an overall social benefit. Humankind survived for hundreds of thousands of years without surveillance devices. Have we created a situation where we're incapable of living without them?

There isn't space in this text to enumerate every kind of surveillance system, but examples will be given to illustrate common systems that are used in a variety of applications such as

- public safety systems,
- industrial, corporate,[50] and residential security,
- national security,
- global financial monitoring, and
- personal devices that can be reappropriated for surveillance.

Surveillance devices are common in North American and Europe, but given that most are manufactured in Asia, it is not surprising that surveillance is spreading throughout Asia, as well. Laws aimed at mandating or regulating surveillance overseas are sometimes based on U.S. law.

Surveillance need not always be high-tech. Peepholes are surprisingly effective and windows or mirrors have long been used in law enforcement and mental health institutions to observe subjects brought in for observation or questioning, but the trend is toward smaller and more powerful devices that can function autonomously and intercommunicate with other devices to perform cooperative tasks.

PREVALENCE OF SENSORS OTHER THAN CAMERAS

Motion Sensors

Motion sensors are based on a number of technologies, including infrared, radar, acoustic sensing, and vibration sensing. They are commonly used to open doors, to monitor gateway traffic, turn on lights, and detect intruders.

In industrial settings, motion sensors are built into production lines and equipment where safety may be an issue. Motion sensors are also used to manage livestock and study wildlife. Newer smartphones and other PDAs are being equipped with motion- and tilt-sensing components to enable a wider range of applications to run on them.

Infrared motion sensors are especially prevalent, with the components selling for less than a dollar. This enables them to be incorporated into a wide range of products. Passive-infrared detection is common in devices that trigger security lights.

Robotics is a growing industry and motion sensors are integral components that enable a robot arm to locate nearby components or for a moving robot to interact with its environment and with other nearby robots.

In 2012, Market Research released a report that global demand for motion sensors will likely increase to $4.8 billion by 2016.

---

[50]Privacy Protection Study Commission (PPSC) convened pursuant to the *Privacy Act of 1974*, report, 1977.

Chemical Sensors

Chemical sensors are common industrial and laboratory safety components and broadly sold for medical diagnostics. Customs and Border Protection uses them to locate illegal narcotics or smuggled items and also to determine quality control in goods scheduled for import or export.

The ATF uses chemical sensors to identify explosives and to investigate possible arson cases. The EPA and other organizations use chemical sensors to monitor water quality and the integrity of oil and gas pipelines. They are also essential for detecting environmental contamination and tracking the progress of cleanups.

Chemical sensors are widely installed in military land and marine vehicles to provide warnings of chemical contamination and the Department of Homeland Security is providing support and funding for the development of handheld chemical sensors and chemical-sensing systems for premises security.

Chemical sensors may even be used in the future to detect and identify individuals and to market products to specific demographics.

In 2010, Global Industry Analysts, Inc., estimated that the global market for chemical sensors would reach $17 billion by 2015. A market report by BCC Research indicates that micro-electromechanical systems (MEMS), components that can integrate sensors and actuators with mechanical elements, will reach approximately $12 billion by the end of 2014.

PREVALENCE OF CAMERAS

Digital cameras are everywhere and the market for small cameras is booming.

Manufacturers regularly add cameras to cell phones, PDAs, netbooks, and other handheld devices to increase their market appeal. Micro-cameras are also incorporated into keychains, pens, sunglasses, and lapel pins and many include both audio and video capabilities.

Cameras squeezed into small multifunction devices are sometimes of lesser quality, but quality is steadily increasing and may soon rival larger digital cameras. New technologies for stabilizing cameras and capturing the image faster, to prevent blur, are quickly advancing. Batteries are improving, as well.

Videocams designed to interface with personal computers are inexpensive and easy to adapt for surveillance purposes. They can be used as local devices to capture video within a room or aimed through a window at a bird feeder. They can be used remotely through a computer network to observe a child at a daycare center and are popular devices for waving at grandma several states (or countries) away. Video chatrooms are a mainstay of social networks and online dating. They are invaluable tools for online tutoring and college-sponsored distance education.

A decade ago, a high-resolution, color, audio-equipped cam the size of a golf ball sold for around $99. Now cameras with similar or better capabilities sell for under $15. Split-screen software can be used to display video from several locations on one screen. With so many economical options, some buildings are installing hundreds or even thousands of cameras.

Left: Tiny keychain and thumb-sized video cameras are capable of capturing medium and high-resolution video at speeds of 20 to 30 frames per second. Right: Tiny video cameras capturing high-resolution color images can be interfaced with computers through USB ports or via Ethernet cables or wireless routers. In the 1990s, a PC card to convert analog video to digital signals for input into a computer sold in the $400 to $1,000 range. Now, since most cameras capture digital data and no conversion is necessary, it is easy to get the video from the camera to a computer and, from there, onto computer networks. [Photos courtesy of Classic Concepts © 2012, used with permission.]

In the U.K., in 2009, it was estimated that about 4.2 million CCTV cameras were in use in public locations. More recent studies suggest earlier numbers may be inflated (not all cameras are operational and some are faux cameras). Nevertheless, the number of cameras is still significant.

Bank machine and alley cams have become commonplace in many cities. They monitor both vehicle and human traffic in an effort to deter crime at ATMs and in less-visible public places. [Photos courtesy of Classic Concepts © 2012, used with permission.]

Cameras specifically targeted at the surveillance market may have additional capabilities such as water resistance, motion sensors, automatic night vision, pan and tilt, "smart camera" image processing, biometric identification characteristics, and Ethernet connectors.

There are many beneficial ways to use surveillance cameras, but some applications are controversial.

Major sports and political events routinely install cameras and scanners for event security, but there are concerns about whether they will be de-installed after the event. In the case of the Republican convention scheduled for August 2012, the decision on how the cameras would be used after the convention was scheduled for September—after the event. Privacy advocates hoped that post-event issues would be decided before the money was spent and cameras installed.

The ubiquity of cameras has provoked recent clashes between the public and law enforcement. Many tweens and teens have cameras in their netbooks, phones, and PDAs, and are used to photographing everything that occurs in their daily lives without a second thought.

Since law is not taught in middle school, nor in most high schools, many youths are unaware of video voyeurism laws, or regulations limiting the use of audio-equipped cameras. Numerous controversial arrests of young people have recently occurred because they recorded law enforcement officers or others without their express permission. If the recordings excluded audio, some may have been legal.

Another concern about cameras is their increasing use in law enforcement activities that are combined with entertainment. Patrol car cams have been used for some time now, but they are less objectionable than news crews that follow police officers into businesses and residences to capture footage for entertainment purposes. This form of reality TV may be educational in some ways, but is confusing to those apprehended, especially if the news crews are wearing police vests that appear to give them the same authority as law enforcement officers. Reality-show footage doesn't always follow up the cases and report whether the people apprehended were released without charges or found innocent.

Expansion of Access for Aerial Surveillance Cameras

One of the more significant developments is Congressional approval, in February 2012, for the expansion of aerial vehicles (UAVs) into domestic airspace. Since uninhabited aerial vehicles have traditionally been developed for the surveillance market (other than very small ones used by hobbyists), most are equipped with cameras and other sensors.

Expanding Private UAVs into National Airspace

"...the Secretary of Transportation, in consultation with representatives of the aviation industry, Federal agencies that employ unmanned aircraft systems technology in the national airspace system, and the unmanned aircraft systems industry, shall develop a comprehensive plan to safely accelerate the integration of civil unmanned aircraft systems into the national airspace system... as soon as practicable, but not later than September 30, 2015."

According to the terms of the *FAA Modernization and Reform Act of 2012*, the FAA must give UAVs (or, as the act calls them, "unmanned aircraft systems") access to commercial airspace and evaluate (and possibly certify) those capable of operating safely with existing air traffic. Opening up airspace to UAVs will likely provoke a significant expansion of domestic surveillance. Some police departments are already testing surveillance UAVs in community neighborhoods. Entrepreneurs will be looking for ways to launch new UAV-based commercial ventures when the FAA regulations are finalized.

The FAA modernization act also requires that tracking technology be transitioned from traditional radar to GPS—a provision that will make vessel tracking more precise and may improve air traffic safety.

Prior to passage, there was little public information on how these changes (and possible broadening of runways and flight paths) would affect the safety of existing air traffic (which will have to share runways and airspace with UAVs), or how the increase in aircraft would affect air quality, visual pollution, residential noise levels, or the health and safety of wildlife such as birds. There was also little discussion of how an increase in UAVs would affect backyard privacy.

The new legislation allocates $63.4 billion to update aircraft access and monitoring capabilities.

In the past UAVs were mainly used in coastal patrol, war zones, or military installations. Now analysts suggest as many as 30,000 may be using domestic airspace by 2020.

The FAA modernization act also directs that areas of the Arctic be designated for 24/7 permanent surveillance by UAVs for research and commercial purposes.

> *Not later than 180 days after the date of enactment of this Act, the Secretary shall develop a plan and initiate a process to work with relevant Federal agencies and national and international communities to designate permanent areas in the Arctic where small unmanned aircraft may operate 24 hours per day for research and commercial purposes. The plan for operations in these permanent areas shall include the development of processes to facilitate the safe operation of unmanned aircraft beyond line of sight. Such areas shall enable over-water flights from the surface to at least 2,000 feet in altitude, with ingress and egress routes from selected coastal launch sites.*

> H.R. 658–65

Large portions of the Arctic belong to Norway, Sweden, Denmark, Canada, Russia, and Mongolia. Most countries probably wouldn't object to collaborative Arctic research with a prior approval process, but many may object to commercial UAV exploitation of the Arctic beyond Alaskan borders—a situation that could harm U.S. international relations and provoke more terrorist attacks from idealists who oppose U.S. expansionism. Commercial exploitation would also require environmental impact studies. The Arctic is a harsh environment and the survival of plants, animals, and indigenous people is tenuous at the best of times.

UAVs are often equipped with multiple imaging sensors, including cameras that can see in the dark or through fog and light cloud. Recent improvements in image stabilization and electronic zoom make it possible for sensors to pick up small details from great distances and facial recognition software is increasingly built into camera surveillance systems.

The main proponents of many of these technologies are not members of the general public seeking to feel more secure, but manufacturers eager to expand their markets. Those markets could include companies eager to collect personal demographics on people's hobbies, movements, and consumer buying habits. It is only a short step between facial recognition and computer matchups with personal profiles gleaned from network aggregation sites or government databases.

Expansion of Cameras in the Hands of Little Brothers

Given the human propensity for voyeurism and to buy gadgets that are cute and convenient, micro-cameras have found their way (via pockets, pens, backpacks, and skateboards) into board meetings, locker rooms, steam baths, and other private places where people prefer not to be photographed.

Incidents of *video voyeurism* are becoming more frequent and the legal uses of consumer gadgets is being tested in courts.

- In February 2012, a thief who stole a laptop a few months earlier, faced additional charges when half-naked photos of a girl, taken through her bedroom window, were found on the laptop.

- Similar charges were leveled against a Louisiana man when video images of teenage girls and one adult female were taped in a bathroom without their consent.

- In Kentucky, nude video images were taken of a minor in a high school locker room when his towel fell off. The cell-phone video may have been transmitted to as many as 30 people offsite. (With current technology, images are not confined to the device that shoots them. With wireless or Internet Protocol [IP], they can be viewed by others.)

- A male corrections officer and his female accomplice led two juveniles to undress in front of a video camera in his home, while he watched.

- An Idaho man was charged with video voyeurism for secretly videotaping sexual relations with his former girlfriend and later placing copies of the images in her boyfriend's vehicle.

- A traveling preacher who had used a pen equipped with a camera and flashcard was caught photographing an Arkansas woman in the shower and apparently had previously recorded two women in Mississippi.

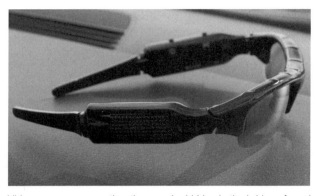

Video cameras are so tiny, they can be hidden in the bridge of a pair of sunglasses and interfaced with controls and a compact memory stick in the stems for recording a video stream. There's no longer a need to have a wireless transmitter (which can be detected up to about 300 feet) or a recording device hidden in a pocket or briefcase. Put on the glasses and, if something significant happens, pretend to adjust the stem while pressing a button to start recording. If the memory stick fills up, insert another one. Expect to see these at the beach. [Photos courtesy of Classic Concepts © 2012, used with permission.]

These examples bring up some questions related to the prosecution of privacy violations.

- Is there a difference between someone deliberately and covertly trying to lure someone or capture someone undressing, and a teammate transmitting nude pictures when a student unintentionally loses a privacy towel in a high school locker room?

- If a high school student is charged with video voyeurism for broadcasting nude images of a teammate in a locker room, should a company, like Google, which broadcast a fully nude woman pouring water in a porch alcove over the Internet Streeview™ service, be charged with a similar offense?

Street-Viewing Cams

There are specialized cameras, such as vehicle-mounted cameras, that record a wide-angle or rotating view of the adjacent area. These can be used to monitor intersections or to record the view along public streets.

The most well-known is Google Streetview™, which uses a sophisticated imaging system that has been integrated with Google maps to provide terrestrial views of major streets throughout the world from almost every viewing angle. The imagery isn't restricted to traffic and storefronts. The high-resolution pictures enable users to zoom in through windows of buildings that line city streets. Snowmobile and human-powered tricycle versions of the Streetview™ camera have also been developed that can be ridden in places that are inaccessible to cars.

It is even possible to invite the trike to visit a premises to take images on private property, a service a real estate agent or owner selling a property would welcome, as would someone trying to promote a sports or entertainment center like a golf course or amusement park. If it is residential property, however, the subsequent owner might be concerned about privacy and security if the images have been uploaded to the Web.

Mini versions of street-viewing cameras can be purchased for under $200. A vehicle camera the size of a tennis ball can be mounted on a dashboard or under the rear-view mirror to take images of the surrounding traffic and landscape. Some of them can rotate and there are models available with more than one lens.

With technology improving at a race-track pace, street-viewing cams similar to the multi-lens Google camera may soon be offered as an option on new vehicles. Some people buy vehicle cams to gather video evidence in case of an accident but whether the evidence is admissible in court depends on many factors and the camera has to be on and pointing in the right direction for the system to be effective. Newer car cams include a shock sensor that senses the impact force of a collision and GPS, so the exact location of an accident can be recorded along with video footage.

Most car cams have one or two cameras, but companies have developed periscope-style cameras that mount on a vehicle's roof rack to capture surroundings from a higher perspective. Periscope cameras with 360-degree views designed for use on marine vessels could be mounted on land vehicles, as well. Some have suggested periscope cams could look over a traffic jam to see what's happening ahead so the driver could decide whether to take the off-ramp.

A few years ago, windshield- or dashboard-mounted videocams were bulky and expensive. Now, for less than $200, it is possible to mount a high-resolution camera anywhere inside a car within reach of a power source, such as the cigarette lighter, and begin recording events to a compact memory card. Often vehicle cams are mounted so truck drivers can see their blind spots on the sides or back of a vehicle, but several vehicle cams were invented because drivers were involved in accidents and wished, in retrospect, they had a visual record of events. Law enforcement officers in patrol cars now routinely record their on-duty activities with vehicle-mounted cameras. [Photos courtesy of Classic Concepts © 2012, used with permission.]

A vehicle cam can be used for other kinds of surveillance in addition to monitoring accidents or traffic infractions, including journalism, marketing, real estate assessment, or locating unauthorized construction projects. Trucks with blind spots are sometimes equipped with cameras to show the sides and rear of the vehicle to protect nearby objects and pedestrians.

Car cams aren't yet prevalent, but new cars will eventually be equipped with them, just as many now include video entertainment systems.

Fake Surveillance

Often surveillance systems are installed to deter or detect crime and faux surveillance cameras have become almost as popular as the real ones. A faux system looks like a real surveillance device, but doesn't include any sensing or recording mechanisms. Sometimes companies that sell surveillance cameras will make the housing available, without internal electronics, for a budget price.

Other surveillance systems are fully functional but are designed to look like something else, such as a light fixture, sprinkler system, or smoke detector.

 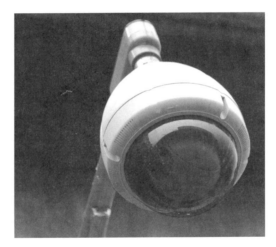

Some security cameras are almost indistinguishable from light fixtures (and may include lights). Only a close inspection reveals whether there is a camera inside and even this doesn't always make it clear whether the camera is operational. It is difficult to determine how many functional video cameras are operational in any given area at any given time. Estimates are usually made on the basis of units sold rather than trying to count the cameras (many of which are hidden), but cam-counts have been attempted, as well, with mixed success. [Photos courtesy of Classic Concepts © 2012, used with permission.]

MILITARY SURVEILLANCE PREVALENCE AND EXPENDITURES

Surveillance is an essential aspect of military life. Sensors and imaging devices are used in every division of the armed services. Some of them help navigate and control vehicles, aircraft, and marine vessels, some are for defense-related reconnaissance and intelligence gathering. Others, such as UAVs, long-range binoculars and cameras, night-vision goggles, and infrared sensing devices, are for directly surveilling other nations or hostile forces in combat zones.

Satellites, piloted aircraft, UAVs, and marine vessels are all equipped with a variety of sensing mechanisms. The most common are optical cameras, infrared/

thermal detectors, ultraviolet imagers, radar, sonar, GPS, chemical sensors. There are also "tilt" sensors based on a combination of attitude (yaw, pitch, and roll) and acceleration.

There isn't space to enumerate all the different systems in use by the military, and there's no way to count them since some of the information is classified, but a few examples will be given.

- In 2005, it was announced that a Raytheon subsidiary had been awarded a $20 million contract to produce shared reconnaissance pod systems that could carry a variety of sensing and imaging technologies on the wings of carrier-based aircraft.

- In January 2011, construction began on a $1.2 billion data storage center at Camp Williams, Utah, for use by the Intelligence Community. The *Community Comprehensive National Cyber-Security Initiative Data Center* (the "Spook Center" is easier to say) sits on acreage more than ten times the size of the White House grounds. Considering that a 2-terabyte hard drive is about the size of a pocketbook, the 1.5-million-square-foot Data Center could potentially have unimaginable capacity for digital information.

- Lockheed Martin was awarded a $47 million contract in 2011 to modernize U.S. Air Force long-range radars that are used for air traffic control surveillance and advanced warning systems. The systems are upgraded every decade or so.

- The *Economic Times* reported that the Air Force and Navy would be spending about $23 billion on the purchase of 55 Global Hawk UAVs to replace the U-2 spy planes that have been in use for decades.

- The payloads that are incorporated into UAVs, which include many forms of imaging sensors, are expected to increase to about $5.6 billion by 2020.

- In 2011, the Brazilian army reported their surveillance needs for monitoring the northwest border and the region stretching across the ocean to offshore oil fields at about $10 billion. The System for the Surveillance of the Frontier (SISFRON) will cost about $6.5 billion and the navy's Blue Amazon (SISGAAZ) surveillance program, up to about $4 billion. Satellites and UAVs are included in the proposed systems. The surveillance is primarily aimed at security threats from criminal gangs, smugglers, and pirates.

- ASD Media BV estimates military spending on full-motion video worldwide at approximately $8.8 billion for 2012. The research was based on sources such as opinions of leading figures in military video-surveillance systems markets together with government and corporate policy documents, and procurement or spending announcements.

- VisionGain released a report that estimated the global market for UAVs, UGVs, and perimeter surveillance systems in 2010 was about $16 billion and expected to grow. The change in FAA regulations that will open civilian airways to UAVs wasn't public knowledge at the time the report was written, but it will significantly increase future estimates.

MONITORING TRAFFIC INTO AND OUT OF THE COUNTRY

Customs and Border Protection (CBP) and the U.S. Coast Guard use a broad range of surveillance technologies and techniques to monitor and protect an extremely diverse landscape that includes land crossings through forests, mountains, rivers, and coastlines, as well as traffic coming and going through the air, not only through land-based terminals, but seaplane landings in lakes and coves.

CBP Overview

The CBP operates one of the most varied and active surveillance systems in the world. It is responsible for more than just monitoring border traffic and smuggling, it also monitors the quality and authenticity of goods going in and out of the country and prevents harmful agricultural pests from entering.

Virtually every aspect of CBP operations involves surveillance of one kind or another.

Surveillance Resources and Equipment

CBP employs more than 40,000 border agents and officers (and 19,000 additional personnel in other capacities) to secure borders from unlawful entry or exit of people or goods, or from harmful intrusion. To accomplish these tasks, it uses border checkpoints, coastal marine and air patrols, ground-based mobile surveillance units, thermal imaging systems, and a wide variety of inspection technologies, including X-ray, gamma-ray, infrared, ultraviolet, density meters, fiber-optic scopes, lab tests, and trained sniffing dogs.

Large-scale cargo imaging systems have led to the discovery of human smuggling, weapons, explosives, and narcotics.

As of Fiscal Year 2011, CBP maintained 46 air units, 71 marine units, 299 large-scale imaging systems, 41 mobile surveillance systems (MSSs) and a mobile processing center (MPC), ~28,000 radiation detection and identification devices, and processed 29.5 million entries of trade goods into the country. More than 11 million marine cargo containers were off-loaded and processed. By studying manifests that are submitted 24 hours before the cargo sets sail for American seaports, those more likely to pose terrorist threats are identified before arrival.

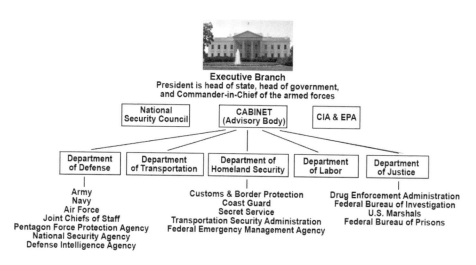

Organization of government departments particularly relevant to surveillance.

CBP Surveillance and Intelligence Departments

The CBP has many departments that collect and process intelligence about people traveling in and out of the country, and about enrollees in expedited travel programs. This is a sampling of some of the departments.

- The CBP maintains the *National Targeting Center-Cargo* (NTC-C) and the *National Targeting Center-Passenger* (NTC-P). The NTC-P is staffed with analysts who carry out research to identify individuals who might pose security threats at border checkpoints using automated data-processing systems to prescreen and identify those who might pose a terrorist threat. It maintains the *Terrorist Watch List* in collaboration with the DHS/National Counterterrorism Center.

- The NTC-P works in cooperation with agencies such as the FBI, the CIA, and the TSA's Office of Intelligence.

- *COMPSTAT* is a desktop database/statistical software application that compiles and processes data from multiple sources, including *Enforcement Case Tracking*, the *Border Patrol Enforcement Tracking System*, and *Intelligent Computer-Assisted Detection* (ICAD) system, and creates reports based on the results of the findings. ICAD enables monitoring of the location and activities of each border agent in realtime and keeps a record of sensor detections and alerts. CBP also makes extensive use of digital mapping services.

- *Border Intelligence Centers* (BICs) are tasked with preventing entry of terrorists, terrorist weapons, and illegal aliens. BICs collect, analyze, and disseminate tactical intelligence. BIC agents utilize operational statistics, interviews, reports from other agencies, and databases and share information with the DHS and law enforcement agencies.

- The *CBP Fraudulent Document Analysis Unit* (FDAU) is a centralized forensic facility that collects and analyzes fraudulent and suspected fraudulent travel documents and gets them out of circulation. The repository has received almost a quarter of a million fraudulent documents. Fraudulent documents are those that have been fabricated or altered or, if valid, have been presented by imposters.

- The NTC-C administrates Agricultural/Biological Terrorism Countermeasures operations to identify and target people and shipments that may pose risks via biological materials.

Collecting data and developing intelligence information is not sufficient by itself. The Integrated Border Enforcement Teams (IBETs) provide practical follow-up. IBETs are comprised of U.S. and Canadian law enforcement personnel that coordinate responses and carry out enforcement operations. CBP also cooperates with Mexican authorities.

Expedited Travel and Watch-List Prescreening

The CBP expedites border crossings through their NEXUS and SENTRI programs. Low-risk travelers who apply for the program, and who pass biographical screening and a personal interview, are given a card with RFID information that matches them to information in the CBP database. Possession of the card enables travelers to use priority "fast" lanes. For the NEXUS program, iris scans provide access to self-service kiosks. More than 1 million card-holders participate in the various CBP expedited-travel programs.

The Advance Passenger Information System (APIS) receives biographical data on passengers and crew traveling into or out of the U.S. by ship or plane, and sometimes by bus or rail, prior to boarding. This information is checked against *no fly* lists and *watch* lists, and cross-referenced with law-enforcement databases to check for alerts. It was reported that about a million people were on the Terrorist Watch List by 2009. About 50,000 of these were U.S. residents/citizens.

AUDIO SURVEILLANCE

Microphones

Just as cameras are incorporated into most PDAs, audio recording is a common feature, one that may accompany video recording or be turned on independently of video.

Dedicated audio surveillance devices exist, as well. In addition to room bugs, there are products that look like doctors' stethoscopes that can be pressed against a wall or door to amplify the conversation on the other side. Newer audio bugs have electronics to suppress ambient noise so the main conversation is more clear. This is important when gathering audio evidence, since tampering with the original raw recording renders it inadmissible in court.

In most places, eavesdropping laws limit audio recording, yet it still occurs often. Millions of tiny audio recorders are sold through spy shops and online auction sites, in addition to being available from companies that limit their market to law enforcement clients.

When a customer calls a company for help or information, the message, "This call may be monitored for quality assurance" may be played at the beginning of the conversation. This is a way of obtaining tacit approval for audio recording. Customers rarely have the option of saying no to help-desk audio recording—it is a recorded statement, not a human asking a question. The company may justify this by saying the caller has the option of hanging up, but then they don't get the help they need.

Commercial firms monitoring customer calls almost never offer information on how the recordings are archived or distributed. Since the caller's voice is linked with account information, the recordings are, in essence, taking identifiable voice prints, even if it is not the original intended purpose. These could be subpoenaed by law enforcement or sold to third parties.

As with cameras, it is difficult to know how many audio devices are in active use for surveillance, but monitoring of business meetings, customer service calls, and phone or chat conversations is common. There is a brisk business in tiny audio recorders hidden inside pens, calculators, and other common business accessories. They are certainly sold in the tens of millions, possibly in the hundreds of millions.

Wiretaps

The U.S. Wiretap Report became an annual requirement with the *Omnibus Crime Control and Safe Streets Act of 1968,* to help prevent wiretap abuse by providing a public record of wiretapping within the United States.

The report for 2010 states that 3194 authorized intercepts were completed in 2010, an increase of 34% over 2009, with 84% of taps related to drug-crime investigations. The installed taps operated for an average of 40 days and 26% of those taps were found to be incriminating.

The report indicates it is rare for intercepted communications to be encrypted. In 2010, there were only six reported cases of encryption and plaintext versions of the communications were obtained by law enforcement.[51] The average cost of each wiretap is about $55,000.

---

[51] United States Courts, Report of the Director of the Administrative Office of the United States Courts on Applications for Orders Authorizing or Approving the Interception of Wire, Oral, or Electronic Communications for 2010.

How much tapping is carried out by the Intelligence Community isn't known. The *Foreign Intelligence Surveillance Act* (FISA) requires that the National Security Agency apply for warrants, but the *Patriot Act* provided broader leeway for monitoring suspected foreign terrorist communications and it is difficult, given the changes in the way people communicate, to monitor suspected terrorists without monitoring everyone else.

There has been considerable lengthy and complex controversy over allegations that the NSA obtained or tried to obtain extensive data mining information from ISPs. There have been many lawsuits related to this issue, so it is suggested the reader consult online resources for further information.

VoIP Monitoring

People use services like Meebo™, AIM™, Skype™, and SiteSpeed™ to exchange audio/video or audio-only messages over the Internet. There are also thousands of chatrooms that support audio and video, including dating sites and gaming clubs.

In addition to legal talk venues, there are many pornography sites that provide private chat services for a fee. Whether these chats and their associated IP addresses are recorded isn't known. Those who use private services usually pay with credit cards, which leave an audit trail.

It is not known how many online digital voice conversations are clandestinely monitored. It might be thousands or tens of millions. They may be live-monitored or recorded for later viewing. It is very difficult for a user to know if "private" is really private.

As mentioned in the Audio Surveillance chapter, Microsoft was awarded a patent for technology that facilitates the monitoring of voice conversations over digital networks. Since millions of users worldwide now converse this way, audio surveillance over digital networks may dramatically increase if there's technology to make it easier, despite wiretapping laws.

GPS SURVEILLANCE

GPS surveillance used to be uncommon but now that GPS is built into most handheld devices and many forms of vehicles, it is becoming a very common way to locate a person, vehicle, or GPS-capable device.

It was reported in 2010 that law enforcement agencies had allegedly obtained millions of GPS location points from ISPs. Many ISPs have electronic surveillance staff that communicate regularly with law enforcement.

GPS location is, in some ways, permissible. If a single location is determined or given to law enforcement upon request by ISPs, this may be lawful and may be useful for locating a suspect or responding to an emergency call.

However, if a series of GPS coordinates is mapped to follow a trail, this may not be lawful—it may constitute tracking which is, in some instances, illegal search and seizure.

BIOMETRICS SURVEILLANCE

Almost every aspect of biometrics surveillance is seen as a growing market, especially since the technological options are increasing and the devices are harder to fool. Biometric scanners that work at a distance are more available than ever and are likely to fuel sales even further.

Biometrics detectors are built into many access-control systems, to verify a person's identity before using a bank machine, computer, or doorway. Even casinos were implementing facial recognition systems by the mid-2000s and the CBP now uses fingerprint scanners.

Biometrics such as DNA, fingerprints, and dental patterns are used for military identification and widely used in forensics to identify suspects, witnesses to a crime, and victims.

Hospitals use biometrics to identify newborns and patients are sometimes now identified by palm scans, in addition to wearing the traditional ID bracelet. In some hospitals, biometrics are used to access medical records.

Biometric photos and sometimes fingerprints are incorporated into ID cards and passports. The Clear Card, a card that includes a fingerprint, an iris print, and quite a bit of additional private information, is used by frequent flyers to expedite travel at airports.

Many programs that offer cards with sensitive personal information to streamline travel are voluntary, but given the long waits in transportation terminals these days—sometimes hours—it is not surprising that more than 40,000 frequent flyers have paid $100 to $200 to purchase them, despite concerns about privacy.

In some countries, fingerprints are now used at supermarkets to verify identity when customers pay.

Large-scale events now use facial recognition and voice surveillance systems, and facial analysis cameras on the backs of airline seats have been tested for possible deployment in commercial airlines.

By the mid-2000s, government spending on biometrics technologies was already about $10 billion per year. In 2008, the DHS Under-Secretary for Science & Technology reported that biometrics was the largest growth area in homeland security technologies.

It was estimated that spending on biometrics technologies in the U.K. would increase to about $5 billion by 2013.

DRUG SURVEILLANCE

Drug testing is surveillance of illegal or restricted chemicals that a person may have smoked, ingested, injected, or otherwise introduced into the body in quantities sufficient to show up on tests. Drug testing is common in competitive sports and mandated by some employers, especially in workplaces where employees are required to operate transportation systems or dangerous equipment.

The Drug Abuse Warning Network System (DAWN) is a public health surveillance system that gathers information from drug-related hospital emergency visits and provides it to policymakers. The program is being superseded by the CDC National Hospital Care Survey, which will collect essentially the same kinds of data at a lower cost.

Drug screening is often conducted through a written questionnaire, but most drug testing is biometric in nature. Body fluids are commonly used, such as blood or urine. Hair samples are sometimes used as they make it possible to check for drug use over a longer period of time. In DUI screening, breath tests are commonly requested, with possible follow-up blood tests, depending on circumstances.

Due to its personal nature, drug testing constitutes a form of search, so it is not permitted in all situations.

Driving-Related Breath Tests

About 38 states have roadside sobriety checkpoints in which vehicles are stopped and checked for anything suspicious that might warrant further investigation. If the driver appears impaired, he or she may be asked to step out and take a sobriety test.

Breath tests combined with behavioral tests are used by police officers to determine if someone is driving under the influence (DUI), as a means to reduce traffic accidents and fatalities.

In 2009, ~10,839 fatalities resulted from crashes in which the driver of a vehicle had a blood-alcohol concentration (BAC) of 0.08 or higher. A BAC of 0.15 or higher was found in 70% of those tested. At least that many drivers were tested for their BAC, in addition to those in accidents who tested below the limit and those tested at roadside checks who were detained or released.

In a four-hour roadside check in Evanston, IL, in 2010, police issued 69 citations, six of which were for driving under the influence (8.7%).

In 2009, the FBI reported that 1.4 million drivers were arrested for DUI for alcohol or narcotics. Surveys indicate this was about 1% of the number of people who self-reported having driven under the influence.[52]

If 9% of those cited in roadside checks result in 1.4 million DUI arrests, it can be estimated that at least 16 million checks are made annually in the U.S.

In 2012, the federal government encouraged the 37 states that didn't already have *per se* ("zero tolerance") DUI policies to establish them. It also recommended development of more reliable and standardized drug testing technologies for checking drivers (and federal employees in the workplace), with an emphasis on oral-fluids testing to complement urine testing.[53] Thus, the development of additional testing technologies and the prevalence of roadside drug testing is expected to increase.

---

[52]R.A. Shults, L. Beck, A.M. Dellinger, "Self-reported alcohol-impaired driving among adults in the United States, 2006 and 2008," *Safety 2010 World Conference, Sept. 2010,* Sept. 21–24, London.

[53]*National Drug Control Strategy,* Executive Office of the President of the United States, 2012.

Drug Testing in Schools

In the 1990s, random drug testing in schools was not common, with only a few hundred schools having explicit testing programs. At that time, random tests were only permitted on school athletes. There was, however, enough testing by the late 1990s that an effort was made to determine if it was effective in reducing drug use.

Between 1998 and 2001, research data from 76,000 schools was evaluated. The report concluded that drug testing "did not have an impact on illicit drug use among students, including athletes." Other ways to use funds that might reduce drug use in schools were listed.

> *"The current push to increase drug testing comes from the drug testing industry as well as well-intentioned educators and parents frustrated by the lack of success of drug prevention programs such as Drug Abuse Resistance Education (DARE)."*[54]

In 2002, the Supreme Court broadened the authority of public schools to test for illegal drugs in students participating in extracurricular activities.

In 2005, under the Bush administration, federal grants of $7.2 million were announced to encourage drug testing, with plans to increase the amount in subsequent years. By 2007, the CDC reported an increase in drug testing programs to more than 4000 schools.

> *More recently, in* Bd. of Educ. v. Earls, *536 U.S. 822 (2002), the United States Supreme Court ruled, in a 5-4 decision, that a School District policy requiring all middle and high school students to consent [sic] to urinalysis testing for drugs in order to participate in any extracurricular activity is a reasonable means of furthering the School District's important interest in preventing and deterring drug use among its schoolchildren and does not violate the Fourth Amendment.*
>
> Guidelines Concerning Student Drug Testing in Virginia Public Schools, Adopted by the Board of Education, June 23, 2004

With government funding, the trend is for increased drug surveillance, but studies of its effectiveness in reducing drug use in schools show mixed results. In some cases, students who knew they would be tested reduced their drug use by about 5% but self-reported they would use drugs again in the future.

Drug Testing Welfare Applicants and Recipients

More than half of U.S. states have proposed that welfare recipients be tested for drugs. Some have suggested those applying for unemployment benefits should be tested, as well, even though unemployment is a form of insurance of limited duration, a fund that workers pay into, as opposed to being strictly welfare.

---

[54] J. Kern, J. F. Gunja, et al., "Making sense of student drug testing: why educators are saying no," Drug Law Reform Project and Drug Policy Alliance, January 2006.

Welfare-related drug tests typically detect common illicit drugs such as cocaine, opiates and marijuana, and certain prescription drugs. They are not always able to distinguish between Tylenol-codeine and opiates. Most tests, if taken once, do not distinguish between those who occasionally take drugs and those who have a substance-abuse problem. The most commonly detected drug, in tests that look for it, is alcohol.

Proposals to test welfare applicants and recipients are being challenged, as they were in 1999, when a Michigan court ruled it was unconstitutional to subject all applicants to drug tests without cause for suspicion of drug use. Pre-screening has been in place in Arizona since 2009.

> *It is "unfair for Florida taxpayers to subsidize drug addiction..."*
>
> Governor Rick Scott, May 2011, upon signing state legislation for mandatory drug testing for TANF recipients.

In spring 2011, after transferring ownership of shares in an urgent-care chain that carries out drug-testing to his wife, the Governor of Florida spear-headed legislation for drug testing of Temporary Assistance for Needy Families (TANF) applicants.

As a result, in May 2011, Florida became the first state to implement mandatory drug testing for the TANF program. The impoverished applicants had to pay for the tests and were reimbursed later if the results were negative.

In July 2011, the Missouri governor similarly approved HB 73 to require TANF drug testing for recipients and certain applicants.

In October 2011, a federal court ruled the Florida legislation as unconstitutional.

In 2012, the state of Utah mandated drug questionnaire screening for the Family Employment Program, with follow-up testing if the screening aroused suspicion. Drug treatment and remaining clean would be a requirement of remaining in the program. This is more in line with constitutional policies, since the biometric testing only occurs if screening shows a reason for follow-up.

While there isn't enough long-term data to fully evaluate the situation, it appears from the Florida statistics that drug testing didn't significantly lower welfare applications and a very low percentage tested positive, contradicting the governor's assertion that TANF was subsidizing drug addiction in Florida. The cost of the testing was found to be more than the expected savings since 98% of those tested had to be reimbursed by the state for the cost of the test and there were associated administrative costs and legal costs related to the program.

The state of Georgia introduced a similar proposal to drug-test TANF applicants. As with the Florida governor, Georgia representative Spencer cited the cost to taxpayers of funding drug addiction as motivation for the testing. Applicants would have to pay for their own drug tests (with reimbursement if results were negative).

"Georgia taxpayers have a vested interest in making sure that their hard-earned tax dollars are not being used to subsidize drug addiction..."

Georgia State Rep. Jason Spencer, November 2011

The proposal was signed into law in April 2012 by the Georgia governor as HB 861.

As a result of the various moves to legislate, drug testing programs were analyzed and ASPE released a brief on the prevalence of drug use and possible costs associated with testing programs.

During 2010 and the first half of 2011, 82 bills on this subject were proposed in 31 State legislatures and the U.S. Congress...

...None of the State cost estimates identified for this paper showed net savings resulting from proposed drug testing programs, though these are all legislative cost estimates rather than rigorous cost-benefit analyses. Also, none of the State cost estimates identified described anticipated unit costs of drug testing programs...

... An evaluation of a Florida drug testing pilot found that those welfare recipients who tested positive for drugs had similar employment outcomes as others on TANF.

Radel, L.; Joyce, K. and Wulff, C. , "Drug testing welfare recipients: recent proposals and continuing controversies," Office of Human Services Policy, *ASPE Issue Brief,* October 2011

Most state proposals were to suspend benefits for a year, if a positive test were found, some for two years, and some proposals would suspend them indefinitely if the test-taker repeatedly tested positive or didn't actively participate in a drug treatment program.

Two of the proposed federal House bills would require all recipients to be tested, with a positive test conferring ineligibility. No provisions for substance abuse treatment or child well-being were addressed in these two federal bills.

One state proposal was for all applicants to be tested, including children under the age of 12.

Civil liberties organizations and individuals spoke strongly against the legislative proposals.

Legality issues aside, drug testing mandates are oppressive, imprecise and largely ineffective. Welfare recipients rarely garner much public sympathy, but stigmatizing poverty by associating it with drug abuse is patently malicious.

Marc Anderson, "Mandatory drug testing of welfare recipients is unconstitutional," *The Daily Cougar.com,* Oct. 19, 2011

In Florida, about 15% of those living in poverty receive TANF benefits.

In Oklahoma, several bills were submitted for drug testing of TANF recipients. The Oklahoma Policy Institute pointed out that the average monthly benefit is $207 and applicants in need of social assistance would have to pay for the drug tests. Most TANF recipients in Oklahoma are children. The proposals are being opposed on constitutional grounds and because applicants are already screened with other measures that have shown good results in the past.

In most areas, recipients are only on the program for a few months. Many TANF recipients are children and mothers trying to raise those children. Some are individuals with psychiatric disorders who may not be capable of holding a job.

It is also a financial burden on the poorest of the poor, even if they are later reimbursed, to pay the equivalent of a week's-worth of food up front for the tests. Children, some of whom are orphans or have parents with mental or physical disabilities, have no money for this kind of expense.

It takes funding to administrate welfare-related drug-surveillance programs and studies thus far show that it may cost taxpayers more than it saves. Constitutional issues will be worked out in the courts.

WORK-RELATED DRUG SURVEILLANCE

General Regulations

Drug testing in the workplace is fairly widespread. A number of professions require it by law. It is estimated that more than half of all accidents on the job are related to drugs (including alcohol), which is part of the reason for drug surveillance. The other motivation for drug testing is loss of productivity due to substance abuse.

Procedures for drug testing have been established by a number of agencies, including the Substance Abuse and Mental Health Services Administration (SAMHSA), which standardizes drug testing in federal agencies. Private employers do not have to follow SAMHSA guidelines, but they are usually on stronger legal grounds if they do.

The Department of Labor points out that drug testing is not a requirement of the federal *Drug-Free Workplace Act of 1988*. Private employers do, however, have the right to test if there is sufficient suspicion to warrant it.

Drug testing of union employees is not permitted without first negotiating with the union.

The *Omnibus Transportation Employee Testing Act* (OTETA) requires drug tests for all personnel who operate aircraft, trains, buses, trolleys, and commercial motor vehicles.

Sometimes drug testing is required as a condition of receiving government contracts. For example, the Department of Defense requires defense contractors to implement procedures for identifying drug users, including random testing.

The Department of Transportation (DoT) has a similar policy for testing workers in safety-related positions. Employees are also required to be tested after an accident to provide forensic evidence.

Examples of Workplace-Related Drug Testing

In 2007, the Oregon state police randomly selected truck drivers for drug testing and found almost 10% tested positive. This was higher than numbers reported for truck drivers by the federal government.

It is estimated that about 9% of employees abuse illegal drugs, based on a sample population that was urine tested. The Department of Justice noted in a 2010 report that increases in workplace drug abuse are primarily prescription narcotics (a similar trend was found in public school children and the two may be related since many children get drugs from their parents' medicine cabinets).

Jobs involving public safety sometimes have mandatory drug-screening requirements but it is difficult to know how many workers are screened since many are working for private companies.

Some companies drug-test job applicants. Job applicants have difficulty saying no because, in most cases, they need the job. The practice has generated controversy and is not yet widespread. Background checks based on Internet information are common, however, and data on a person's social habits, which may include pictures or posts about drinking and drug use, may surface on forums or social-networking sites. Even if the person using drugs or alcohol doesn't post them, friends at the same party might, especially now that every smartphone has a camera.

In March 2012, in Florida, the governor proposed mandatory drug testing of state employees, pre-employment testing of prospective workers, and random testing of all employees in the state, but the moves are being challenged and policies may be revised. The Florida *Drug Free Workplace Act* authorizes random testing every quarter-year and permits workers to be fired if they test positive. An amendment to the bill, which would require the governor himself and elected officials to submit to a similar testing policy was rejected.

In 2012, the Office of National Drug Control Policy released the *National Drug Control Strategy* in which crime, health, and lost productivity costs due to illicit drugs were reported to be $193 billion in 2007.

When is Drug Testing Surveillance?

Is there a difference between drug testing and drug surveillance?

One-time drug testing for a job applicant who isn't hired and whose test results and biographical data are immediately deleted would probably not be considered surveillance.

If the information is kept on file, however, if tests are repeated, or if the information is kept and used for statistical analysis along with other test results and any demographic information that might accompany them, it would be considered surveillance, especially if the data makes its way into national databases accessed by other agencies, is sold for extra revenue, or is transferred to another company.

If the samples provided for testing are used for other purposes, such as the detection of pregnancy, diabetes, or genetic anomalies that could show up on a DNA test, it would be considered surveillance, especially if the test results are retained or compared to those of others. It would not be unreasonable for a woman applying for a job to worry that a positive result for pregnancy might result in the job being denied.

It may also be considered surveillance if the results of a drug test influence promotions, bonuses, responsibilities, manager-subordinate relationships, or specific positions sought by an employee.

LAW ENFORCEMENT SURVEILLANCE

Every aspect of law enforcement is related to surveillance. The police monitor compliance and maintain statistics on crimes and apprehended criminals. They carry out the work on foot, from the air, and with a variety of ground and marine vehicles. Law enforcement also makes extensive use of databases with personal information such as photos and fingerprints, as well as databases for checking backgrounds, ballistics, sex crimes, and fraud. Police officers use radar detectors, cameras, audio listening devices, wiretaps, subpoenaed data, tracing beacons, and chemical sensors for compliance monitoring and forensic investigations.

In 2008, the FBI awarded almost $1 billion to Lockheed Martin to help create a database of biometrics information to extend the current fingerprint system to include other identifiers such as iris prints, facial features, palm prints, scars, and tattoos. The goal of the project is to make the database open-ended and extendable so that new biometrics developed in the future could be "plugged in" to the system without significantly changing the underlying structure.

In February 2011, the DHS announced a solicitation for about $1 billion worth of audio and video surveillance devices for investigative surveillance for various federal agencies. Equipment sought included easily concealed wireless systems appropriate for use by law enforcement agencies.

Welfare Fraud Surveillance

The Public Assistance Reporting Information System (PARIS) is an information-sharing database matching system that was initiated in 1993, sponsored on Department of Defense computer systems. It takes personal data and social security numbers and checks whether any duplication occurs in other states to determine whether individuals are receiving taxpayer-funded social assistance, such as Medicaid, food stamps, or other benefits from more than one source.

Some of the results of police surveillance are shared with the public. City crime statistics are publicly posted in newspapers and on the Internet. Some are combining the information with Internet mapping services to provide a visual picture of local crime patterns and locations.

INFORMATION SURVEILLANCE

Intelligence isn't always sought with cameras and microphones. A great deal of surveillance occurs over the Internet. Almost everyone is seeking information on others or gathering content to attract people to their websites.

Predators prowling the Net look for information on potential victims.

The Intelligence Community and law enforcement agencies make regular use of Internet data. Employers regularly look online for the profiles and histories of job seekers. Marketing agencies make extensive use of information found on the Web.

Businesses seeking venture capital or setting up a business plan use market demographics and projections to plan ventures and seek financing.

Searching out data on consumer profiles, consumer spending, census statistics, and other market-related information is a large part of business surveillance, especially when developing a business plan or seeking additional financing.

In 2011, Miranda Mizen projected that market surveillance spending by financial services brokers in the U.S. and Europe would be approximately $268 million in 2013.[55]

Email Surveillance

Email is rarely private. ISPs may search email transmitted through their systems for a number of reasons, including the determination of compliance with terms of service, and to provide assistance to law enforcement.

Since many ISPs explicitly state in their terms of service that there can be "no expectation of privacy" while using their systems, users are agreeing to waive privacy claims when they sign on. Thus, there may be widespread divulgence of email content behind the scenes without user knowledge or authorization, but the prevalence is impossible to determine.

It is also possible for email on local computers or on network services to be hacked. Breaches of email privacy through hacking or malware are not uncommon. Email can also be intercepted en route by packet sniffers or other means.

Email can be intercepted via court-approved wiretaps or subpoenas, provided the subpoena request is of a specific nature.

Comments about the privacy of email, or lack of privacy, apply to many chat services, as well.

Economic Factors in Surveillance Development and Installation

Spending reports can give some idea of the prevalence of surveillance systems, but not all money allocated to surveillance is effectively used. Some systems are shelved, damaged, stolen, or left unattended for lack of operational funds. No one knows exactly how many systems, or how many dollars, have been lost, but taken

---

[55]M. Mizen, "Dynamic surveillance: detection, prevention and deterrence," TABB Group, Feb. 22, 2011.

together with obsolescence and replacement costs, the proportion may be as high as 30%.

As an example, the Space Radar program in development in the mid-2000s was terminated in spring 2008 due to schedule delays, technological problems, and concern about costs. The system was intended for global reconnaissance and surveillance by the military and the Intelligence Community. An estimate of spending up to the time of termination was about $450 million. It was hoped the first satellite would be launched in 2016, but it appeared unlikely that schedule could be met. Alternatives were being sought after the announcement. A program to launch blimps over Afghanistan was experiencing similar technical and cost-related problems.

Most democratic nations have a requirement that contract awards and governmental fiscal year spending be announced, since they are funded with taxpayer dollars. Even with this information, it is difficult to measure surveillance expenditures. Not all purchases are broken into categories; some allocations are classified, and maintenance and continued-deployment costs are more difficult to determine than initial procurement costs. Some guesstimates can be made, however, based on fiscal budgets.

For Fiscal Year 2013, the U.S. government budgeted $851 billion for security spending,[56] a large portion of which is for the development, acquisition, and maintenance of surveillance systems and associated operational personnel.

Given that only about 75 million American taxpayers earn enough that they are required to pay federal income taxes (about 46% of those filing), this is an average of about $11,347 per year per taxpayer devoted just to defense spending—it doesn't include the additional $15,000 or so paid toward Social Security and Medicare. Some of the money comes from corporations, but a surprising number of profitable Fortune 500 companies paid no income tax in recent years.[57] Some comes from high income earners, but more than 1000 households with earnings of more than $1 million per year paid no federal income tax.

Surveillance spending is in the hundreds of billions of dollars, so it is hard to know where all the funding is coming from, unless U.S. defense spending is based partly on money borrowed from other countries like China, Japan,[58] and Brazil.

---

[56]Office of Management and Budget, *Fiscal Year 2013 Budget of the U.S. Government*, U.S. Government Printing Office, February 2012.

[57]Study on federal income taxes paid or not paid by 280 profitable Fortune 500 corporations in 2010, published by the Citizens for Tax Justice and the Institute of Taxation and Economic Policy, November 2011.

[58]Current debt to China in Treasury bills reported by the U.S. Treasury as of April 16, 2012 is $1,178,900,000,000 ($1.2 trillion) and to Japan is $1,095,000,000,000 ($1.1 trillion).

EMERGING TECHNOLOGIES

Terahertz Vision

One area of surveillance is so new, it doesn't have a chapter in this book. It has been mentioned since the 1990s in scientific research papers, but wasn't yet practical and convenient for small-scale applications.

In April 2012, the University of Texas, Dallas, announced that researchers under the direction of Dr. O had designed imaging technology that can see through things in much the same way that X-radiation enables us to image inside certain objects or tissues in the body.

The new technology uses electromagnetic radiation that falls between microwaves and infrared, in the terahertz frequency. While it was possible to access these wavelengths in the past, there were no readily available micro-components for the consumer market. Now it is possible to fabricate terahertz-capable components so small they could be added to cellphones.

Terahertz technology has great potential for medical diagnostics and treatment and the researchers explicitly mentioned they will be concentrating their efforts on proximity sensing (within about 4 inches) rather than long-range sensing characteristic of surveillance applications.

The New York Police Department, in cooperation with the DoD, are testing a terahertz scanner that could possibly be mounted on police vehicles to detect objects like guns, knives, or anything that might block the heat that radiates from human bodies and show up on a scan.

The Terahertz Imaging Detection (TID) system is receiving criticism from civil liberty groups who liken it to everyone walking down the street being given a pat-down search without a search warrant. Others say it makes people safer. A third group asks whether cutting back youth group programs and increasing surveillance funding decreases or increases crime and suggests there are better ways to spend the money so that society is enriched rather than enslaved.

The technology is not yet prevalent, but will no doubt find its way into other surveillance applications and represent a growing market.

Shift From Technology to Data Management and Analysis

Tech devices aren't even the most significant aspect of surveillance any more. The technology has matured to the point where devices are so powerful, they not only do the job, but we haven't discovered all of their uses yet. Data accumulation and analysis is becoming the greater concern.

Current technology enables far more information to be gathered than can be reasonably filtered and analyzed. In the past, the focus was on improving surveillance technologies, and making devices smaller and easier to use (and this is still true), but the emphasis is shifting to improving and optimizing information management so that the burgeoning volume of surveillance data can be effectively utilized.

EFFECTIVENESS OF SURVEILLANCE

KEY CONCEPTS

- *cost/benefit considerations*
- *inconsistent results from empirical studies*
- *controversy about the effectiveness of surveillance*

Introduction

The effectiveness of surveillance depends partly on technical capability and appropriate installation, and partly on the administration and operation of a system. Sometimes public acceptance and cooperation is required, as well. When governments use taxpayer money to surveil taxpayers, there may be discontent or significant opposition. When corporate surveillance moves toward intrusive monitoring, such as implanting chips on employees to track them 24/7, most citizens complain to lawmakers.

Surveillance devices are often installed based on marketing promises or subjective hopes. Rarely are they installed after exhaustive studies on whether they will be effective. Assessment of the cost of monitoring and maintaining surveillance systems is often underestimated. Once they are in place, unless good data has been gathered prior to installation, it is difficult to compare effectiveness before and after, especially when other factors change. In many areas, cameras that have been used to monitor traffic for several years have been turned off following public controversy or studies questioning the effectiveness of the systems.

Law enforcement organizations make widespread use of surveillance—but does it reduce crime? Schools and individuals have ready access to surveillance devices but does it enforce, encourage, or discourage legal or moral behavior? Corporations make widespread use of surveillance to protect property, monitor employees, and target potential consumers—does this have a positive effect on society in general?

It will take time to answer these questions. Consumer-level technological surveillance is recent—very recent if you focus on micro-electronics and "smart" devices that go beyond recording events by adding image processing or expert-system analysis to recorded events. Studies are needed because intuitive off-the-hip answers often contradict research. Most people assume, intuitively, that cameras will stop crime. Often, they don't.

There are many short-term studies, but the results of long-term studies aren't in because many surveillance technologies are less than a decade old and the technology changes so fast, by the time a study is underway, the devices have been upgraded or replaced.

This chapter gives a few examples that illustrate whether surveillance is effective for specific tasks and looks at the trends that might illuminate the bigger picture, one that may change as research continues.

ANECDOTAL COMMUNITY EXPERIENCE AND RESEARCH

Sometimes requests for increased surveillance come from citizens concerned about crime in their neighborhoods. Other times, it is promoted by local politicians or law enforcement agencies.

Many of the local surveillance systems installed in the last few years were financed largely, or in part, by federal grants. Local politicians often reassure their constituents that "no state money was used for this," but taxpayers pay for it nonetheless. Money paid in federal income taxes by people in all states are funding local surveillance technologies via DHS grants.

Many communities have reported discontent with surveillance systems because installation was never completed, the systems were inadequately monitored, deterrent effect was negligible, or the position of the surveillance devices had little relationship to areas of high crime or vandalism.

One example is cameras installed in patrol cars in Detroit which, it turned out, were not programmed to record continuously, raising questions about whether officers would selectively turn them off or on. Surveillance isn't always about law enforcement monitoring the public; sometimes law enforcement itself is scrutinized by administrative officials or the public.

Privacy is also cited as a reason to reject or abandon local projects, and citizens are concerned that some neighborhoods might be unfairly targeted. Furthermore, there have been complaints about incremental extension of law enforcement and targeting of citizens who have no criminal record.

In 2001, a proposal to equip public streets of Ybor City, Florida, with face-recognition systems met with opposition in city council meetings. The main proponents of the system were the mayor and people in the commercial security products field. The main opponents were citizens concerned about extension of police powers. Despite public resistance, several dozen cameras interfaced with face-recognition software were installed on a main avenue and entertainment complex. The software enables camera images to be compared to a database of wanted felons and sex offenders so that law enforcement may follow up with appropriate action. There can be mistakes, however. Sometimes people's faces are released to the news before the person has been notified and the case investigated, as happened with one case of mistaken identity.

A decade later, a similar controversy over surveillance cameras occurred in nearby St. Petersburg. It illustrates several common concerns with city-based surveillance.

A camera surveillance pilot project was instituted in 2009 and 2010. Given limited funds, 10 cameras with zoom and scan capabilities were installed, mostly in city parks. As operational funds were not yet available and no one had been officially assigned responsibility for monitoring, the active video feeds were initially not watched by law enforcement officers (IT staff sometimes watched them) and there were liability concerns over what would happen if incidents were recorded on video but missed by the staff. Whether the cameras had any deterrent effect after more than a year of installation was unknown—no funds were allocated for studies. Community members also questioned why the cameras weren't monitoring higher crime areas.

This kind of situation is not limited to Ybor or St. Petersburg. The idea of using cameras, and the reality of putting together an effective program of administration and operations, are two different things. It takes funding, planning, studies, evaluation, and decisions about responsibility and liability. Even if policy and deployment steps are documented, they aren't necessarily implemented, as in this case.

Research is important because public acceptance or controversy over surveillance is often based on personal experience, anecdotal "evidence," or observations in a limited geographical area or time-frame. Some is based on political leanings.

While long-term studies on the effectiveness of surveillance systems are still needed, there have been many short-term studies. One area of surveillance that is frequently debated is whether traffic surveillance reduces accidents or is merely intended to line city coffers with more revenue from fines. Another is whether surveillance deters crime or merely displaces it.

TRANSPORTATION SURVEILLANCE

Transportation surveillance isn't restricted to any one particular technology—many devices are used—cameras, radar, X-rays—but visual surveillance of traffic is common, whether it is officers in patrol cars watching a highway from an overpass, or cameras in tunnels and subways.

Visual recording of license plates is now commonplace at many border crossings, but the technology is not limited to border protection—license-plate cameras can catch those who break traffic laws in intersections or, as is the case in London, U.K., hundreds of cameras intended to reduce congestion take photos of cars entering the city so that fines may be levied against those who don't pay an access fee by the end of the day. Once the cameras were in place, law enforcement agencies saw opportunities to appropriate them for other purposes, such as adding facial-recognition software to locate criminals.[59]

Even toll booths can be a source of surveillance information, if they accept electronic payments or access cards that are logged or traced back to their owners. A murder story that made the news in New York City in 1997 was solved partly with information from an electronic tag that enables quick passage through tolls.

Traffic Surveillance

Technology-based traffic surveillance began as a way to manage large loads that were transported along public highways, but it gradually spread to law enforcement tasks such as monitoring border crossings, and the apprehension of speeding drivers and those who ignored red lights or made illegal right turns on red without first stopping and checking for pedestrians or cyclists. The National Highway Traffic Safety Administration has reported that nearly a third of all traffic fatalities are speed-related, hence the use of radar detectors and other speed traps.

Researchers in Australia have reported that cameras to detect speeding have reduced traffic-related injuries and fatalities, and a study of speed cameras in urban areas reported in *Injury Prevention* (Summer 2011) shows that property damage and medical costs were significantly reduced, as well.

But research on events at intersections is harder to interpret than speed violations. Traffic intersection cameras (TIC)s illustrate how surveillance-as-deterrent can have unanticipated consequences.

A number of studies from the late 1990s and early 2000s suggest red-light cameras may increase crashes at intersections. The methodology of some of these studies was questioned however, so a national study was conducted by the Federal Highway Administration (FHA). In 2005, the FHA published results of a safety evaluation of red-light cameras (RLCs)[60]. The intent of the project was to determine the economic effect of RLCs and their effectiveness in reducing crashes. The study found a statistically significant decrease in right-angle collisions but also a significant increase in rear-end collisions.

In an Insurance Institute study, released in June 2011, it was reported that fatalities involving red light crashes in selected cities were 24% lower in those without cameras and that people may react to the presence of cameras by causing more rear-end collisions.

Was there an overall benefit or did the negative outcome override the deterrent effect? Apparently there was a modest benefit overall, in economic terms, of having the cameras. The FHA evaluation also suggested that cameras may be more effective if installed at intersections where right-angle crashes were common and rear-end crashes infrequent.

---

[59] Asher Price, "Technology: surveillance and the city," *The Next American City,* June 2003.

[60] Federal Highway Administration, "Safety evaluation of red-light cameras," FHWA-HRT-05-048, April 2005.

There are still many who cite studies that suggest the economic cost of RLCs might override the deterrent effect. Thus, there is a need for more research, not only because of contradictory results from different studies, but also because it appears that some kinds of surveillance may be more effective than others in reducing injuries and fatalities.

Transportation surveillance is not always for traffic infractions, sometimes it is to prevent theft or vandalism of property or unauthorized entry into restricted areas. It may also be installed to prevent theft of rail services (access without a ticket). Access surveillance is very location-dependent, however. Many parts of the world operate transit ticketing on the honors system, with no surveillance at all or with occasional spot checks.

In the U.S., surveillance of trains and subways is prevalent, particularly since the WTC attack, as there is concern about bombs or other terrorist threats against transportation systems.

Many bus and train systems are equipped with CCTV cameras that have been there for some time, but the trend is to gradually replace them with IP-based analog or digital cameras, cameras that can be interfaced with computer networks so they can be monitored from remote locations. Cameras in transportation corridors are usually mounted high enough to deter tampering and may have rugged casings for weather-resistance.

Smaller cameras in moving vehicles can't always be interfaced with wireless services, and may use compact tapes, compact drives, or memory cards to store information. These can be swapped out if longer recording times are desired.

While traffic surveillance has the potential to reduce death, injuries, property damage, and medical costs, it also has a downside in that law-abiding citizens are being monitored and recorded on an almost continuous basis as they travel to work, shopping malls, or social activities. Surveillance cameras are indiscriminate—they don't blink or look the other way when people are minding their own business.

A large portion of ground-based traffic surveillance is visual in nature but X-rays may be used to look inside container trucks, and radar may be used to detect speeding.

Traffic surveillance isn't always by commercial or government transportation providers or law enforcement agents. Some people have installed cameras in their cars to record what is happening around them in case of accidents or if they are stopped by a police officer.

Intelligent Transportation Systems

Increased population and industrialization are resulting in severe traffic congestion in urban centers worldwide. Highways are expensive to build and maintain and almost always result in destruction of valuable agricultural land and sensitive ecosystems. Instead of adding more highways, planners are increasingly seeking other solutions to improve traffic flow and efficiency. As a side effect of these efforts, vehicle traffic surveillance has increased substantially.

The Evolution of ITS and Traffic Surveillance

It is a well-documented fact that highway traffic congestion in the metropolitan areas of the world is becoming a major concern.... However, due to the high financial, social, and environmental costs of such projects, that is no longer seen as a viable option. Many innovations have come to use in a more efficient way the existing infrastructure, such as: improved traffic-signal controllers, changeable highway signs, rerouting rush-hour traffic, creation of traffic-control centers that monitor and display gross traffic conditions, use of preplanned alternative traffic solutions based on repeated daily traffic patterns, etc.

Martin, Alberto, et al., "Intelligent vehicle/highway system: a survey," *Florida International University, Dept. of Mechanical Engineering,* Miami, Florida.

A 2010 study by the Department of Transportation (DoT) reports that the collection of realtime data from cameras and sensors has changed the focus of traffic surveillance from planning support to operations support, and that communication between various agencies has significantly improved to the point where operational coordination is possible.

The DoT also reports that technical advancements have changed the character of traffic surveillance from passive monitoring to monitoring that can send alerts and commands to various other devices. Automation of toll gates with electronic collection systems is now widespread—a situation that law enforcement agents have leveraged to solve crimes.

Sensors and cameras are an intrinsic aspect of traffic congestion monitoring and mitigation. Cameras are also used to deter and detect traffic violations. In prototype systems, sensors are further used as control components for autonomous vehicles—which may lead to fully automated transportation systems in the future.

Thus, traffic surveillance has evolved far beyond monitoring traffic congestion, speeders, or those who run red lights. It is becoming a comprehensive and sophisticated way to oversee an entire transportation ecosystem as though it were a symbiotic organism. Intelligent Transportation Systems (ITSs) use a broad range of technologies, including pavement sensors, traffic light sensors, ramp meters, traffic counters (and other statistics-gathering devices), together with autonomous and remotely controlled cameras.

Control centers reminiscent of a command console on a starship now monitor thousands of cameras interfaced with banks of video displays that are jointly operated by software and human operators.

As a spin-off of technology developed through the NASA space program, this traffic control center in San Antonio, Texas was developed as an alternative to adding more lanes. Studies suggested better management of the existing system, through technology, could be a better solution than constantly paving over more land. The first phase became operational in 1995 and the system now oversees nearly 200 miles of freeway networks. [Photos courtesy of NASA Spinoff, released.]

San Antonio has been developing an intelligent traffic surveillance system since the early 1990s. Some of the benefits include better traffic flow, reduced fuel consumption and emissions, and faster incidence response. Each operator has four video monitors at a desk, including a mapping system for displaying the location of incident alerts, plus a view of a 60-foot video wall. The system includes both incoming and outgoing communications—vehicle detectors and cameras are installed at half- and one-mile intervals—and operators can remotely aim cameras and post messages on fiber-optic electronic signs.

A literature review of ITSs published in 2005 concludes that ITS can be effective in reducing incident delays, travel times, accidents, and transportation costs of freight vehicles.[61] ITS communications can also improve emergency response times.

Most ITS alerts to drivers are conveyed by roadside message boards, but ITS concepts can be taken a step further by equipping cars with networked access to information from the ITS software or command console operators. Imagine images of a traffic accident being broadcast to nearby cars, along with GPS-based location information and a map to show the best detour to take to avoid traffic congestion (and the long wait that might occur if drivers enter the crash site).

Integrating Intelligent Vehicles with Intelligent Transportation Systems

Intelligent Vehicle Systems (IVSs) are a growing aspect of ITS. Cars that can check their own electronic systems, sense nearby obstacles and signal crash alerts

---

[61] Robert L. Bertini. Christopher M. Monsere, Thareth Yin, "Benefits of intelligent transportation systems technologies in urban areas: a literature review," *Portland State University Department of Civil & Environmental Engineering,* April 2005.

call up GPS-based mapping systems, or video-monitor vehicle blind spots, already exist. Japan has been working on dynamic routing systems since the 1970s.

The next step is to integrate intelligent vehicles into intelligent transportation systems. Software programs could help drivers plan the best routes based not just on raw distance calculations, but also on time of day, traffic flow, temporary obstacles such as construction or accidents, and perhaps even on driver preferences in terms of scenery. If an architect is interested in viewing buildings along the way, or an artist prefers to drive past parks and marine vistas, the system can customize those preferences into the route recommendations. It would be possible to maintain lists of favorite restaurants and rest stops in a vehicle database, as well.

When vehicles are designed to intercommunicate within an ITS, they become like computer terminals within a larger computer network, capable of two-way communication and remote monitoring or control. Data can be uploaded, downloaded, or analyzed locally or off-site, and computerized components, such as the engine, air conditioning, lights, or even an entertainment system could be checked, tweaked, or tuned remotely by expert mechanics or entertainment content providers.

It is only a small step before biometric monitors, such as cameras or body sensors, could be installed, as the EU is planning to do on airplanes in Europe. Biometric safety systems could analyze faces to see if drivers are exhibiting signs of *road rage* or of *highway hypnosis* that could presage falling asleep at the wheel.

Biometric monitoring may prevent accidents, but also raises significant privacy issues. If cars are equipped with driver preferences, location information, and two-way communication between the vehicle and the ITS, then logs of a person's whereabouts and driving patterns are easily stored and analyzed. If vehicles are further equipped with biometric monitoring, then moods, conversations, outward signs of thoughts, and logs of bodily functions could theoretically be maintained.

If you don't think this will happen, consider that private companies managing fleets of freight trucks have greater leeway than governments in installing surveillance systems to safeguard their assets. Many companies already mandate RFID cards to monitor employee movements throughout the workplace and trucking services already use tracking systems on fleet vehicles, so it is not a stretch to extend the concept to monitoring drivers for signs of road fatigue. Once the technology exists, people find ways to use it.

Eventually, automation may replace drivers altogether. With roadside and vehicle sensors, rails, and other monitor-and-control devices (MCDs), cars may drive themselves.

Integrating Vehicles with the World Wide Web

Even in places where ITSs are not yet in place, technology is changing the way we use and view cars. Driver services such as GM's OnStar™ and Ford's

SYNC®, designed for theft protection and accident/emergency service communi-cations, are moving into wider integration of vehicles with computer networks.

In 2010, OnStar announced wireless access as an added feature of their services so users could tune into Internet radio, podcasts, and possibly also control Web-related resources from their vehicles with voice commands. Taking this one step further, drivers could potentially communicate with other drivers through the service, by voice or text messages.

In 2012, a new RelayRides™ partnership was announced, which would enable car owners to rent out their vehicles when not in use. Services to bring video chat into the rear seats of cars were announced as well.

The Ford SYNC services include hands-free phone, GPS navigation maps with turn-by-turn directions, and voice-controlled phone apps.

Cars are rapidly turning into network service and media entertainment "peripheral devices."

Privacy, again, becomes an issue. If a driver requests Web searches, video chat, and other communications that may be logged by service providers, what will the companies do with the information? And what if those companies are subpoenaed to give up the information to law enforcement upon request if, for example, a driver is in the vicinity of a robbery or car accident and the police want to see what they were doing, who they were talking to, and what they were listening to at the time of the incident, even if the driver wasn't directly involved?

Also, what happens if service providers want to sell information gathered from vehicle-network subscribers to other companies?

These questions have already prompted criticism of privacy policies by lawmakers and consumers, leading to a change by OnStar, in September 2011. The company had been planning to collect and share various information, such as GPS-tracking data, diagnostic codes, odometer readings, and crash information, even if a subscriber had stopped using OnStar services. The proposed change was reversed after complaints about privacy invasion. Consumers also made it clear they wanted potentially intrusive services to be "opt in" rather than "opt out" so that users aren't unwittingly subscribed to something they may not understand or want.

The company justified the information-gathering by saying it would provide opportunities for the company to gather statistics for the development of future services, and to provide former subscribers with urgent alerts, but many felt the disadvantages for consumers might override the advantages.

Global Positioning System (GPS)

GPS-based services are very effective in providing maps, tracking, and loca-tions for navigation and search-and-rescue operations. The technology does a good job of providing increasingly accurate latitude, longitude, and altitude information. GPS-equipped devices are becoming common, both as handheld geolocators and feature-rich devices such as multifunction phones.

GPS information is also incorporated into many Web applications such as street-viewing, mapping, aerial imaging, and directory services. Some devices even learn a person's favorite routes and haunts and incorporate the information, along with any scheduled alerts (like a visit to the eye doctor or dentist) into suggested routes and information about local transportation options.

Some of these new applications are helpful and some are intrusive. Mapping programs greatly benefit from GPS information that users can coordinate with handheld or vehicle-mounted GPS locators but there is also controversy over programs that tell people to avoid certain areas. Those who live and work in those areas could be unfairly targeted and marginalized and a technological boycott may only worsen problems for poverty-stricken neighborhoods.

Another aspect of GPS-service multifunction calendars that skirts the line between convenience and privacy-invasion are services that want access to an individual's preferences and personal calendar so they can broadcast targeted ads to the consumer. Imagine a person walking to a coffee shop the day before a relative's birthday. The software checks the calendar, sees the birthday, and chooses a sponsoring store in the neighborhood, alerting the user to shop there to find an appropriate gift.

For the most part, despite potential for intrusive tracking and targeting, GPS is a highly effective technology that helps navigation and saves lives in search-and-rescue operations.

Border and Seacoast Surveillance

Border security is one of the most prevalent forms of surveillance. Traffic across international borders takes many forms, including

- tourists taking vacations and students studying abroad,

- container trucks transporting food and goods,

- import/export of items through federal postal services,

- smuggling,

- protestors attempting to prevent illegal or excessive harvesting of marine life,

- legal immigrants and refugees, and

- undocumented migrants crossing borders to flee oppression or legal issues, or to locate relatives or jobs.

Border surveillance is an extensive topic, worthy of several books—there isn't space in this text to discuss it at length—but a few examples of new technologies and border surveillance issues will be given.

The Wescam MX-20 (left) is a long-range high-resolution, rotating, digital surveillance camera that can be joystick operated. It is gyro-stabilized for use on aircraft and the turret can accommodate up to seven sensor systems. The system illustrates how contemporary surveillance cameras go far beyond taking a picture. Multiple sensors enable capabilities such as high resolution, thermal imaging, night vision, and haze penetration, while "smart camera" features include realtime image enhancement, motion scanning, and target tracking. One of its applications is in U.S. naval research onboard a modified Orion aircraft, as illustrated by Rear Admiral M. Klunder (right), with Dr. M. Pollack, surface and aerospace surveillance program manager, observing. [News photo by Wescam, released. March 2012 U.S. Navy news photo by John F. Williams, released.]

SeeCoast[62] is being developed as part of the Coast Guard Port Security and Monitoring system that uses technology to detect, track, and classify vessel traffic, using electro-optical and infrared cameras, as well as analytical systems that can create alerts if any unusual vessel transportation patterns are detected.

The U.S. Coast Guard uses the MX-20 in a Hercules long-range aircraft and a smaller version of the camera is installed in a U.S. Air Force fixed-wing aircraft. The MX-20 has also been evaluated as part of the *Advanced Integrated Multi-sensor Surveillance* study by the Defence Research Agency of Canada, where it was used for ergonomic observations within a CP-140 trainer vehicle.

Just as a person can be observed for signs of nervousness or abnormal behavior, so can marine traffic. This process is known as *automated scene understanding* (ASU) and is a combination of stored information on familiar patterns combined with analysis of sensed scenarios. The goal of most ASU systems is to free staff from sitting all day watching camera feeds. If the computer can stand watch and send alerts when something happens, the staff can be engaged in more productive work.

Marine, aerial, and terrestrial war-zone surveillance have special challenges not always found in ground surveillance. In urban environments, video analytics generally involves separating a target of interest from surrounding trees, bushes, traffic, and crowds—the camera may rotate, but the base generally is stable. In contrast, on water or in the air, the camera is often jiggling and changing altitude,

---

[62]Michael Seibert, Bradley J. Rhodes, et al., "SeeCoast port surveillance," *Proceedings of SPIE Vol. 6204: Photonics for Port and Harbor Security II*, Orlando, April 18–19 2006.

and may be mounted on a very fast-moving vessel. Separating the background from the target is a special challenge at high speeds where the visible field changes rapidly.

Vessel identification also involves estimating how large it might be and if it is friend or foe. Camera zoom features can, in part, help the software estimate size, which is particularly helpful in marine applications, where there may not be physical references in the background like buildings or cars.

The Internet, more than any other historic milestone, is serving to globalize the planet and some have suggested getting rid of international borders entirely. Despite this acceptance of more open communication and melding of diverse cultures, most countries still have a desire to protect local resources (such as gas, oil, mineral deposits, and jobs) and to preserve their cultural heritages. Thus, border surveillance not only is ongoing, but has increased.

Forty years ago, the Canada/U.S. border was minimally monitored, other than a few border checkpoints and occasional patrols, the border "fence," in many areas, consisted of scattered trees and shallow ditches—people on either side could walk up and shake hands, and could pass through checkpoints with minimal identification without their ID being recorded. Now it is increasingly monitored with sensors, vehicle license plate scanners, computer databases to log individual vehicles, fingerprint scanners, and swivel cameras on tall poles scanning the regions between checkpoints.

Close to home, most marine surveillance is handled by Customs and Border Protection, but in international waters, national security surveillance objectives are carried out by submarines, patrol boats, aircraft, and aircraft carriers. Here, Airman Erin Johnson is stationed in the launch and recovery television surveillance room aboard an aircraft carrier responsible for maritime security. [U.S. Navy Dec. 2011 news photo by Kenneth Abbate, released.]

The Canadian and U.S. economies are very similar in terms of quality of life and access to social services, so movement back and forth across the border tends to be orderly and uncontentious. The Mexican/U.S. border is a different situation, however.

Mexico's economy has had many upheavals in the last century, from one in which most land was owned by a minority, to a post-WWII boom in which resources were more generally distributed and the country experienced various forms of prosperity. Since then, the economy has had problems, including lower revenues for oil resources, slower growth, devaluation of the currency, political unrest, and rapid inflation.

All these events, plus a greater discrepancy between U.S. and Mexican economies than is found between Canada and the U.S., have provoked a flow of illegal migrants across U.S. borders. Since Americans are concerned about losing jobs to undocumented aliens (even though many of the jobs have substandard wages and working conditions), there has been considerable public and governmental attention to securing the southern border.

Surveillance of Marine and Airspace Territories

Protection of oceans and airspace is more contentious than terrestrial protection (they're harder to monitor), so they have been heavily surveilled for at least the last century. A large proportion of marine and aircraft surveillance is international in character since boats and planes can easily move (or stray) into foreign territory. This aspect of border protection has been a problem for some time. Fishing boats are often caught in foreign waters as they chase after diminishing resources, illegal prey (such as whales and dolphins), and fleeing schools of fish.

Spy planes are acknowledged as being on regular patrol in national and shared airspace, but they are also caught, from time-to-time, in foreign airspace. No one admits they do this, but incidents occasionally make the news to reveal that it happens.

U.S. Customs and Border Protection (CBP) is heavily involved in the surveillance of terrestrial, marine, and airborne craft (and their passengers) that cross national borders, and uses a wide variety of surveillance technologies to carry out the work.

The technologies used include, but are not limited to, sensors, light towers, mobile night vision scopes, remote video surveillance systems, directional listening devices, various database systems, and unmanned aerial vehicles (UAVs).[63]

---

[63]Chad C. Haddal and Jeremiah Gertler, "Homeland security: unmanned aerial vehicles and border surveillance," *Congressional Research Service,* July 2010.

To this list, we can add sniffing dogs that search for illegal drugs or explosives, and forensic departments that test goods to see if they meet import/export standards or are evidence of poaching.

Results of CBP Surveillance

In 2011, CBP surveillance apprehended 340,252 illegal aliens within the country and blocked admission at ports of entry to almost a quarter million more. More than 1 million prohibited vegetable and animal materials were seized at border crossings, along with almost 200,000 agricultural plants and pests that might pose a danger to domestic crops. Almost 5 million pounds of drugs were seized, including heroin, methamphetamine, cocaine, and marijuana.

Marine Surveillance

Marine surveillance involves a great variety of technologies including telescopes, periscopes, sonar, and interception of wireless communications. Often, those that are not inherently visual in nature are converted to some visual form that humans can recognize. Sonar pings used to be interpreted aurally, but now that most sonar is outside of human hearing ranges, sonar signals are tracked with visual representations of the sonar echoes.

Marine surveillance used to be limited to what one could see with a spyglass, but new monitoring devices and stealth technologies create a Catch-22 cycle—better surveillance devices provoke the development of better stealth technologies.

Sonar and radar were originally designed to prevent collisions at sea and to aid rescue workers in finding distressed vessels. Over time, they were developed to detect hostile ships and submarines. More recently, thermal and magnetic detectors were added to the surveillance arsenal. Detection and homing capabilities were also added to torpedoes and missiles to help them find a target.

In turn, ships and submarines were redesigned to make them harder to see and find. Ships have been equipped with lights to make them appear as thermal patterns or "mirages" on open water. When radar became prevalent as a detection technology, they were outfitted with chaff—reflective streamers that confuse radar signals by scattering them in all directions. Both ships and submarines were given new shapes and materials to make them difficult to recognize. Special foam surfaces or coatings could absorb or convert signals so they didn't return to the probing party, or returned in a way that wasn't recognizable as a marine vessel.

Aircraft Surveillance

Air traffic controllers are one of the most common and most welcome professionals monitoring airborne traffic. They track incoming and outgoing planes and prevent them from flying into each other or colliding on the runway, but there are also many kinds of aircraft surveillance that are not so apparent to the public, including tracking of foreign aircraft through the Northern Shield and marine vessels, stealth planes, and a fleet of UAVs that patrol the borders.

UAV programs are expanding at the border and at local police precincts. UAVs are perceived as a lower-cost solution to inhabited aircraft and the smaller ones can fly nearer to the ground than helicopters and planes.

There is controversy, however. In addition to noise, local residents are concerned that UAVs could crash into their homes and cause damage to people or property, or fly over their yards when they are sunbathing or hot tubbing. Fences do nothing to protect home-owners from eyes in the sky. They may also provoke increased barking of dogs if they see the UAVs as intruders.

Cost/benefits are not always clear-cut either. UAVs tend to crash more often than piloted aircraft and, depending on size and configuration, sometimes cost more to operate.

Despite public resistance and questions about cost benefits, UAV programs are increasing. The U.S. Congress has expressed interest in UAVs for border security and for expanding CBP fleets.

Transportation Security Checkpoints

By spring 2012, about 1000 backscatter radiation scanners (which measure ionizing radiation from low-energy X-rays) had been installed in U.S. airports to scan passengers (at considerable cost to taxpayers).

Millimeter wave scanners, which come in both active and passive varieties, are also used, as an alternative or adjunct to backscatter systems, and may have advantages in terms of safety. Passive millimeter wave scanners do not project radiation at the target—the image is created by interpreting naturally emitted radiation in wavelengths that enable the technology to "see through" clothing.

Unfortunately, airport security inconveniences travelers in more ways than just asking them to pass through a radiant imaging device. Coats and jackets have to be stowed in cargo baggage or placed in a bin that is sent through the X-ray machine along with any metal items such as mobile phones or PDAs. With the exception of young children and certain frequent flyers, all passengers must remove their shoes.

The Transportation Security Administration (TSA) is also telling travelers how to dress. Some examples of clothing recommendations include instructions to avoid wearing

- underwire bras,
- clothing with metal buttons, snaps, or studs,
- belts with buckles,
- hair accessories with metal attachments, and
- heavy jewelry.

In addition to these recommendations, the TSA has announced that body piercings may lead to a second screening or pat-down search (with the option to remove the piercings in private). It is a sad comment on our times when a 1/4-inch navel ring can be perceived as a potential terrorist threat.

On a lighter note, if the terrorist "shoe bomber" incident of 2001 can provoke a ruling to remove shoes at transportation checkpoints, should there be a similar ruling as a result of the "underwear bomber" trying to detonate plastic explosives in his underwear on a flight from Amsterdam to Detroit in 2009?

Where is the line between consistency and common sense? Perhaps this illustrates that surveillance doesn't necessarily stop crime, but may inspire criminals to greater ingenuity. There is evidence in Iraq that efforts to thwart car bombings have resulted in the rapid development of more sophisticated bombs.

Military Surveillance

As mentioned previously, not all military spending results in fruitful surveillance data. It is not uncommon for projects in the millions of dollars to be shelved before vehicles or cameras are deployed and many aerial vessels have crashed and need to be replaced. Some of the recent blimp programs have been plagued with problems and may not go forward. Sometimes technical problems cause active surveillance to become dysfunctional or nonfunctional. Satellites sometimes are damaged while being launched or don't work once they are in orbit. The U.S. GPS system has spares in orbit, in case some of the satellites fail.

Another form of "loss" is in the form of unusable data. Even if every aspect of a surveillance system works perfectly, there's no guarantee a camera or microphone is aimed in the right direction or even whether there's anything worth observing. The operational costs of launching drones, blimps, and installing every vehicle with cameras and chemical sensors is considerable. If $5 billion is spent and there's one incident of chemical detection, the cost/benefit ratio leads to questions about whether there's a better way to spend the money.

CRIME PREVENTION

Cameras don't always deter crime. Sometimes crime is displaced a few feet (or a few blocks) from the cameras, and sometimes crime continues, despite the cameras. Even if assault, theft, or vandalism remain undiminished, however, it may be easier to prosecute cases if video evidence is available. Then the question becomes whether the cost of installing, maintaining, and monitoring the cameras is worth the occasional lost case. Cost/benefit studies are still needed.

Does surveillance stop crime?

In the broadest sense, surveillance encompasses most of the activities of law enforcement agencies, from a police officer walking a beat, to DNA swabs of criminal suspects. It also includes corporate and residential surveillance, as well as ecological surveillance to prevent unlawful habitat destruction, off-season hunting, and poaching. Surveillance is so broad and varied that most research aiming to determine its effectiveness is targeted at specific aspects and sometimes specific technologies such as certain devices or methods, rather than overall trends.

Crime prevention often involves the use of video cameras—some clearly displayed, some hidden, some "dummy" devices that look like cameras but have no imaging sensors. Unlike studies where test and control subjects can be paired for similar traits, surveillance sites and systems are difficult to match. One bank or retail store may border a dark alley, another is near a busy street or police station. Cameras or sensors may be different makes and models and installed in different ways. Monitoring may be real-time streaming video, recorded video, or still shots taken every few seconds or minutes. With so many variables, it is difficult to set up a valid experiment and draw statistical comparisons.

It is important to look at what is happening beyond the boundaries of a regional study. It may appear that installing surveillance cameras near a brothel reduces crime, until another study determines that crime didn't stop, but was deflected a few blocks down.

Most people intuitively accept that a video camera is a crime deterrent but statistics and anecdotal observations don't always support that assumption. A group of nannies suspected of abusing young children were told they would be surveilled by cameras, yet they abused the children anyway. Bank robbers know every bank and convenience store is equipped with surveillance systems and yet they continue to rob them. Motorists are aware that there are red light cameras at many urban intersections and yet they still run red lights. Shoplifters know department stores and supermarkets are installed with dozens, sometimes hundreds of cameras, and yet they steal. Statistics for violent crimes such as murder and rape showed very little decline between 2000 and 2009 despite the increase in surveillance cameras.

Banks are more consistently equipped with sensors than most institutions. Cameras are aimed at doorways, tellers, vaults, and alleyways adjoining the buildings. They also aim at ATMs (and often are installed inside the ATM as well). An FBI report of bank crime statistics for the third quarter 2010 showed a total of 1323 robberies and burglaries. Victim institutions reported more than 1289 alarm systems and 1306 surveillance cameras, almost all of which were activated during the crimes. There was a significant reduction in bank crimes since 2003, but how would it compare to the 1970s or 1960s when cameras were less common? Are cameras responsible for recent declines?

The Bureau of Justice Statistics has reported that homicide rates declined in the late 1990s, so it is tempting to assume the decline is related to surveillance, except that homicide rates are the same, per 100,000 population, as they were in the early 1960s, when surveillance cameras were rare. Robberies showed a similar decline in the late 1990s and 2000s such that 2009 statistics match robbery rates from 1969. Thus, 10- or 20-year studies may not adequately characterize statistical correlations.

If current crime rates match those from 40 years ago, based on equivalent population samples, can we say crime has declined? Or was there an upswing in the 1980s that has now settled back to previous levels regardless of the presence of cameras or motion detectors?

Effectiveness of Operator-Controlled Cameras

When cameras are selectively aimed or turned on only for suspicious events, operator bias may skew the objectivity of the recorded information. Security personnel may also make biased decisions when responding to live streaming video. As an example, if a camera operator assumes one gender, ethnic group, or age group steals more than another, a camera might be preferentially trained at certain shoplifters, thus providing faulty evidence to support the bias and unfairly targeting one group over another. There are statistical studies that indicate adults shoplift more than teens, but that doesn't mean an adult is more likely to be stealing in any given situation. Profiling is a means to try to guess who is more likely to commit a crime to effectively use limited resources, but generally tends to introduce operator bias.

How can these problems be overcome to increase the effectiveness of security surveillance? One way is to train operators to be aware of biases and ways to avoid them. Another is to have objective oversight and quality assurance measures, such as multiple operators and multiple cameras, but this approach is costly.

In the end, the solution may come from even more pervasive (and potentially invasive) technology. Three-hundred-sixty-degree cameras, like those used for street mapping, may provide enough context and a broad enough view so operators are not selectively aiming the camera or even involved. But even this may be subject to bias when evaluated by law enforcement personnel or a jury.[64]

Another problem with operator-controlled surveillance cameras is inappropriate use. There are frequent reports of operators training cameras on attractive pedestrians or shoppers rather than potential shoplifters or crimes in progress.

Implied-surveillance systems are sometimes installed as a cost-reduced means to prevent crime. There is brisk trade in dummy security cameras and signs that indicate the existence of a security system when there may not be one. The effectiveness of implied surveillance is difficult to assess without long-term controlled experiments. The prior discussion on property damage pointed out that property crimes are down, compared to the 1980s, but about the same as the mid-1960s. So, have surveillance cameras had any impact on crime at all? Was the recent reduction in property crime due to increased surveillance, other social factors, or simply a balancing out over time?

Effectiveness of Community-Policing Cameras

The effectiveness of cameras in deterring crime has been mentioned in several chapters, with the results being mixed. Sometimes, after cameras are installed, crime goes up; sometimes crime goes down. Sometimes crime changes character; sometimes crime shifts a few blocks away.

One thing that appears fairly certain is that crime isn't going to stop because of cameras, particularly crimes of passion that occur when emotions are highly

---

[64]Justin D. Levinson and Danielle Young, "Different shades of bias: skin tone, implicit racial bias, and judgments of ambiguous evidence," *West Virginia Law Review,* Winter 2010, 112 (2), 307–350.

pitched and the person committing the crime doesn't care who sees what he or she is doing.

When crime rates are studied over the short term, what appears to be an increase or decrease may be a cyclical fluctuation or even errors in the design of a study. This is why researchers come back to a situation time and again with more experiments, to try to replicate results or get better ones.

Decades-long studies are needed to draw better conclusions and most of the more sophisticated forms of surveillance are very recent.

A four-year study of D.C.'s Metropolitan Police Department by the Urban Institute found no reduction in crime from a citywide surveillance system, but this could be because the number of cameras was low, or because they were not live-monitored feeds. In other areas with differently configured systems, there appeared to be some reduction in crime.

Effectiveness of Biometric, Radio-Frequency, and Other ID Cards

It is hard to say how effective cards with radio-frequency IDs or biometric information such as fingerprints, iris prints, and other personal information are in identifying people at access points. The cards are commonly used to access workspaces and individual computers, or resources at commercial copy service companies. They are also presented at airports and border checkpoints by frequent travelers, with tens of thousands of them purchased for this purpose.

No matter how much encryption data is built into radio-frequency cards, or how much biometric data is added to other cards, there are still limitations to how much security they provide. Here are some of the limitations.

- *Even if everyone were carrying them, they may not be checked.* A majority of cashiers don't check the signatures on the back of credit cards when customers make purchases, and people in the justice system sometimes forget to check the IDs of inmates coming before a judge for release. A few inmates have escaped this way, by impersonating others. Even if the justice system switches from cards to iris scans, it is still necessary for the chaperon to remember to take a scan to verify a person's identity.

- *Theft is always a problem.* The cards themselves can be stolen and sometimes reprogrammed. Bank cards, student cards, and door-access cards are sometimes stolen and repro- grammed or swiped and impersonated, and the computers storing the data can disappear, as well. In 2008, the TSA announced that a laptop containing the data associated with 33,000 Clear Cards, cards used by frequent flyers, had been lost at the San Francisco airport. At the time it disappeared, the system was in use at 17 airports worldwide. Theft of credit-card data is not uncommon, either, with tens of thousands of names and numbers disappearing over the last decade.

- *Companies can go bankrupt or close their doors for other reasons.* In the past, biometric data was only in the hands of government agencies such as customs and the police, and a few private institutions like hospitals and research labs. With the advent of frequent travel cards, commercial ventures like Clear are storing fingerprints, iris prints, facial features, social security numbers, employer information, email addresses, date and place of birth, gender, and height. In 2009, Clear abruptly announced it was closing its doors and customers were left not knowing what would happen to their sensitive data. The company's privacy policy did not explicitly explain what would happen if the company were bought out. When asked about the fate of the data, the company announced that it would be deleted, but there may have been legal restrictions, since the data was a company asset and investors might make claims for keeping it. In 2010, AlClear acquired assets of the company along with a database of 160,000 subscribers. Apparently the data wasn't deleted, as promised. The new company claimed that the data of those who chose not to renew their subscriptions under the reborn company would be deleted.

- *Data integrity can be compromised.* If the data is compromised, if an employee or unauthorized person were to access the database and change one or more entries, it could be possible to create a false identity or reprogram a valid one sufficient to gain entry. Strong security is necessary for databases that contain such sensitive information. Even strong security can be breached, as hackers have proven time and again by breaking "unbreakable" encryption schemes.

- *Not every crime can be predicted by profiling people or prevented by denying cards to those with criminal histories.* Every person in the criminal justice system has a first crime which occurred when they had no prior criminal record. Many terrorist acts are committed by university-educated idealists with no previous criminal history. Community leaders sometimes commit domestic terrorism out of idealistic frustration, after years of trying to go through regular channels. There will always be incidents regardless of how much information is on an ID card or how carefully people are screened before handing them out.

Some card-service providers have dropped cards altogether and switched to biometric identifiers for financial transactions. In 2006, Citigroup, Inc., Singapore announced the introduction of a payment service based on fingerprint scans.

As soon as a service like this is announced, thieves determined to thwart the system look for ways to create fake fingerprints, as do researchers and students curious about the effectiveness of the security.

Sometimes you can fool a fingerprint system with a Gummi Bear™ or, more correctly, a wad of gelatin. In 2002, a Japanese cryptographer demonstrated how gelatin and a latent fingerprint lifted from a glass could be etched onto a PC board and combined to create a fake finger that could fool 11 commercial detectors about 80% of the time.

In 2002, the Biomedical Signal Analysis Lab in West Virginia used dental impression mold materials and casts of Play-Doh™ and clay and were able to spoof fingerprints successfully. Then they went a step farther and used cadaver fingers, on the possibility that criminals might try to use dismembered fingers to gain access, with a verification rate, once enrolled, of 40–94%.

In follow-up tests in 2006, using off-the-shelf products, students at the Washington & Jefferson College found that Gummi Bears were better as food than as fingerprint spoofers. After trying various materials and methods, they found that a wax cast that was frozen to harden the imprint and then pressed with Crayola modeling material could successfully register as a fingerprint on one of the commercial scanners about a third of the time. On another brand of scanner, with a more sensitive sensor, Hasbro Play-Doh was found to be more successful.

Various combinations of paraffin and bee's-wax (which are used by sculptors), silicon, and other movie-makeup materials can be molded into very lifelike prosthetics that might fool some machines, as well.

A fingerprint alone might not be enough to ensure security. It may be necessary to combine it with other biometrics, like pressure-sensing or iris scans, or to use palm-vein scans instead, which can't be as easily impersonated.

The research reports released in the early 2000s demonstrated that it was possible to spoof fingerprints, so the National Science Foundation, with support from DHS and the DoD, funded research on spoofing vulnerability in biometric systems and possible strategies to prevent them in a project called *ITR: Biometrics: Performance, Security and Societal Impact*. The project was initiated to explore the technical, legal, and privacy issues in biometric identification systems.

One of the outcomes of the research was an algorithm for including perspiration patterns as part of the verification process, so that cadaver and Play-Doh fingers, which don't sweat, would be excluded as valid prints.

Report on Camera Surveillance

In 2009, a study funded by the Privacy Commissioner and the Social Sciences and Humanities Research Council of Canada was undertaken by university researchers to see if there were overall benefits of camera surveillance systems.

The document also presents some examples of cameras that were installed and later removed, partly due to lack of evidence for their effectiveness, partly because of privacy concerns.

Most funding of public CCTV in Canada is by local law enforcement agencies or contributions. There is no widescale federal grant funding specifically for surveillance systems as in the U.S.

Results of the study, which show consistent public perception that cameras will deter crime, but few systematic studies supporting the assumption, can be summarized in a few quotes.

> *"Current public opinion research, whether from marketing firms, political organizations and media outlets or from the social sciences, consistently shows strong support for the use of camera surveillance in public and in private spaces.... Regardless of the setting, cameras are seen as useful against crime, though their usefulness has not been proven in quantitative evaluations. The public largely presumes, or even hopes for usefulness: cameras are seen as worth installing even if they will generally not be useful, on the hope that they might prove useful eventually..."*

> *"A 2003 study by Welsh and Farrington performed a meta-analysis on twenty-two CCTV evaluations in the UK and North America. They found that the overall reduction in crime averaged approximately 4%, and that in the five North American studies in particular, none demonstrated evidence of a reduction in crime..."*

> *"...one striking fact drawn from these findings is that the majority of respondents have [sic] little knowledge of camera surveillance in their own neighbourhood..."*

The Canadian study pointed out that the capability of surveillance cameras to deter crime may vary with the public's awareness of the cameras (many hadn't noticed them), the location, and the form of crime.

In San Francisco, research by King et al focused on the crime-deterrent effects of the city's Crime Camera Program. The researchers carried out an empirical assessment of cameras installed at 19 sites throughout the city based on data supplied by the police on all reported incidences between January 1, 2005 and January 28, 2008 within 1000 feet of the cameras, a total of 76,390 incidences.[65]

Crime rates in the 200-day period preceding camera installation were compared with those after installation.

---

[65] Jennifer King, et al., "Preliminary findings of the statistical evaluation of the crime-deterrent effects of the San Francisco Crime Camera Program," Center for Information Technology Research in the Interest of Society, UC Berkeley, March 17, 2008.

Preliminary findings indicated that crime deterrence was dependent on the distance from the cameras, with a 22% decline in property crime within half a block of the cameras. Results were summarized as follows:

"In an analysis of the change in crime occurring within 500 feet of a camera site, we find a statistically significant decline in property crime occurring within 100 feet of camera locations, but no statistically significant changes in crime 100 to 200 feet, 200 to 300 feet, 300 to 400 feet, or 400 to 500 feet from the site."

Declines in property crime were very specific, however. There was no significant reduction in burglary or auto theft. Most of the reduction was in larceny/theft.

Violent crimes did not appear to decline. Homicides declined within 250 feet of the cameras, but increased in the area 250 to 500 feet from the cameras—thus canceling out any reduction.

"When we extend the analysis to the areas that are 500 to 1,000 feet from crime camera locations, we find no significant changes in these areas for violent crime and no overall significant changes for property crime."

In a document published by the U.S. Department of Justice in 1999, guidelines are laid out for security in U.S. schools, including video surveillance, metal detection, entry control, and duress alarm devices. Despite research to the contrary, the document describes the deterrent effect of cameras. It also encourages video cameras as engendering "peace of mind of both students and faculty" thus supporting the findings of the Canadian study that there is a strong *perception* of security associated with cameras even if empirical evidence does not bear out an actual deterrent effect.

The DoJ document acknowledges that misbehavior or crime on school premises may be displaced rather than prevented.

Cameras are like smoke detectors. Many get installed, but how many are properly maintained? Cameras and other surveillance devices have little surveillance value if they are not monitored, checked, aimed in effective directions, and the data stored and retrieved in ways that are effective when needed. Adequate training on their use, and knowledge of the possible liabilities associated with incorrect use, are also important, as are followup studies to determine if they are cost-effective and effective in providing a more secure environment.

The Perception of Security and Control

Many studies indicate that people have hope or confidence that the installation of cameras or other surveillance systems will reduce crime, but additional studies suggest that the presence of cameras may instill a concern about loss of control over the installation of the technologies or, in specific instances, may create the perception of a threat.

Research by Brooks[66] suggested that subjects of the study perceived a social benefit from having surveillance cameras, but expressed concern over their ability, as a community, to ensure appropriate control over public-street surveillance.

A study published by Williams and Ahmed[67] in 2009 described how 120 shoppers were presented with a photo of a fictional urban street scene. When the scene included a CCTV camera together with a skinhead or woman, participants reported raised concern about walking in the scene, as compared to the same image of a skinhead or woman without the CCTV camera. Thus, visible surveillance devices, in some circumstances, may create a sense of unease or of imminent danger rather than a sense of security.

---

[66] David J. Brooks, "Is CCTV a social benefit? a psychometric study of perceived social risk," *Security Journal*, Vol. 18, 2005.

[67] D. Williams, D. and J. Ahmed, "The relationship between antisocial stereotypes and public CCTV systems: exploring fear of crime in the modern surveillance society," *Psychology, Crime & Law*, 15 (8), 743-758, 2009.

CONSTITUTIONAL FREEDOMS

LEGISLATION

CONSTITUTIONAL FREEDOMS

- *constitutional rights*
- *ubiquity of surveillance devices*
- *balance between the needs of law enforcement and the privacy of law-abiding citizens*

Introduction

Freedom is more than the right to speak one's opinion and move about freely—it also includes the right to be left alone. Privacy is a fundamental aspect of human dignity that was well articulated by Supreme Court Justice Brandeis in *Olmstead vs. the United States* when he stated

> "The makers of our Constitution undertook to secure conditions favorable to the pursuit of happiness. They recognized the significance of man's spiritual nature, of his feelings and of his intellect.... They sought to protect Americans in their beliefs, their thoughts, their emotions, and their sensations. They conferred as against the government the right to be left alone—the most comprehensive of rights and the right most valued by civilized men."

The right to be left alone, a right that was probably taken for granted in the days before high population and micro-electronics, is now under severe assault by burgeoning numbers of surveillance devices. Buildings, personal tech gadgets, even cars, are equipped with motion sensors, GPS location capabilities, and sophisticated, high-resolution cameras.

In March 2012, the European Union announced plans to protect an aspect of privacy related to the right to be left alone, that is, the "right to be forgotten." This, too, is an important aspect of freedom that is eroding due to all the private information that is collected and retained in electronic databanks without the knowledge or consent of the individuals who are the focus of the information.

DOCUMENTS ACKNOWLEDGING AND ASSERTING CIVIL LIBERTIES

A few key milestones in the establishment of American liberties will be mentioned that have a bearing on surveillance or its relation to basic freedoms, including the Declaration of Independence, the American Constitution, and the Bill of Rights.

The Declaration of Independence

The presentation of the Declaration of Independence by the committee that drafted the document, including Thomas Jefferson, John Adams, Roger Sherman, and Benjamin Franklin with recipient John Hancock seated on the right. The painting by John Trumbull was commissioned in 1817 and placed in the Rotunda in 1826. The document asserted independence from the political leadership of the British King.

When the American *Declaration of Independence* was written in 1776, it was primarily to assert independent sovereignty from England. When taken together with the Constitution, which includes the Bill of Rights, it served to confirm basic "common sense" human rights, as opposed to "American" rights. The purpose of the Declaration, as articulated by Thomas Jefferson, the chief architect, was to assert self-evident rights or, as Jefferson termed it, "the common sense of the subject."

> *"... This was the object of the Declaration of Independence. Not to find out new principles or new arguments, never before thought of... but to place before mankind the common sense of the subject..."*

Thomas Jefferson letter to Henry Lee, 1825

In the text of the Declaration itself, it states

> *"That whenever any Form of Government becomes destructive of these ends, it is the Right of the People to alter or to abolish it, and to institute new Government, laying its foundation on such principles and organizing its powers in such form, as to them shall seem most likely to effect their Safety and Happiness...."*

American Declaration of Independence, July 4, 1776

Contributors to the Office of Technology (OTA) report also pointed out that many forms of communication, such as digital phone calls and digital graphic communications over networks, were not protected by existing statutes.

Two significant pieces of American legislation enacted to protect privacy and freedom, the *Wiretap Act of 1949* and the *Omnibus Crime Control and Safe Streets Act of 1968 Title III,* came into existence to counteract perceived abuses of personal privacy. They governed a number of new technologies for a while, but they are inadequate to address the explosion of surveillance occurring now, particularly the use of surveillance devices that incorporate "smart" processing—software capable of analyzing the contents of an image or communication and making decisions on what is relevant and what is not.

Laws have been drafted to prevent intrusive uses of surveillance, but so far results have been inconsistent. Fair Information Practices (FIPs), principles for storing, handling, and archiving information, have been laid out by a number of countries, but adherence to these practices is voluntary and individuals and for-profit firms often disregard them, especially when new, potentially profitable, technologies arise that make it easier to do so.

Left: J. Edgar Hoover, Director of the FBI, fingerprinting Vice President Garner. Right: James Lawrence Fly, Chairman of the Federal Communications Commission, at a meeting with journalists. [Photos courtesy of the Library of Congress, public domain.]

The long battle between J. Edgar Hoover and J.L. Fly exemplifies the constant struggle to balance use of surveillance by law enforcement agencies versus public disclosure and regulations to govern eavesdropping. Fly was a vocal proponent of constitutional rights and critical of what he called the "FBI's illegal practice of wiretapping." He opposed ungoverned institutional eavesdropping and advocated public disclosure of wiretapping. He wasn't opposed to law enforcement seeking information to solve crimes, he just wanted sufficient oversight to prevent abuse.

One of Fly's recommendations was that information be gathered via subpoenas of communications carriers rather than by wiretapping. In addition to wiretapping laws, term limits for FBI directors were instituted in response to Hoover's controversial surveillance activities.

The Hoover/Fly debate seventy years ago may seem like old news, but the issues raised are still surprisingly relevant to digital communications. When the *Communications Assistance for Law Enforcement Act of 1994 (CALEA)* was drafted to require communications carriers to set up their systems so that "wire-tapping" for law enforcement purposes could be facilitated, there was outcry from carriers regarding the cost of compliance and privacy rights. Despite controversy, the regulations were implemented and further expanded in 2005 to include private networks such as those on university campuses.

Since then, CALEA's influence continues to grow and has moved beyond U.S. borders. In 2010, India proposed a law permitting interception of voice communications, email, and Internet data, based on CALEA.

When Practical Obscurity Becomes Public Accessibility

The Internet hosts thousands of companies that monetize personal information for commercial gain. Some is gathered through covert surveillance. Some is gathered from a wide variety of public sources and presented on websites such that it takes on a new character.

The law does not yet make a distinction between personal information held in separate local repositories and that which is aggregated and uploaded to the global Internet. Personal information that has been compartmentalized is like a series of small peepholes in a very large fence. You see a glimpse here, travel a couple of miles, catch another glimpse there, travel a mile, catch another glimpse. It takes effort to drive around to local sources to present ID and ask to see all the files on a particular individual. This acts as a deterrent to large-scale data aggregation. On the Internet, the fence has been removed, and all the "glimpses" moved closer together to provide an instant and very personal picture that can significantly impact a person's life.

In February 2012, the U.S. government proposed a privacy "bill of rights" to increase consumer control over private information, but the way it was written, compliance would depend heavily on the voluntary commitment of companies, many of whom already openly disregard personal privacy.

The Federal Trade Commission (FTC) and Department of Commerce have responded to a number of concerns about the way companies use information.

As an example, the chairman of the FTC, Jon Leibowitz, has stated that a failure to follow through on privacy policies could be considered a deceptive practice violation, provoking action by the FTC.

Most of the early initiatives to protect privacy through legislation were aimed at groups that are particularly vulnerable, such as children, or they targeted specific kinds of data, such as credit reports. No broad protection for the general public is yet in place.

Online Privacy

It is difficult to determine whether the greatest intrusions into personal privacy come from cameras and listening devices, or the dissemination of personal data on the Web.

There are programmers who actively develop spyware and other malware to infect local computers. The software snoops around on people's hard drives, seeking out personal information such as passwords, email contacts, and financial data. A large amount of identity theft, credit card theft, and hacking of online banking services occurs this way.

Sometimes privacy violations are committed by large, popular companies offering a wide range of services. Google, Inc., provides a premiere search engine, free translation services, white-board document sharing, and the incredibly successful Youtube™ video-sharing site. Popularity does not guarantee, however, that every aspect of the company's operations are fully compliant with the law.

In 2011, Google was charged by the FTC for using "deceptive privacy practices" when launching Google Buzz.

> *"Google Inc. has agreed to settle Federal Trade Commission charges that it used deceptive tactics and violated its own privacy promises to consumers when it launched its social network, Google Buzz, in 2010. The agency alleges the practices violate the FTC Act."*
>
> Federal Trade Commission news release, March 30, 2011

With the groundswell of concern over privacy, responses by companies like Google suggest a realignment of privacy policies is underway. It also creates an entrepreneurial window for companies to create products to prevent intrusive tracking and data collection. The International Association of Privacy Professionals (IAPP) had grown to almost 10,000 by 2011.

It was revealed in a Indian blog in 2010, and announced more broadly in the news in 2012, that a Stanford researcher had discovered code that enabled Google and three other companies to use software to bypass privacy settings within the popular Safari Web browser so they could track user movements on the Web. This practice contradicted information on Google's site telling users how to prevent tracking. When the companies were contacted, they had various justifications for their actions, but it illustrates how consumer confidence in browser settings may provide a false sense of security when surfing the Internet.

Whether new products can adequately protect privacy as surveillance technology advances remains to be seen. Obsolescence is common these days.

Private Data Aggregation

Aggregated data doesn't have to be on the Internet to impinge on privacy.

One of the primary ways in which data mining erodes privacy is by the consolidation of information from different sources in closed company vaults. A user might divulge a few private facts to Company A and others to Company B, on the promise that private information "is never sold to third parties." Then, a year down the road, Company A may be bought out by Company B, or enter into a cooperative agreement, and the puzzle pieces combined.

The Liquid Nature of Surveillance Data

Once it gets established, surveillance data spreads like weeds, sending out roots and popping up in unexpected places. Surveillance data is considered a significant asset by government agencies and companies that collect it. As such, it is often sold for immediate profit, or transferred to other agencies or companies as institutions are reorganized or bought out.

When the company that managed Clear Card biometric expedited travel cards went out of business, the firm promised to delete personal information on file. This apparently never happened. The company buying the assets announced a large repository of data as part of the acquisition. As mentioned in a previous chapter, it may not be legal to delete surveillance data if it is owned by a public company with shareholder stakes even if privacy is at stake or promises were made by prior management.

Convenience and Awareness

A large proportion of surveillance data is given up voluntarily for reasons of convenience or because individuals receive something in return, like a green card or Social Security check.

Those giving up personal data for convenience often aren't aware of how the data is stored or traded with other companies. They enjoy faster service or bigger discounts and don't have time to think about what happens behind the scenes.

Individually, personal biometrics such as voice or facial recognition systems may not seem overly intrusive, especially when used to monitor activities like financial transactions at an outdoor teller machine, but many companies are selling or licensing their data repositories to other companies. If you combine fingerprint data from one organization, facial information from another, and GPS location from a third (such as a cell phone company), you have a composite portrait of a person's financial transactions, buying patterns, and patterns of movement that is highly revealing.

Internet World Stats for 2011 reports that more than 20% of North Americans weren't using the Internet. Other surveys show that many people use it only occasionally, primarily for email, and yet data is being compiled on everyone. Those with limited Internet skills, or Internet access, are probably unaware that their lives are being monetized and globally distributed.

One of the findings of studies on camera surveillance that surprised researchers was that many people were unaware of the cameras. Similarly, many people are unaware that credit cards now include RFID chips. They can be recognized by the small speaker icon that is printed on the card, but even if a consumer sees the icon, he or she may not understand the symbol or the nature of RFID.

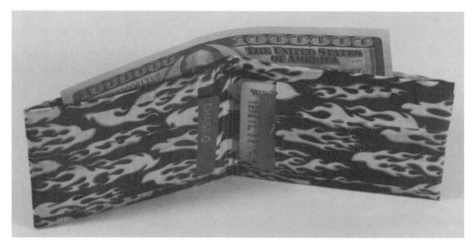

Some consumers are unaware that RFID chips are now built into many credit cards and identity cards. Those who are have begun buying products that protect the cards from electronic snooping. The small speaker symbol on the credit card (top left) identifies an RFID-transmitting card. To prevent RFID cards from broadcasting IDs (or other information that might be included on a "smart RFID" card), people are shielding them with metal-lined sleeves, metal cases, and wallets, such as these Tanya Cartwright billfolds crafted with shielded duct tape (bottom). When the shielding is built right into the billfold, individual shielding sleeves become unnecessary. [Photos © 2012 Classic Concepts, used with permission.]

If the populace is unaware of the technology that underlies the surveillance, people are unlikely to consider how data might be abused.

Urban Planning and Residential Policing

Up to now, the decision of whether to install security cameras in private buildings has been an individual choice, but this appears to be changing.

Building permits for financial institutions and ATM installations require that security cameras provide a full view of any lobby areas as well as both interior and exterior ATMs. Most would support this kind of surveillance to protect financial-transaction privacy and to discourage muggings at ATMs. The permit requirement is spreading to other areas, however. Some cities are requiring a commitment to install surveillance cameras as a requirement of getting a building permit. This isn't only for commercial or industrial developments; it includes residential developments, as well.

While some feel this is a small concession to security, surveillance is often developed incrementally, which raises a question—if surveillance cameras were required in all new buildings, how long would it be before the police department asked the city to require that all building feeds be accessible by central law enforcement?

Perhaps examples from recent history can provide an answer.

Congress supported law enforcement requests to require ISPs to build remote tapping capabilities into their communications systems so law enforcement agencies could stop criminals and fight terrorism. In principle, this is a reasonable request.

Unfortunately, with today's powerful technologies, listening for suspects and criminals doesn't mean tapping just their phones, it sometimes means monitoring *all* the conversations that transpire through a digital link.

Having police eyes and ears in every corner of every building might make us safer and it might not. Research results are very mixed on whether cameras reduce crime. In some areas they appear to have no effect at all. In others, crime appears less. In others, crime has increased despite the cameras. Very specific cameras appear to have some positive effect, like those aimed at ATMs or the entrances of parking lots, but more general surveillance doesn't seem to deter crime.

Providing police access to all security camera feeds would extend the capacity of law enforcement to monitor and archive every aspect of people's lives without a warrant. Wiretapping laws were traditionally established to regulate audio conversations, not video feeds, since tiny, powerful video cameras as we know them today didn't exist when the laws were originally framed.

All this supposition may seem alarmist, but reality is often ahead of our imaginations. The idea of putting police security cameras on private property has already been suggested by Harold Hurtt, Houston's Chief of Police.

"Houston's police chief proposed... placing surveillance cameras in apartment complexes, downtown streets, shopping malls, and even private homes as a way of combating crime with a shortage of police officers...

...Hurtt said if a homeowner requires repeated police response, he thinks it is reasonable to require cameras surveillance of that property."[50]

Harold Hurtt as reported by Pam Easton, Associated Press, February 2006

Part of Hurtt's justification for security cameras at a person's home was because he felt they were putting an unfair burden on the justice system and "cheating the

---

[50]Pam Easton, Associated Press, "Houston police chief wants cameras across city to help fight crime," *Victoria Advocate*, February 16, 2006.

other residents of Houston..." This is, in essence, saying that the victim is responsible for taking more police time. Why Hurtt would feel it was the victim's fault rather than the fault of whoever carried out the theft, property damage, abuse, or other illegal act was not explained.

Additional surveillance is not always initiated by law enforcement authorities. Sometimes concerned neighborhoods request cameras or take it into their own hands to install cameras on private property.

In Georgetown, D.C., in Feb. 2012, Georgetown community members acted as a voluntary arm of the law with plans to set up cameras in places that could not be installed by police, including private yards. The concept is to tape events in the area and turn the tapes over to the police if any are felt to be related to a crime. As stated by Diane Colasanto, a member of the public safety committee, their primary concern was deterrence. "We just want crime not to happen here."

Community policing is not new. It has been implemented successfully in some areas, and has been termed vigilantism in others. As plans progress, there are questions about privacy and who would be permitted to access the video "and under what circumstances."[51] Since community members may not have training in law or a hierarchy by which to provide checks and balances on the integrity and completeness of the information and how and for how long it is stored, it may be an experiment that takes some time to work out.

Once the surveillance footage exists, it can be legally subpoenaed as evidence.

Most law enforcement agencies are now linked to national databases, which means if law enforcement eventually obtains local access to private camera feeds, that information could end up in federal archives, as well—perhaps indefinitely.

Since there's no guarantee that crime will stop with increased camera coverage, and many examples from history suggest that concentrating too much information in too few hands upsets the balance of power between institutions and individuals, society will have to establish where the line is between the perception of security, the right of privacy, and the right to prevent intrusion on the lives of ordinary citizens.

Surveillance doesn't only chip away privacy, it also erodes the expectation of privacy because humans are adaptable—it is our greatest asset and maybe our greatest curse.

Background Checks

It was noted in the chapter on computer surveillance that employee background checks are prevalent and those who offer contract services to do the research on behalf of companies are amassing large storehouses of information gleaned from the Net and other sources. This data is not restricted to current job hunters; the search begins at birth. The more complete a picture the contractor can provide 10 or 30 years hence, the more competitive the company is in relation to other firms.

---

[51] Andrea Nobel, "Public surveillance from private property questioned," *The Washington Times*, February 5, 2012.

"While some people are not concerned about background investigations, others are uncomfortable with the idea of an investigator poking around in their personal history. In-depth background checks could unearth information that is irrelevant, taken out of context, or just plain wrong."

"Fact Sheet 16: Employment Background Checks: A Jobseeker's Guide," *Privacy Rights Clearinghouse,* November 2011 revision

The Fact Sheet brings up one of the biggest problems with background checks—those who are subjected to checks rarely have an opportunity to see the information to verify its accuracy or to judge whether it includes information that may be legally discriminatory. They may not even know there's been a background check.

There is always the possibility that the wrong person has been investigated, or that data from different people with the same name have been combined.

There may also be information that's irrelevant to the job. If a background check unearths information that suggests an unemployed job applicant paid some bills late in the last few months, it might be irrelevant to the job. If the job is an accounting position, it might matter. If it is laying optical cable or doing data entry, it may not. There's no way for the applicant to know if irrelevant data was included.

Banks may also use background checks for loan applicants, especially for larger loans such as mortgages or business loans. In addition to requesting a credit report, the company may check employment history and ask to see tax returns.

RESTRICTIONS ON SURVEILLANCE

Restrictions on surveillance are partly federal law and partly governed by individual states. As an example, most states conduct roadside checks for traffic infractions and signs of driving under the influence but some have a legal mandate restricting them.

Overt Versus Covert Observation

The legality of audio/video recording varies from state to state.

Cameras recording from a hidden location may be legal as long as the recording is video-only. In some places a notice must be posted. A security device with streaming audio, as opposed to recorded audio, may be legal in some states.

In most places, video recording is permissible but not in places where people have an expectation of privacy, such as bathrooms and change rooms.

Despite restrictions, video cameras are now widely incorporated into personal devices and audio/video recordings of activities in public places and the workplace, and occur millions of times a day in schools, parks, parking lots, offices, cafés, streets, parties, and public entertainment events.

Surveillance for Titillation or Sexual Exploitation

There are many ways in which surveillance is used to acquire or deliver sex or sexually explicit imagery.

It is arguable that procedures have to be followed, but children are sternly taught not to go to strangers, not to leave their mother's or father's side when they are traveling, and not to let strangers compromise their privacy by touching their bodies. Procedures for adults may not be effective for children, or appropriate.

There have been a number of lawsuits for privacy invasion tendered against the TSA, but so far most judgments are in favor of the transportation authority having leeway to carry out scans and pat-down inspections.

Intellectual Property

Legal violations related to surveillance do not stop with invasions on personal privacy. Video recording encompasses copyright issues, as well. Google Streetview includes images of artwork such as large sculptures that are permitted to be photographed as tourist images for personal use, but not for republication for commercial gain. Since Google is a for-profit company that uses information to build traffic to websites and monetize its advertising and other revenue streams, it has been argued that the company's imagery is for commercial purposes.

In Oregon, a tourist taking pictures of downtown Portland in 2009 was told by a police officer that he was not permitted to photograph buildings and yet Streetview pictures clearly show buildings around all the major streets in the same area.

These seeming inequities raise a recurring question in discussions about surveillance. Are corporations entitled to greater leeway than individuals when it comes to conforming to surveillance laws and matters of physical privacy or intellectual privacy?

Laws to Protect Privacy on Public Communication Venues

Faltering efforts to draft laws to protect people from online exploitation, such as sexual exploitation, include the 1996 *Communications Decency Act* (CDA) which was passed as part of the *Telecommunications Act*. Many people supported the spirit of CDA, but advocates of personal freedom challenged that it could be used to censor constitutionally protected rights such as freedom of speech.

Lawmakers followed up by proposing the 1998 *Child Online Protection Act* (COPA), which, if passed, would make it unlawful for commercial entities to make obscene or sexually explicit materials available to minors. The model for most of these laws has been conformance with television and radio broadcast decency standards, but many freedom advocates argue for taking a different approach to regulating the Internet due to its different nature.

In today's more liberal Internet climate, one might wonder why all the fuss, but in the late 1990s and early 2000s, almost every search on any search engine led to pornography sites and they often filled the top 200 slots in any search due to the way pornography sites embedded pages and pages of keywords related to common searches. It was impossible to avoid them and parents were particularly concerned about children being exposed to porn without parental supervision.

At that time, search engines also led directly to thousands of "upskirt" sites, pornography sites that obtained photos by planting tiny spy cameras in the toilet bowls of public washrooms or the stalls of change rooms. Cameras were even hidden in skateboards or baseball caps, so that the board could be placed under the skirt of an unwary pedestrian or the cap dropped on the sidewalk without the victim knowing she was photographed. The practice continues, but the Internet red light district no longer crowds out searches for other sites on major search engines.

Some attempts were made to create software to filter out controversial content for use in homes and schools, but it was difficult, not only because of high search engine placement of objectionable sites, but also because low-cost computers were flooding the market, giving children easy access to computers that weren't installed with filters.

Why are so many opposed to children seeing explicit content of natural acts? Some children are frightened by such things, but of greater concern is the propensity of children to mimic.

Children are eager to copy adults, but are not mature enough to understand the risks and responsibilities of adult sex. In addition, they do not comprehend that porn sites depict fantasy sex rather than responsible real-life sex. Online porn is geared primarily toward male fantasies and the videos rarely include condoms. Internet porn does not show consequences such as sexually transmitted diseases (STDs), pregnancy, the emotional and physical consequences of rape or the ire and frustration of unsatisfied sexual partners. It makes sex appear as uncomplicated as a game of checkers and thus may lull children to become more vulnerable to sexual exploitation.

Thus, laws to protect children generally prohibit the collection of intimate images, with the definition of "minor" varying region-to-region, and also restrict public display of pornographic content.

Video Voyeurism

Video voyeurism is observing or recording in a way that violates a person's reasonable expectation of privacy. In some states the definition is more specific, specifying a lewd purpose.

Lawmakers have struggled to draft laws that protect people's privacy while not being overly broad and censoring nudity in art or journalism. In 2000, the Wisconsin Supreme Court struck down a state voyeurism law as unconstitutional when a man appealed a charge of video voyeurism after taking nude photos of his ex-girlfriend through a window.

By 2008, a little more than half the states had laws specifically addressing video voyeurism via the use of hidden cameras. Some laws are specific to video voyeurism and some are included within broader eavesdropping, privacy, and "peeping Tom" laws. A few states address video voyeurism within their stalking laws. By 2011, all states had addressed this issue in one way or another and some had further strengthened the laws by changing video voyeurism from a misdemeanor to a felony so law enforcement could issue subpoenas and search warrants to investigate reported cases.

The increase in lawsuits for video voyeurism is not surprising, considering the availability of tiny cameras that are easily hidden in shoes, garbage cans, and bathroom fixtures. This may not be the only reason, however. Some have suggested that sensationalist television talk shows and popular reality shows stoke the public's appetite for voyeuristic entertainment and have dulled sensitivity to the toll it takes on the persons being watched. The popularity of voyeuristic broadcasts also implies public acceptance of this practice. A Canadian study on television programming between 1993 and 2001 indicates significant increases in both physical and emotional acts of violence during prime-time hours. Video voyeurism can be seen as a form of emotional abuse.

As a response to these incidents, Public Law 108-495, the *Video Voyeurism Prevention Act of 2004* was enacted, which states, in part,

> *"Whoever, in the special maritime and territorial jurisdiction of the United States, has the intent to capture an image of a private area of an individual without their consent, and knowingly does so under circumstances in which the individual has a reasonable expectation of privacy, shall be fined under this title or imprisoned not more than one year, or both...."*

It further exempts "any lawful law enforcement, correctional, or intelligence activity." The law restricts acts of voyeurism on federal property such as embassies and military bases, but leaves the regulation of other offenses under the jurisdiction of individual states.

In February 2012, the Florida Senate unanimously passed a bill to raise video voyeurism from a misdemeanor to a third-class felony. This distinction is now common to most states. The Senate also passed the *Protect Our Children Act* requiring convicted violators to be registered as sex offenders if there is more than one video voyeurism offense against minors. Thus, as law tries to catch up with technology, efforts to protect individual privacy have increased since the early 2000s.

An important issue in this debate is whether live video should be regulated in the same way as recorded video. Most lawmakers agree that it should, but a lawsuit with no recorded evidence may be harder to prosecute.

A second issue is whether parents and guardians have greater leeway in video recordings than strangers. If a family tapes a child's birth and uploads the event to a video-sharing site, most people would not consider this video voyeurism, but what if 5-year-olds are recorded naked by a cousin while splashing in a wading pool and uploaded in the same way? At what point does it change from innocent play that does no apparent harm to video voyeurism?

Voyeurism and UAVs

As has been mentioned, the FAA is taking steps to enable UAVs to share commercial airspace, a move that will result in commercial firms producing more UAVs to satisfy demand. The new UAVs are smaller, more powerful, and much easier to fly than remote-controlled planes of the past. The cameras are more powerful, as well, taking high-resolution shots from higher altitudes than ever before. It wouldn't be surprising if many of them were flying over backyards (and perhaps into backyards) in the near future.

Voyeuristic sites on the Net may soon be loading up with UAV "down-shirt" images. Many of the newer UAVs are priced so that a hobbyist could buy one for the express purpose of building a website full of voyeuristic backyard shots, provoking even more lawsuits related to privacy.

Even law enforcement precincts that may benefit from UAV technology are divided on whether aerial camera surveillance should be deployed. Others are going ahead with training and testing. Concerns from the public are partly about privacy and partly about whether the surveillance of criminals may gradually change to routine surveillance of everyone by Big Brother air patrols.

Balance between Security and Personal Freedom

"No one is challenging the notion that a police officer does not have the right to trail a suspect for a couple of blocks.... But we know from history that this technology will expand into routine surveillance. We need to be wary when proponents suggest only the benefits of a new technology."

Jay Stanley as quoted by Kenneth Artz, Dallas

The dangers of excessive visual surveillance are not limited to governments or commercial interests. Individuals are actively using cameras, and law enforcement agencies sometimes entreat them to provide video evidence if they have witnessed events related to police investigations.

Biometrics Surveillance

The use of biometrics is increasing as rapidly as the use of smaller, more powerful cameras.

In August 2010, the Department of Homeland Security released a Privacy Impact Assessment for the Iris and Face Technology Demonstration and Evaluation (IFTDE). In the Privacy Impact Analysis, three privacy risks were identified.

- That data could be used for purposes "outside the originally stated purposes (i.e., for testing and evaluation purposes)."

- That inaccurate information might negatively impact the individual.

- That there might be unauthorized access or use of the iris prints.

To respond to these concerns, it was stated that a standalone, unnetworked system would be used, that no identifying information would be linked to the iris images or biographical information, and that access to the iris-image data would be restricted to authorized NIST and U.S. Naval Academy employees and contractors on the research team, who must sign a "Rules of Behavior" agreement. It wasn't specified whether the same rules would apply when the pilot study was concluded and the actual program begun.

Keystroke and Email Monitoring

It has become common for businesses to monitor employee email and sometimes even the keystrokes and mouse clicks used on work-based computers. For the most part, businesses are permitted to surveil computer use, but the monitoring of email may be limited to email programs originating on the work terminal or server. Web-based email may not be monitored in some areas without the consent of the user.

Privacy and Investigations of Illegal Surveillance Activities

When government surveillance of electronic communications and consumer access to the Internet became more prevalent in the 1990s, issues of who is responsible for network content came to the fore. There were some instances of ISPs being charged due to content, such as pornographic images or pirated files uploaded by Internet customers. ISPs claimed that it was impossible to police every piece of data 24 hours a day, and mounted a defense against charges for acts they had not committed.

Over time, the pendulum swung in the other direction and the *Communications Decency Act* (CDA), part of the *Telecommunications Reform Act of 1996*, somewhat settled the debate by stating that "No provider or user of an interactive computer service shall be treated as the publisher or speaker of any information provided by another information content provider."

This act has a significant side effect. It illustrates a divide between the laws for printed content and those for online content. If information is printed in a magazine or newspaper, the publisher may be held liable for content distribution. If the same information is published online, the publisher may be exempt.

Another issue emerged from the CDA. If content is owned by the publisher or speaker, is it protected under the Fourth Amendment?

For the most part, third-party records of a person's use of services, such as Internet services, library usage, or retail purchases may be subpoenaed and are not protected under Fourth Amendment rights. Thus, a person's pattern of usage or purchase or loan of specific items may be requested by law enforcement agents. Search engine patterns might also be requested, along with logs of online chat sessions.

Purchases of surveillance equipment, network transmission of surveillance data, logs of phone conversations, or sharing of information over email or file-sharing services may not be private if requested by authorities. Protection for a reasonable expectation of privacy and the requirement of probable cause does not apply to customer records regarding patterns of behavior and may not apply to content even if the records are somehow "sealed" in a way that is analogous to sealing a postal letter.

Economic Surveillance

Surveillance isn't always about cameras and microphones. Financial transactions are monitored, as well. It is difficult to discuss civil liberties without including economic surveillance. Without the ability to make personal transactions and economic decisions, people have few options and very little freedom to choose.

Economic surveillance is not limited to individuals or local businesses—it is conducted globally. Some of the surveillance is to benefit people in need. Human rights watchers conduct visual and financial surveillance to locate areas where villagers live in poverty or are being evicted from their homes by stealth or by force.

Some financial surveillance is to study and try to influence global financial stability. Not everyone agrees on the best way to do this. Much of commerce is based on inequities—it opens entrepreneurial windows—but a certain level of stability is necessary or there may be crashes similar to the Great Depression, but on a global scale. If no one has money to buy goods and services, no one can sell—products will rot on the shelf.

The International Monetary Fund (IMF) monitors the economic and financial policies of almost 200 member countries both individually and in relation to their global relationships by sending out economists to meet with governments and labor organizations. As part of this economic surveillance, the IMF reports on situations that might threaten economic stability and publishes the *World Economic Outlook* and the *Global Financial Stability Report*.

Global financial surveillance is also conducted for multinational empirical purposes. Countries with shaky economies are more vulnerable to exploitation and more likely to enter into loan agreements that may not favor them long term—a situation that is often exploited by those concerned only with making money.

The transition from cash to digital payments makes it easier to track the financial status and expenditures of institutions and individuals. There's a high likelihood that cash may be phased out—many places no longer take personal checks.

With computer networks and the high number of mergers within the banking system, a perfect audit trail of every penny is theoretically possible. Banks would have significant power over a person's money in an all-digital economic system—an issue that relates to both freedom and privacy.

Some lending institutions are already requiring that loanees agree to automatic account deductions as part of the loan agreement. This takes financial control out of the hands of the individual. If a man loses his job and needs to juggle limited funds for baby food or an extra tank of gas to find a new job, it was once possible for him to set temporary priorities—to let a mortgage slide for a few weeks until new work is found. Banks are entitled to be paid, but the choice of what is the highest priority in a person's life used to lie with the individual.

With auto deductions (which are often difficult to cancel or delay), the option of short-term reallocation all but disappears. In an electronic world, there is no humanistic leeway. It is done according to numbers and big institutions dip first.

Not only does digital currency make it more difficult for a person to choose how his or her money is managed, but financial surveillance is at an all-time high. Credit reports and economic status are eagerly posted by data miners on directory sites and a person's "financial health" is often assessed, rated, and posted on data aggregation sites.

Many firms use the Web to research loan applicants and potential employees and their credit histories, so the gradual loss of control over one's finances is compounded by greater visibility. Bankruptcies used to be taken off the books after 10 years to give people a fresh start. Since they are publicly announced at the time of the procedure, and data miners now harvest public announcements and post them on the Web, bankruptcies are listed in perpetuity.

New online payment systems have significant control over money, as well. A number of online sales venues hold funds according to company policies that may be a surprise to a seller and are as yet loosely regulated.

Online financial systems also enable a company to know what competitors are doing. eBay and PayPal are associated companies—one is an online auction site, the other a payment system. Since PayPal processes payments for competing auction sites, eBay has first-hand information on its competitors' financial transactions.

Every form of surveillance has the potential to affect privacy and individual freedoms. The next chapter examines some of the ways in which legislation is used to regulate surveillance systems and safeguard people's rights.

PRIVACY RESOURCES

A number of organizations strive to protect privacy, promote free speech, and encourage the ethical use of technology, including

- American Civil Liberties Union (ACLU)
- Electronic Privacy Information Center (EPIC)
- Privacy International
- Privacy Rights Clearinghouse
- U.S. Privacy Council
- Electronic Frontier Foundation (EFF)
- Global Internet Liberty Campaign (GILC)
- Computer Professionals for Social Responsibility (CPSR)
- Commission for Liberties and Informatics
- Consumers Against Supermarket Privacy Invasion and Numbering (CASPIAN)
- Institute for Ethics & Emerging Technologies (IEET)
- Canadian Office of the Privacy Commissioner
- Cyber Rights and Cyber Liberties (U.K.-based)
- Hong Kong Privacy Commissioner

A sampling of print and online publications devoted to privacy includes

- *Privacy Journal*
- *Privacy Law Adviser, Privacy Times*
- *Privacy and American Business*
- *Privacy Laws and Business*
- *Privacy Law and Policy Reporter*
- *Computer Privacy Digest*

LEGISLATION

KEY CONCEPTS

- *the lawmaking process*
- *laws related to privacy and freedom*
- *the balance between surveillance, security, and individual rights*

Introduction

Privacy and freedom are not easily protected. People with personal or commercial agendas are often quick to impose them on others. The long history of emancipation from slavery, and various gender and ethnic inequities (many of which were formerly sanctioned by law) illustrates that freedom isn't always freely given—it must be carefully monitored and protected. There have been many recent proposals to protect freedoms that are at risk due to the proliferation of surveillance devices— some have been tabled, others enacted into law.

Federal laws and guidelines are proposed by citizen groups, individual petitioners, corporations, and Congressmembers, and debated in the Senate and House of Representatives. If an issue receives sufficient interest and votes, it is passed by lawmakers.[52] Individual states often frame local laws after federal laws and some U.S. federal surveillance laws have been the model for recent laws enacted in other countries.

The lawmaking process is similar in Canada. New laws are proposed in the form of a bill and introduced to the House of Commons, or sometimes the Senate. Following debate, there is a vote and a process of reviews by committee and members of Parliament as it moves through the House of Commons and the Senate, until there is agreement on wording, at which point it passes to the Governor General for approval.

Since society changes and governments change, in both the U.S. and Canada, bills may be adjusted with amendments, either during or after they are enacted in law.

U.S. Congressional bills may be public or private. While most surveillance issues become public bills, surveillance also affects immigration security and thus may impact certain private bills.

---

[52] John V. Sullivan, U.S. House of Representatives, "How our laws are made," *110th Congress, 1st Session, Document 110–49,* July 2007.

It is outside the scope of this text to provide a detailed list of all the laws relating to surveillance, but a brief overview of how surveillance is regulated, with a sampling of significant laws as examples, is included here.

It should be noted that the author is not an attorney. The information gathered is informational research to summarize some of the main issues. For legal interpretation or legal assistance, the reader is encouraged to consult an attorney specialized in surveillance and the Internet.

More detailed information on specific Congressional bills and the lawmaking process can be found in the Thomas archives (named after U.S. founding father Thomas Jefferson) at the Library of Congress (*loc.gov*), and at the Canadian parliamentary information site *(www.parl.gc.ca/)*.

Legislative Process

"In legislating the appropriate uses of electronic surveillance, Congress attempts to strike a balance between civil liberties—especially those embodied in the first, fourth, and fifth amendments to the U.S. Constitution—and the needs of domestic law enforcement..."

> *"Electronic surveillance and civil liberties," Federal Government Information Technology, Congress of the United States, Office of Technology Assessment, Washington, D.C., 1985*

Some laws related to surveillance are enacted so the FBI, local law enforcement agencies, the CIA, NSA, and military forces can more effectively do their jobs. Some are more general bills designed to facilitate commerce in balance with the needs of law enforcement.

Many government bureaus and regulatory bodies have been established and reorganized to govern the use of new technologies such as motorized vehicles, radio, television, and the Internet.

Internet law is still evolving. Some laws are being explored on a case-by-case basis, with decisions that have surveillance implications that apply beyond Internet usage.

GPS Vehicle Tracking

In June 2011, a bill entitled the *Geolocation Privacy and Surveillance Act (GPS Act)* was introduced to establish a legal framework for the access and use of GPS devices with regard to privacy. If passed, such a bill would require authorities to cite probable cause and obtain warrants before conducting GPS-enabled surveillance. The bill could have wide-ranging implications for companies selling GPS trackers and, indirectly, LPR systems that monitor vehicles through multiple checkpoints. While not all LPRs are based on GPS data, scanning a vehicle license plate through several checkpoints yields a tracking result similar to the "triangulation" location calculations used in GPS.

In June 2011, the *Location Privacy Protection Act* was introduced to Congress to address voluntary location-tracking of electronic communications devices. If enacted, the bill would amend the federal criminal code to prohibit nongovernmental entities or individuals that offer electronic communications services from knowingly collecting, obtaining, or disclosing geolocation information to nongovernmental recipients. The bill illustrates concern over unregulated use of the geolocation information that is now built into most popular PDAs and smartphones.

In January 2012, before the location privacy bill reached committee stage, a related issue came before the courts. In *United States v. Antoine Jones,* the U.S. Supreme Court announced a unanimous decision that a warrant must be obtained prior to attaching a device to a vehicle designed to track its movements. The decision was based on FBI and D.C. actions in which a GPS-transmitting device was installed in a suspect's vehicle in a parking lot and used to monitor its movements. The Court ruled that the tracking activity constituted a warrantless search, a violation of Fourth Amendment rights, and inadmissibility as evidence.

Cookie Tracking and "Respawning"

Many online sites have been accused of using network "cookies" (unique software identifiers) to track user movements, not only on their sites, but on visits to subsequent sites, sometimes even if they have logged off the originating site. This is primarily accomplished with browser cookies, but there are also software applications that enable cookie tracking across sites and may even be capable of rebuilding supposedly deleted cookies.

Consumer Privacy Rights

"Consumers have a right to know what data is being collected about them and have a right to decide whether they want to share that information and when..."

Senator Al Franken, Chairman of the Senate Judiciary Subcommittee on Privacy, Technology and the Law, 2010

Pressure from the White House and the FTC to discontinue practises that include intrusive tracking that might involve employment, credit, and medical insurance or healthcare resulted, in 2010, in calls for a "do not track" option in Internet browsers. The Digital Advertising Alliance (DAA) and U.S. Department of Commerce were also involved in the effort to encourage browser developers to build an easy opt-out of cookie tracking. Support from several major companies such as Mozilla and Google resulted in changes to browsers.

Nevertheless, tracking continued, sometimes by circumventing browser tracking by using other applications to follow users from site to site.

Microsoft, McDonald's, and Interclick were charged with using Adobe Flash cookies to track users for extended periods as they surfed the Internet, but the suit was dismissed in August 2011. The lawsuit alleged that tracking violated the federal *Computer Fraud and Abuse Act,* but the judge dismissed the majority of claims because injury to users could not be quantified in monetary terms. In other words, privacy damages are hard to enumerate in dollars and cents and thus not as easily protected by law as financial damages.

The judge stated in part, "Advertising on the Internet is no different from advertising on television or in newspapers." This statement can be debated on several levels.

- Internet advertising differs from television advertising in both quantity and character: (1) TV advertisers aren't secretly following you around your house logging everything you do and (2) Internet ads have a more specific and intrusive nature than television ads.

- Television and newspaper ads are not targeted at one specific person, but on broad-based demographics. Internet advertising is often based on highly individualized information that may be sensitive and assumed to be private by the user.

- Online tracking goes far beyond harvesting demographics for advertising. Anyone who has been denied access to a job, medical insurance, a social group, or an elite club knows that people and companies track personal information for many purposes other than posting ads.

The debate over cookies and privacy did not die with the federal suit. Those concerned with privacy were determined to seek redress from other companies that might be tracking their Internet usage and patterns, and more suits ensued.

In September 2011, Facebook was accused of a number of practices opposed by privacy advocates, including cookie tracking after a user logs off the site, in a suit filed in Illinois. The FTC was also asked to investigate whether privacy might be compromised by the company combining biographical information with tracking data.

A couple of weeks later, a federal suit was filed in Kansas against Facebook, charging that tracking cookies violated state consumer protection laws and federal wiretap laws and that cookies allegedly continue to track users after they leave the site.

Soon after, a federal suit was filed in the Northern District of California including allegations that the company violated the *California Internet Privacy Requirements Act* and *California Unfair Competition Law.*

Facebook asserted that user cookies are removed when they log off, a statement that was contradicted on September 26, 2011. Since then, Facebook has apparently made changes to a process that broadcasts information about a user to plugin partner sites. A Facebook spokesperson responded to the Kansas suit that privacy allegations were "without merit."

Then, in March 2012, two class action lawsuits were filed against Google, in New York and California, alleging that the company inserted code into Google ads that would place tracking cookies on Mac-related products such as OS X computers and iPhones. The search engine giant allegedly violated the *Federal Wiretap Act,* the *Computer Fraud and Abuse Act,* and the *Stored Electronics Communications Act.* The lawsuits came about as a result of a Stanford researcher's observations that browser privacy settings were being circumvented to data mine information about Net searches without the consumer's permission.

Spyware for smartphones that includes call tracking, message tracing, GPS location tracking, and phone book access from a remote location through the Internet is openly sold on the Net. Since the sellers of such software claim it is for users to remotely track their own phones if they are stolen, or for parents to track their children, they may be exempt from liability, but ads for these kinds of products almost always imply that the software is for covert surveillance of a broader nature.

Email, IP, and Internet Activity Access

For the most part, ISPs are permitted to access the content and processing information for customer email messages. The terms of service may expressly stipulate that email privacy is not granted. Lack of privacy protection applies to both the sending ISP (the customer's provider) and the receiving ISP (which may be the same company or any other that provides Internet services).

Thus, law enforcement agents are at liberty to subpoena email messages, provided they follow due process.

There are limits to what attorneys and law enforcement agents may request, however. A decision in 2003 ruled that an attorney could not seek "all copies of e-mails sent or received by anyone" and cautioned that "those who invoke [subpoena authority] have a grave responsibility to ensure it is not abused..." Thus, while targeted requests for emails may be granted, wide-net requests are discouraged.

It should be noted that cases of mistaken identity can occur. Many Internet users are not technically literate and do not understand how to secure their systems. Their firewalls and routers may be open for other people to access the Internet through their connection. Wireless router accounts are sometimes "hijacked" by neighbors in nearby houses or apartments for general Web surfing or illegal activities.

If the network-connection hijacker engages in Internet-related crime, a subpoena issued to access emails, Web-surfing logs, and the IP number assigned to the client by the ISP may point to the ISP's authorized customer rather than the hijacker, and the person being charged may have difficulty proving he or she wasn't involved. A nontechnical user may not even understand a connection can be hijacked.

Another area of Web activity that may provoke subpoenas is incidents of libel—situations where defamatory remarks have been made about a person or company. This is not uncommon on forums, chat sites, and sites that post product reviews.

Libel is prohibited by law, but there are situations where the information posted is factual and thus may not be considered libelous, and the situation can only be sorted out by legal professionals arguing the case in court.

Many companies conduct regular Internet surveillance to see what people are saying about their products and their executives. There may be situations where a company feels that negative remarks cause economic harm but the line between free speech and negative remarks may not be clear without legal investigation. Thus, subpoenas for the identity of the people making the remarks may be denied.

Given the open climate and exchange of opinions and information on the Internet, there are many situations where trade secrets, transcripts or videos of internal meetings and company policies have been leaked to friends or colleagues through email, or openly posted on the Web. In situations like this, the court tends to side with the company expressing the grievance and permits issuance of a subpoena to determine the identity of the person obtaining or leaking the information, since the disclosure of trade secrets is not permitted and could do economic harm.

Monitoring of Mobile Services

Tracking is not limited to Internet surfing. It is possible to monitor user actions, preferences, and movements on mobile communications devices.

One diagnostic tool, called Carrier IQ, which is widely installed on mobile phones, has come under particular fire, especially after a class action suit was filed in December 2011 against a number of companies using the software on their systems.

An Illinois suit claims that Carrier IQ and HTC intercepted and recorded private data. A California suit claims that Carrier IQ, Inc., T-Mobile USA, Inc., et al., violated *California's Invasion of Privacy Act* and *Unfair Competition Law* by surreptitiously intercepting communications of mobile phone users without informing them of this activity. The suit also alleges keystroke monitoring, which could reveal the content of text messages and private passwords.

A federal suit filed in Delaware claims that use of the software violates the *Federal Wiretap Act*, the *Stored Electronic Communications Act*, and the *Computer Fraud and Abuse Act.*

Carrier IQ responded by saying the software is intended to help providers improve wireless services, not violate privacy, and that a video purporting to show keystroke monitoring was incorrectly interpreted. It also briefly published a document explaining what the software can and can't do, but this is no longer available at its original Internet location. Other analysts have explained that ISPs have wide latitude to use monitoring tools to manage their services.

Whether the companies named in the lawsuits will be found guilty of privacy violations is yet to be seen, but information and data protection authorities in the

United Kingdom and the European Union who may be affected by the results are watching the debate unfold.

Iris and Retina Scans

In 2005, a bill called the *Iris Scan Security Act* was introduced into Congress. On the surface this might sound like a privacy bill, but it was a surveillance bill introduced to authorize grants to law enforcement agencies so they could add iris prints from convicted criminals to state databases linked to the National Instant Criminal Background Check System (NICS).

It would also require federal firearms dealers to obtain iris prints from anyone seeking to purchase firearms. The bill did not pass through committee, but since it was re-introduction of a bill from 2003, it may resurface again now that iris scans are easy to capture from a distance of up to 30 feet. A rewording of the bill could, for example, require gun-shop owners to install iris scanners that forward information to NICS. Cameras are small and iris scans from a distance are instantaneous. A person could be scanned and never know it, so there would likely be strong opposition, but such a proposal is not impossible.

Radio Chip Monitoring

Employee and Military Tracking

As mentioned in the chapter on radio-wave surveillance, an RFID chip is a form of radio transceiver. It can be made small enough to be implanted in a human being. When the Food and Drug Administration approved the chips as safe for implantation into humans, employers immediately considered implants as a condition of job employment, arguing that it was no different from employees carrying ID cards and more convenient. One company in Cincinnati chipped the forearms of two employees for access to a secured area.

Right away, there were concerns about privacy and analogies to the humiliations suffered by Holocaust victims who were branded with tattoos to inventory them like cattle or articles on a store shelf.

Implanted radio chips are different from ID cards because you can't take them off—they impinge on human dignity by eroding a person's sense of control and self-determination.

The outcry against RFID employee tracking was not universal, however. There are people who voluntarily get chips to use as security devices for access to machinery, computers, or entrances and exits. Wave your wrist and you're in. There are also those who argue that chipping a child or an elderly person with mental instabilities could save their lives if they wander. By fall 2007, only three states had outlawed the practice: North Dakota, Wisconsin, and California.

There are arguments both for and against RFID implantation. Some say there are better ways to raise and protect a child or protect an elderly person than subjecting him or her to electronic tracking. In California, for example, it is illegal for a parent or guardian to force a child or elderly person to be chipped.

Voluntary chipping is a different situation. Few are opposed to a person chipping him or herself in much the same way that people choose artistic tattoos or body piercing. But there are good arguments against an employer requiring an employee to wear a tracking device 24/7 to become eligible for a job. Not only can this damage a person's dignity, but RFID broadcasts could be recognized and used by unauthorized persons or firms to spot individuals as they pass by sensors outside the workplace. There is also a concern that employers or other entities may offer enticements for people to get chipped, like discounts, bonuses, or other financial incentives that could put pressure on those with limited means to get chipped whether they want to or not.

Another problem with chipping people is that selective chipping stratifies society and may marginalize some. If blue-collar workers were chipped and white-collar workers weren't, or if regular staff were chipped and management wasn't, or women were and men weren't, then electronic surveillance becomes a way to magnify inequities.

If chipping of humans starts in the workplace, it sets a precedence for other chipping requirements. For example, some people have interpreted the new HR 3200 health care bill as a legal framework for a mandate for all citizens to be implanted with a "class II device" (e.g., RFID) to determine eligibility and provide other information for access to medical care.

In Canada, the Privacy Commissioner published a consultation paper of "good practice" rules for organizations seeking to use RFID systems[53] based on concerns that workplace surveillance could "very much undermine the dignity and autonomy of employees." The paper directs that RFID technologies "must be respectful of privacy and in conformity with Fair Information Principles."

In the European Union, the EC is developing guidelines for the use of RFID cards and chips that may be included in consumer goods. It is recommended that RFID operators make impact studies before installing RFID systems, and that there be provisions for deactivating tags that may be associated with personal information.

In Mexico, where the political climate is unstable, members of the government have been implanted with chips for access to high-security areas, including a presidential candidate who was the victim of a violent kidnapping. Retailers of RFID devices find fertile ground in regions where crime rates are high and people are frightened, and use incidents such as this as selling points.

Do chips protect us from kidnapping as well as manufacturers claim?

- Being implanted with a chip doesn't prevent a person from being kidnapped, nor does it prevent kidnappers from cutting out the tag and using it as a ransom tool or decoy to lure law enforcement away from, instead of toward, a victim.

---

[53] Office of the Privacy Commissioner of Canada, *Radio Frequency Identification (RFID) in the Workplace: Recommendations for Good Practices,* March 2008.

- There's no way to hide the implant location from kidnappers if it is transmitting radio signals and chips small enough to be implanted may not be strong enough to send a signal more than a few yards—which significantly limits them as tracking beacons.

Rice-sized implants can be effective as access control devices for short-range devices, but are inadequate as emergency beacons—currently only palm-sized devices can send strong enough signals to broadcast the location of a victim.

RFID Privacy and Consumer Protection

Concerns about privacy soon followed implementations of RFID systems in the mid-2000s. There were also concerns about security. It was suggested that thieves equipped with RFID readers could scan the contents of a home or car to "case" a potential target for robbery. In the days when RFIDs were uncommon and only transmitted about 3 feet this was unlikely, but RFID inventory chips are selling by the millions and transmission ranges have increased as much as sevenfold.

With RFID chips embedded in shoes, coats, and backpacks, it is also conceivable that police officers with RFID readers could scan items worn by people near a crime scene and match that with data from the company that sold the items via a legal subpoena. It hasn't happened yet, but the possibility exists.

In order to curb malicious or inappropriate use of RFID, there have been communications with the FTC by individuals and organizations like EPIC, regarding consumer protections, including the following recommendations:

- Those checking RFID tags should signal consumers with an alert when information is drawn from a tag.

- RFID tags should be removable and anonymous.

- Tags should not transmit personal information (as might happen with smart RFID systems).

- Multiple scans of consumer RFIDs at different locations should not be logged or the information could be used for tracking purposes.

To date, most of these recommendations have not found their way into U.S. law. Some consumer protection has been initiated, however.

In the mid-2000s, Washington State announced that intentional scanning of RFID tags for fraudulent purposes without the person's knowledge or consent constituted a Class C felony. This doesn't offer protection for RFID used for commercial purposes, but even this concession to consumer protection took considerable effort by lawmakers, against objections by businesses that prefer RFID to remain unregulated.

California, Rhode Island, and Nevada followed up with similar laws between 2008 and 2010, but as of 2012, most states do not explicitly prohibit improper use of RFID chips.

By 2006, the EC was establishing a policy framework for laws related to RFID.

EMPLOYER SURVEILLANCE

Employers use a wide variety of surveillance methods to monitor job applicants, employees, computer use, fleets of vehicles, equipment, and every aspect of their businesses that may be subject to abuse, theft, or internal spying. They also monitor the safety of employees and equipment.

Job Applicants

Most employers conduct hiring surveillance. Some of them keep a watch on competitors' employees to see if they can entice them away. It is a well-known strategy among businesses of all sizes, and even happens between universities.

Employers also check the backgrounds of contractors, as mentioned in the computer chapter, and will sometimes check people slated for promotions or relocation to another branch. Staff members will search the Web or pay a fee to companies with large data repositories that specialize in work-related background checks. Sometimes private detectives are hired. There are legal restrictions on what may be searched and used in hiring decisions, in order to prevent workplace discrimination, but there are no laws enabling applicants to see and correct background-check information.

Access to Private Information

"In an age where more and more of our personal information—and our private social interactions – are online, it is vital that all individuals be allowed to determine for themselves what personal information they want to make public and protect personal information from their would-be employers. This is especially important during the job-seeking process, when all the power is on one side of the fence."

Chuck Schumer, U.S. Senator from New York

Background checks aren't the result of mere curiosity—an employer could be sued for negligence if an employee were to harm others and was found to have a documented violent history that wasn't noted by the hiring staff. Employers do checks partly to protect themselves from liability.

Employers are also seeking to verify information provided in résumés, since employees sometimes inflate their qualifications or claim credentials they don't have. Some people have purchased fraudulent "degrees" from diploma mills—

companies that provide a short simple course of instruction and then "award" a diploma for a fee. It is a fast-track way to purchase a document that appears reputable on the surface but is deplored by employers in both business and academics.

Even executives are subject to greater scrutiny than in the past, due to insecurity provoked by headline news stories such as Enron, and because of due diligence requirements of public companies.

Politicians are subjected to checks as part of the process of being installed into office. Independent agencies usually seek out additional information about people slated for positions of responsibility and leadership.

Fair Credit Reporting Act Restrictions

There are legal restrictions on what may be searched for a background check, as outlined in the *Fair Credit Reporting Act* (FCRA), which puts restraints and notification requirements on certain kinds of records like bankruptcies, tax liens, and civil suits. Background checks and credit checks are both considered "consumer reports" by the FCRA.

FCRA restrictions do not apply to jobs with annual salaries over $75,000 a year. At one point, criminal records were not reportable after a certain number of years, but this prohibition has been lifted at the federal level. Some states still restrict the reporting of criminal offenses after seven years.

Some documents require hiree permission to obtain, including medical records, education records, and military records.

There are certain employment positions that require background checks by law, such as jobs that involve the handling of pharmaceuticals, the care of children, the transfer of large amounts of money from one location to another, and cleaning staff who need to enter secured facilities outside of working hours.

Government Employee Background Checks

Certain government jobs require background checks, and the background check process has been broadened under a 2004 presidential order for more uniform identification standards for federal employees, as documented in a Supreme Court Syllabus on a decision rendered in 2011.

> *"Respondents [at NASA/JPL] were not subject to Government background checks at the time they were hired, but that changed when the President ordered the adoption of uniform identification standards for both federal civil servants and contractor employees. The Department of Commerce mandated that contract employees with long-term access to federal facilities complete a standard background check....*
>
> *The employee must also sign a release authorizing the Government to obtain personal information from schools, employers, and others during its investigation."*
>
> "National Aeronautics and Space Administration et al. v. Nelson et al.,"
> *Syllabus, Supreme Court of the United States,* decided January 19, 2011

Note that the respondents in this Supreme Court ruling were not employees, but contractors, some of whom had worked at the facility for decades. Independent contractors are considered to be self-employed for most business-license and tax purposes and are legally distinguished from salaried employees in a number of ways.

> *"At the time when respondents were hired, background checks were standard only for federal civil servants.... In some instances, individual contracts required background checks for the employees of federal contractors, but no blanket policy was in place."*
>
> Ibid.

Justices Thomas and Scalia concurred with the judgment and Justice Thomas had this to say:

> *"I agree with JUSTICE SCALIA that the Constitution does not protect a right to informational privacy.... No provision in the Constitution mentions such a right."*
>
> Ibid., Justice Thomas concurring

Government background checks now extend to a larger number of employees, and contractors—even those with long work histories at the same facility.

Certain federal employees are also required to take drug tests.

Credit Checks

Employers sometimes run credit checks within FCRA guidelines, in situations where it can be justified that it is an important aspect of the position to be filled. A survey by the Society for Human Resource Management, in 2010, found that 40% of employers consulted credit reports on potential hirees.

Sometimes employers ask job applicants for direct access to private spaces.

In March 2012, Associated Press reported that job-related agencies were asking job seekers for their personal information on social media sites. Apparently some have gone as far as asking job applicants to supply passwords that unlock private areas of their accounts that may include information that is protected by federal employment anti-discrimination laws.

Alarmed that these requests may be a violation of federal law, two U.S. Senators asked the Attorney General to investigate and alerted the Department of Justice and the Equal Employment Opportunity Commission. Administrators of social media sites have also spoken out against this practice, although for a different reason. Most social networking sites have terms of agreements in which users promise not to share passwords.

Being asked for social network usernames or passwords puts a job applicant in an awkward position. People hunting for jobs are afraid if they say no to employer requests they might be denied the opportunity to be hired even if the employer request impinges on privacy.

Buying Background-Check Services

Companies don't always do their own background checks, partly because of liabilities involved in searching or finding information that could influence hiring decisions through discrimination.

There are many companies that offer employee background checks. One company alone boasts more than 21,000 background checks a day, and there are about 1000 companies offering similar services, which gives a sense of how common it is.

Employers contract information for a number of reasons. Companies specializing in background checks have a significant investment in data storehouses that the employer doesn't have. Reputable companies have expertise in filtering important information and ensuring that the correct person is being searched.

Obtaining the information through another party may shield the employer from charges of discrimination if the contractor provides only the kinds of information permitted by law.

How reputable these companies are is not always known. Nor can it be determined how carefully they fact-check their information or update their databases. If an individual is arrested and the charges dropped because it was found to be a mistake, which information is in the database—the arrest or the dropped charges? And how carefully do they conform to FCRA requirements?

Definition of Personal Information

What does the legislative system consider to be personal information? This may vary from country to country and law to law and also may vary with the age of the people involved.

In the *Children's Online Privacy Protection Act (COPPA) of 1998*, personal information is defined as follows:

"(8) PERSONAL INFORMATION.—The term "personal information" means individually identifiable information about an individual collected online, including—

(A) a first and last name;

(B) a home or other physical address including street name and name of a city or town;

(C) an e-mail address;

(D) a telephone number;

(E) a Social Security number;

(F) any other identifier that the Commission determines permits the physical or online contacting of a specific individual; or

(G) information concerning the child or the parents of that child that the website collects online from the child and combines with an identifier described in this paragraph."

Thus, in COPPA, the definition of personal information is fairly specific—mainly focused on contact information for the child. In contrast, the South African *Electronic Communications and Transactions Act of 2002*, personal information has a broader definition as follows:

"personal information" means information about an identifiable individual, including, but not limited to—

(a) information relating to the race, gender, sex, pregnancy, marital status, national, ethnic or social origin, colour, sexual orientation, age, physical or mental health, well-being, disability, religion, conscience, belief, culture, language and birth of the individual;

(b) information relating to the education or the medical, criminal or employment history of the individual or information relating to financial transactions in which the individual has been involved;

(c) any identifying number, symbol, or other particular assigned to the individual;

(d) the address, fingerprints or blood type of the individual;

(e) the personal opinions, views or preferences of the individual, except where they are about another individual or about a proposal for a grant, an award or a prize to be made to another individual;

(f) correspondence sent by the individual that is implicitly or explicitly of a private or confidential nature or further correspondence that would reveal the contents of the original correspondence;

(g) the views or opinions of another individual about the individual;

(h) the views or opinions of another individual about a proposal for a grant, an award or a prize to be made to the individual, but excluding the name of the other individual where it appears with the views or opinions of the other individual; and

(i) the name of the individual where it appears with other personal information relating to the individual or where the disclosure of the name itself would reveal information about the individual,

but excludes information about an individual who has been dead for more than 20 years..."

In Canada, in the *Personal Information Protection and Electronic Documents Act (PIPEDA) of 2000*, with amendments proposed in 2010, personal information is defined as

"'personal information' means information about an identifiable individual, but does not include the name, title or business address or telephone number of an employee of an organization."

In interpreting PIPEDA, personal information includes a person's behavior, and information that may be contained in the NETBIOS of a personal computer, but does not include business information or the tangible results of a person's work. Items owned by another entity but used by an individual, such as a credit or debit card would be considered personal information. Documents that don't specifically name or otherwise identify a person are not considered personal information. An alternate identifier that can be related to the person's name, such as a code number, would be personal information. Camera coverage and broadcast of public events is not considered personal information because the individuals are not identifiable by name and license plates generally cannot be read. However, the issue of license plates may change in the future when higher-resolution cameras are the norm.[54]

One of the more significant aspects of PIPEDA is the right of access of a person to his or her information.

Legislation Regarding the Use and Storage of DNA

DNA Data and Sample Banks

In most cases, establishment of DNA banks for law enforcement was initially limited to violent offenders, in some cases, a specific subset of offenders—those convicted of violent sex offenses against children. Gradually, other categories have been added both in the U.K. and the U.S. By 2009, almost 6 million criminals and suspects were included in DNA data banks in the U.S.[55] Providing a DNA sample has become mandatory in the armed services.

In Australia, the ethics of human DNA data banking has been taken seriously since before DNA became well-known, and is documented in the *HGSA Guidelines for Human DNA Banking* (July 1990).[56] Eight years later, the *Genetic Privacy and Non-Discrimination Bill* was proposed to the Australian senate.

In the U.K., more than 5% of the population is now included in the national data bank. It was also found that ethnic minorities in the database who had not committed crimes were over-represented in the database almost three-to-one compared to their incidence in the general population. Some have proposed putting everyone in the DNA bank and some have even suggested taking samples from visitors. The main reason given for not including everyone in the DNA bank was not a concern about privacy, but rather one of cost and bureaucratic unwieldiness.

Prior to 2001, in the U.K., DNA samples were required to be destroyed for those who were released without charge or acquitted. Then a ruling permitted police in

---

[54] Commentary and links to relevant abstracts of some legal decisions (www.pipeda.info/).

[55] Karen J. Maschke, "DNA and law enforcement," *From Birth to Death and Bench to Clinic: The Hastings Center Bioethics Briefing Book for Journalists, Poicymakers, and Campaigns,* The Hastings Center, 2008, pp. 45–50.

[56] Deborah Crosbie has documented some of the similarities and differences among DNA banking policies in several countries in a report to the Human Genetics Commission of Australia, *Protection of Genetic Information: An International Comparison,* September 2000.

England, Wales, and Northern Ireland to retain the profiles of those charged with recordable offenses. A few years later, there was considerable public outcry over the fact that DNA samples of more than 51,000 children who had committed no crimes had been included in a police database, blurring the line between criminals and law-abiding citizens. In May 2006, Members of Parliament called for the immediate removal of the children from the archive and in 2008 there was a court ruling by the European Court of Human Rights for innocent citizens to have their profiles removed from the DNA database.

Despite the outcry, the database increased rather than decreasing. By 2009, about 1 million children were included—a 20-fold increase in four years. It is not known how many were involved in crimes, but it is believed by the director of GeneWatch that about half have no criminal record. Some of the DNA from children was taken without parental consent.

Originally, the archive was to include only convicted criminals. Those powers were expanded in 2005 to include suspects, as well, who had not yet been proven guilty.

There are also cases that border on the absurd. Some boys kicked a ball into the neighboring yard of a Derbyshire grandmother. The woman was accused by the boys' father of stealing the ball, which she claims she never saw. Her house was searched and she was taken into custody in 2006 and swabbed. Even though the case was dropped, her DNA was retained.[57]

The response of the Home Office about the archive administrated by the National Policing Improvement Agency (NPIA) was that "Innocent people have nothing to fear from DNA" despite the fact that inclusion in a DNA bank can show up in background checks and prevent a person from getting a job or, as has happened, a person could lose a job either because of suspicion of that person, even if innocent, or because the company is concerned about protecting its image.

Some of the people in the U.K. data bank were witnesses to crimes, not themselves charged or convicted. This too raises problems, as witnesses may not come forward and report what they see if they know they are going to be included in an incriminating-looking database with criminals.

Innocent people may indeed have something to fear, if their personal information is mingled with that of criminals. Throughout the ages, people have lost social status, job opportunities, and immigration opportunities by association, even if those associations were not their fault, so the Home Office statement that there's "nothing to fear," while perhaps true in a technical sense, is not true in a social sense.

It can be very traumatic to victims of crime to be processed in the same way as criminals. They think the police are there to help them and feel abused, confused, and betrayed if they are treated in the same way as the perpetrators. While DNA can potentially help solve crimes, aggressive collection policies that include innocent

---

[57]Ian Drury, "One million innocent people could have their profiles wiped from Britain's 'Orwellian' DNA database after court ruling," *MailOnline*, December 4 2008.

people tend to create a loss of trust and respect in law enforcement agencies that may cancel some of the crime-fighting benefits.

Controversy over DNA banks is not limited to the western world. In Malaysia, there was considerable debate over the introduction of the *DNA Identification Bill of 2008*. Most opponents were not against a law enforcement DNA bank in principle, but were concerned about the form of the proposed legislation.

As a society, we have to ask which is worse—a few thousand crimes going unsolved, or the entire populace living under constant suspicion and thousands, perhaps tens of thousands, of innocent people experiencing humiliation and discrimination due to personal information being held indefinitely in criminal databases at taxpayer expense.

DNA and Nondiscrimination

"It is difficult to imagine information more personal or more private than a person's genetic makeup. It should not be shared by insurers or employers, or be used in making decisions about health coverage or a job. It should only be used by patients and their doctors to make the best diagnostic and treatment decisions they can."

Senator Edward Kennedy, remarks to Congress, January 2007

After a long series of proposed bills by various lawmakers, dating back to 1995, Senator Edward Kennedy introduced a proposal for the *Genetic Information Nondiscrimination Act* (GINA), in 2007, to safeguard individuals from discrimination based on their genetic makeup. He, along with President Bush, recognized the great potential of genetic research for improving medical care but also wanted to protect the public from unfair use of this potent information in matters of employment and insurance eligibility.

Left: Senator Edward Kennedy worked tirelessly to introduce legislation to protect citizens from discrimination based on their genetic makeup. Right: President George W. Bush signed GINA into law in May 2008 so that employers could not discriminate based on genetic predispositions and health insurers could not deny, change, or cancel medical insurance due to a person's DNA. [White House photos by Lawrence Jackson and Eric Draper, public domain.]

In 2008, GINA was enacted into law. It is important to note that while GINA protected the use of genetic information against certain forms of misuse or abuse, it did not prohibit its collection or storage.

Onus of Responsibility

Social attitudes about responsibility and blame go through cycles. In the 1950s, in many regions, people were expected to respect the rights and property of others and cars and houses remained unlocked and property-related crime rates were no higher, per capita, than they are now.

There are still parts of the world where the onus is on the visitor to respect the host and property is minimally protected. In some regions, this attitude has changed or never existed and there are those who want to shift all responsibility to potential victims.

In 2003, there was video voyeurism lawsuit in which an officer had parked by an apartment building and used a video camera to zoom in on a woman's crotch and chest as she sat in her second-floor apartment. The attorney for the defendant stated, in part

> *"If you don't want people looking through an open window, you have a responsibility to put blinds or curtains on the windows."*[58]

Thus, the attorney implied that the onus is on the potential victim to do everything possible to prevent being victimized.

Is there a problem with this argument? In the most literal sense, does it mean we can't leave our houses or carry money or credit cards in case we might be mugged? Should we remove all bus shelters because they might be vandalized, or keep children locked indoors 24/7 so they can't be bullied, molested, or kidnapped? Should we ban all driving to prevent traffic accidents? Shouldn't there be some responsibility on the part of others to respect privacy and property whether curtains are drawn or not?

What does Mississippi law say about situations like this?

> *Any person who with lewd, licentious or indecent intent secretly photographs, films, videotapes, records or otherwise reproduces the image of another person without the permission of such person when such a person is located in a place where a person would intend to be in a state of undress and have a reasonable expectation of privacy, including, but not limited to, private dwellings or any facility, public or private, used as a restroom, bathroom, shower room, tanning booth, locker room, fitting room, dressing room or bedroom shall be guilty of a felony and upon conviction shall be punished by a fine of Five Thousand Dollars ($ 5,000.00) or by imprisonment of not more than five (5) years in the custody of the Department [*8] of Corrections, or both.*

Mississippi Code Ann. § 97-29-63

---

[58]*Eddie Gilmer vs. State of Mississippi*, counsel for appellant J. Epps.

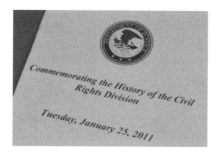

In January 2012, the Department of Justice hosted an event commemorating the history and legacy of the Civil Rights Division, which was established as a result of the Civil Rights Act of 1957 which sought to protect all Americans from discrimination on the basis of religion, familiar status, national origin, gender, disability, or ethnic background. Now discrimination due to information held in surveillance databases, such as images, personal records, DNA banks, along with information about a person's associations and long-past history may also serve as data for discrimination. [Department of Justice News Photo by Craig Crawford, Jan. 2012.]

The fact that it is easier to violate privacy with surveillance devices shouldn't automatically confer greater rights to those who do it, compared to those who are entitled to a reasonable expectation of privacy by law.

Some people feel that privacy has already been diminished to the point that it is impossible to reclaim it. Others feel that a concerted effort can create change and that laws can be adapted to accommodate new technology. Even people within government feel that there is still choice in the matter.

> *"You can have the greatest technology in the world. You can stand behind previous Supreme Court cases on the lawfulness of using it, but if the public— who is really running our government—if they turn against a technology, it is going to be very difficult to implement it."*

> Al Frazier, as reported by Dan Gunderson, *Minnesota Public Radio,* January 2012

U.S. Congressional Acts Related to Privacy and Surveillance

There isn't space in this text to list all the laws related to privacy and surveillance, but a sampling of significant laws is included.

Acts Related to Communication, Law Enforcement, and National Security

- The *Communications Act of 1934* was enacted to organize and regulate interstate and foreign communications for national defense and to promote competitive communications technologies and services. It established the Federal Communications Commission (FCC) to regulate the allocation and use of radio waves.

- The *National Security Act of 1947* put various military agencies under the secretary of defense and created the National Security Council (NSC) and Central Intelligence Agency (CIA). The CIA was not a law enforcement agency, but rather an intelligence-gathering agency with some leeway to conduct covert operations.

- The *Omnibus Crime Control and Safe Streets Act of 1968* regulated how law enforcement agencies could obtain authorization to conduct electronic surveillance. It required telecommunications carriers to provide the "technical assistance necessary to accomplish the interception," a move that was hotly contested by communications carriers, but nevertheless passed into law. It further created the National Institute of Justice (NIJ), a research and technological development agency of the Department of Justice (DoJ).

- The *Foreign Intelligence Surveillance Act (FISA) of 1978* established procedures for authorized government officials to acquire judicial orders to authorize electronic surveillance or physical search of foreign entities. It provided legislative authority for surveillance of foreign powers both within and without the country. It further established the Foreign Intelligence Surveillance Court (FISC) to review and approve surveillance that could be used to monitor U.S. persons. FISA was amended in 1994 to provide limited authority for physical searches. This act was expanded in 1994 and, in 1996, was brought into the digital age with the *Electronic Freedom of Information Act Amendments of 1996* which established requirements for making FOIA-relevant information electronically available.

- *Intelligence Oversight Act of 1980*. Two committees were established in response to allegations of wrongdoing by U.S. intelligence agencies: the Senate Select Committee on Intelligence (SSCI), in 1976, and the House Permanent Select Committee on Intelligence (HPSCI), in 1977. These, along with the Armed Services and Foreign Relations and Foreign Affairs Committees, would oversee and authorize activities of intelligence agencies. The Hughes-Ryan Amendment required that covert action notifications be given only to the two committees (other committees no longer had to be notified). A number of executive orders followed to shape and direct surveillance of foreign powers.

- The *Intelligence Identities Protection Act* of 1982 imposed penalties on those who wrongfully divulged identities of covert intelligence personnel.

- The *Central Intelligence Agency Information Act of 1984* exempted the CIA from search-and-review requirements of the FOIA concerning sensitive files.

- The *Computer Fraud and Abuse Act of 1986* amends Title 18 to provide additional penalties for fraud-related activities in connection with computers and access devices.

- The *Communications Assistance for Law Enforcement Act of 1994 (CALEA)* is sometimes called the Digital Telephony law. It requires telecommunications providers to ensure that law enforcement agents are able to execute court-authorized wiretaps. In many cases, this requires physical changes or upgrades to equipment or, more recently, the option of sending the entire data stream to a third party. It made it possible for law enforcement agencies to remotely implement broad-based taps.

- The *Digital Communications and Privacy Improvement Act of 1994* facilitated court-authorized electronic surveillance.

- The *Comprehensive Counterterrorism Prevention Act of 1996* was a strategy to improve security in federal buildings and aircraft cargo holds. It also authorized relocation of U.S. forces in foreign stations that might be at risk for terrorist attacks.

- The *Communications Decency Act of 1996* was a provision of the *Telecommunications Reform Act* that provoked controversy as to definitions of "lewd" or other materials that were being promoted as criminal because they might be objectionable to the general public, yet were considered acceptable in the more open climate of the Internet. The act was contested and, in 1997, declared unconstitutional and in violation of individual free speech rights.

- The *Telecommunications Act of 1996*, the first substantive overhaul of the *Communications Act of 1934,* was enacted to promote open access to the telecommunications business and to permit competition. Responsibility related to phone and broadcast services shifted from state courts to the FCC, while much of the administrative workload remained with state authorities. It made it possible for

Regional Bell Operating Companies to provide interstate long-distance services and for telephone companies to provide cable television. Cable companies could now provide local telephone services, as well.

- The *Children's Online Privacy Protection Act (COPPA) of 1998* requires providers of Internet services directed to children to obtain verifiable parental consent for the collection, use, or disclosure of personal information from children and to disclose what information is collected from children on the website and how it is used. It also prohibits using enticements, such as games, to encourage a child to disclose more personal information than is reasonably necessary to participate in the activity and requires that the confidentiality, security, and integrity of the personal information be maintained.

- The *Department of National Homeland Security Act of 2001* established the Department of National Homeland Security as a hasty response to the September 2011 attack on the World Trade Centers. Given the speed at which it was drafted and enacted, it is not surprising it went through a number of amendments to become the *Homeland Security Act (HSA) of 2002.*

   As a result of the HSA, almost two dozen federal agencies were brought under one jurisdiction, including agencies such as the Coast Guard, the Immigration and Naturalization Service, the Customs Service, and the Secret Service. One of the justifications for the significant restructuring was to facilitate communication between agencies, which has obvious benefits but, by putting them under the same national security administrative umbrella, and centralizing information gathered by these agencies, a certain amount of specialist autonomy was lost and controversy over "top-heavy administrative power" ensued. The Department of Homeland Security could be characterized as the civilian analog of the Department of Defense. HSA met with public controversy right from the start, not only because of its hasty inception, but because some feared that over-centralized government might threaten democratic freedoms.

   Significant outcomes of the Act directly affecting surveillance are the consolidation of intelligence-gathering and investigative units into Homeland Security Investigations and merging a number of border-related agencies as U.S. Customs and Border Protection (CBP).

- The *USA Patriot Act (Patriot Act) of 2001* was a direct response to the September 2011 attack, enacted less than seven weeks after the collapse of the World Trade Center. It gave broader powers to intelligence-gathering agencies, and those tasked with law enforcement and immigration matters, to facilitate identification of terrorist risks and individuals. Foreign and domestic terrorism were both brought together in this bill. Some portions of the Act became permanent law while others would require periodic renewal.

- *Intelligence Reform and Terrorism Prevention Act (IRTPA) of 2004.* As a continued response to the September 11th attack, IRTPA rearranged national intelligence agencies and established the position of Director of National Intelligence (DNI) as well as the Counterterrorism Center. There was public concern that various anti-terrorism bills could take away civil liberties so the bill also established the Privacy and Civil Liberties Oversight Board as an advisory body to senior officials, including the President.

- The *Video Voyeurism Prevent Act of 2004* amends the federal criminal code to prohibit the deliberate videotaping, filming, photographing, or recording by any means, or broadcasting the private areas of an individual without the person's consent in situations where the person has a reasonable expectation of privacy. There are exemptions for law inforcement- and intelligence-related activities.

- *USA Patriot Act (Patriot Act) extension of 2011.* A four-year extension of key provisions of the 2001 Patriot Act bill was remotely signed (through an autograph pen) by the President in May 2011. At the urging of intelligence community directors, the extension was passed to facilitate intelligence-gathering related to search, surveillance, and roving wiretaps of specific individuals.

Acts Related to Privacy and Protection of Commercial Interests

- The *Wiretap Act* was initially passed as *Title III* of the *Omnibus Crime Control and Safe Streets Act of 1968* act to protect the privacy of spoken and wired communications. It prohibited the deliberate interception or disclosure of communications other than through statutory permissions (e.g., court-approved warrants). A 1986 amendment broadened privacy protection to include electronic communications.

- *U.S. Privacy Act of 1974.* With this act, the federal government acknowledged gathering personal information on individuals and established certain limitations on information practices and the maintenance of information associated with a person's First Amendment rights.

- *Electronic Communications Privacy Act (ECPA) of 1986.* This amends the *Omnibus Crime Control and Safe Streets Act of 1968*.

ECPA extended existing restrictions on unauthorized interception of traditional communications media to cover electronic communications. It does not extend prohibitions in cases where one of the parties consents to the interception, and does not extend the right into work-related communications (there are employer exceptions that permit monitoring). There are also exemptions for communications providers to permit system administrators to manage and troubleshoot systems. A number of state laws are patterned after the ECPA.

- The *Economic Espionage Act of 1996* is a vehicle for prosecuting those committing economic espionage, that is, the stealing of trade secrets and other business information. Similarly, the *National Information Infrastructure Protection Act of 1996*, an amendment to the *Computer Fraud and Abuse Act* and response to computer hacking, protects proprietary economic information.

- The *Electronic Communications and Transactions Act of 2002* facilitates, regulates, and promotes access to electronic communications and transactions, and seeks to prevent abuse of information systems.

- The *Workplace Privacy Act of 2011* was established in response to a significant increase in workplace monitoring and stipulates that workers should be given written notice of workplace surveillance and be made aware of the presence and location of visual sensors such as cameras. It also provided protection for employee access to communications such as email messages.

- National Data Privacy Day, a Senate Resolution to designate January 28, 2012 as National Data Privacy Day to encourage educators and others to increase awareness of data privacy. The resolution was passed two days after the designated day had passed.

A large number of privacy bills was introduced in 2011 and 2012, some of which have been referred to committee. They include topics such as geolocation privacy, consumer privacy, commercial privacy, and financial information privacy. It can take years and sometimes decades for specific bills to work their way through the system and some are tendered in different forms many times before they are enacted or abandoned.

Acts Related to Government Accountability and Public Access to Records

- A measure of government accountability was established in 1966 with the *Freedom of Information Act* (FOIA) which granted individuals the right to request access to records controlled by executive branch agencies, subject to a number of exemptions. Fees are charged, but may be waivered if the information is primarily of benefit to the public rather than intended for commercial use.

Links Relevant to Data Communications, Privacy, and Legislation

This is a sampling of websites with information related to data communications, privacy, and legislation. There are additional links in Section 5, References.

Canadian Standards Association Model Code

http://www.csa.ca/cm?c=Page&childpagename=CSA%2FLayout&cid=1239124810319&packedargs=itemcontext%3Dca%252Fen%252Fnull&pagename=CSA%2FRenderPage

http://laws.justice.gc.ca/eng/P-8.6/20090818/page-1.html

Council of Europe (COE) Convention for the Protection of Individuals with Regard to Automatic Processing of Personal Data (1980).

http://www.privacy.org/pi/intl_orgs/coe/dp_convention_108.txt

DuckDuckGo search engine, which doesn't track searches.

http://www.duckduckgo.com

European Union Data Directive, Directive 95/46/EC of the European Parliament and of the Council of 24 October 1995 on the protection of individuals with regard to the processing of personal data and on the free movement of such data

http://ec.europa.eu/justice_home/fsj/privacy/law/index_en.htm

Germany, *Federal Data Protection Law 1977,* amended in 1990, 1994, 1997, 2002. *Federal Data Protection Act,* 1994 version.

http://www.iuscomp.org/gla/statutes/BDSG.htm

   Federal Data Protection Act current version.

http://www.bfdi.bund.de/EN/DataProtectionActs/DataProtectionActs_node.html

Open Identity Trust Framework (OITF) White Paper

http://openidentityexchange.org/sites/default/files/the-open-identity-trust-framework-model-2010-03.pdf

Organization for Economic Cooperation and Development, OECD Guidelines on the Protection of Privacy and Transborder Flows of Personal Data (1980).

http://www.oecd.org/document/18/0,2340,en_2649_34255_1815186_1_1_1_1,00.html

Report of the Committee on Privacy, chaired by Rt. Hon. Kenneth Younger, Chairman) (1972). See Appendix B of the 1973 HEW Report for a summary of the report.

http://aspe.os.dhhs.gov/datacncl/1973privacy/appenb.htm

Sweden, *The Data Act 1973* as amended in 1989.

http://archive.bild.net/dataprSw.htm

The *Personal Data Act* of 1998 replaced Data Act of 1973 to be in compliance with EU Privacy directives.

http://www.datainspektionen.se/in-english/legislation/The-Personal-Data-Act/

Library of Congress Thomas Register, searchable Congressional bills.

http://thomas.loc.gov/

Trust Framework Provider Adoption Process (TFPAP) for Levels of Assurance 1, 2, and Non-PKI 3, Version 1.0.1, September 4, 2009. PDF file.

http://www.idmanagement.gov/documents/TrustFrameworkProviderAdoptionProcess.pdf

U.S. Department of Homeland Security, *Privacy Policy Guidance Memorandum*, (2008) (Memo. 2008-1). PDF file.

http://www.dhs.gov/xlibrary/assets/privacy/privacy_policyguide_2008-01.pdf

U.S. Identity Credential and Access Management (ICAM) personal identity verification.

http://www.gsa.gov/portal/category/27240

U.S. National Strategy for Trusted Identities in Cyberspace (2010 DRAFT).

http://www.nstic.ideascale.com

U.S. Privacy Protection Study Commission (PPSC), *Protecting Privacy in an Information Society* (Chapter 13).

http://aspe.hhs.gov/datacncl/1977privacy/toc.htm

SECTION 5
RESOURCES

REFERENCES

GLOSSARY

INDEX

REFERENCES

References of interest available on the Web.

Allman, Mark; Kruse, Hans, Ostermann, Shawn, "A history of the improvement of Internet protocols over satellites using ACTS," *Online Journal of Space Communication,* Issue 2, Fall 2002.

"An overview of the federal wiretap act, electronic communications privacy act, and state two-party consent laws of relevance to the NEBUAD system and other uses of internet traffic content from ISPs for behavioral advertising," Center for Democracy & Technology, Washington, D.C., July 8, 2008.

Bauer, Robert; McMasters, Paul, "Survey of advanced applications over ACTS," *Online Journal of Space Communication,* Issue 2, Fall 2002.

Bergamo, Marcos A., Hoder Doug, "Gigabit satellite network using NASA's Advanced Communications Technology Satellite (ACTS): Features capabilities, and operations," NASA, 1995.

Jim Berry, President, Outlier Software, LLC, "Separating static and surveillance data collection," The Information Marketplace: Merging and Exchanging Consumer Data, Public Statement to the Federal Trade Commission, date unspecified.

Biale, Noam, Advocacy Coordinator, "What criminologists and others studying cameras have found," ACLU Expert Findings on Surveillance Cameras, ca 2008.

Bonarini, Andrea: Aliverti, Paolo; Lucioni, Michele, "An omnidirectional vision sensor for fast tracking for mobile robots," *IEEE Transactions on Instrumentation and Measurement,* 49 (3), June 2000.

Brooks, David Jonathan, "Public street surveillance: a psychometric study on the perceived social risk," Thesis for M.S., *Edith Cowan University,* May 12, 2003.

"COSPAS-SARSAT 406 MHz MEOSAR implementation plan," *COSPAS-SARSAT, C/S R.012* Issue 1—Revison 2, October 2006.

da Costa, Beatriz; Schulte, Jamie; Singer, Brooke, "Surveillance creep! new manifestations of data surveillance at the beginning of the twenty-first century," *Radical History Review,* Issue 95, Spring 2006, pp. 70–88.

Diesman, W. and Derby, P., et al., *A Report on Camera Surveillance in Canada, Surveillance Camera Awareness Network,* Contributions Program of the Office of the Privacy Commissioner, and the Social Sciences and Humanities Research Council of Canada, published in two parts January and December 2009.

Duff, James C., *Report of the Director of the Administrative Office of the United States Courts on Applications for Orders Authorizing or Approving the Interception of Wire, Oral, or Electronic Communications,* June 2011.

Farrington, Nathan M.; Nguyen, Hoa G.; Pezeshkian, Narek, "Intelligent behaviors for a convoy of indoor mobile robots operating in unknown environments," SPAWAR Systems Center, *SPIE Proceedings 5609: Mobile Robots XVII,* Philadelphia, PA, October 2004.

GPS.gov includes a variety of information on GPS technology, including legislative bills related to surveillance and court decisions on individual GPS-related cases.

Green, Mary W., "The appropriate and effective use of security technologies in U.S. schools: a guide for schools and law enforcement agencies," National Institute of Justice, U.S. Department of Justice, September 1999.

Hao, HaiYang; Cao, YongCan; Chen, YangQuan, "Autopilots for small unmanned aerial vehicles: a survey," *International Journal of Control, Automation, and Systems,* 8(1), 2010, pp. 36–44.

Olds, Jr., Robert B., "Marine mammal systems in support of force protection, *Intelligence, Surveillance, and Reconnaissance, SSC San Diego Biennial Review,* 2003, pp. 131–135.

Police and Security News, published by Days Communications, Inc.

Rathlev, Frank; Meyer, Benedikt; Juerss, Sven, "Innovative technologies for aerial survey of gas pipes," *New Technology Reports, Gas for Energy,* Issue 1, 2012.

Ruffier, Franck; Franceschini, Nicolas, "Visually guided micro-aerial vehicle: automatic take off, terrain following, landing and wind reaction," *Proceedings of the 2004 IEEE International Conference on Robotics & Automation,* New Orleans, LA, April 2004.

Seeley, Juliette; Richardson, Jonathan M., "Early warning chemical sensing," Lincoln Laboratories, ca. 2007.

Sward, Ricky E.; Cooper, Stephen D., "Unmanned aerial vehicle camera integration," *Institute For Information Technology Applications, United States Air Force Academy,* Colorado, Nov. 2004.

Traynor, Patrick; Carter, Henry, "(sp)iPhone: Decoding vibrations from nearby keyboards using mobile phone accelerometers," *18th ACM Conference on Computer and Communications Security,* Chicago, Oct. 2012.

Tsang, W.M.; Aldworth, Z., et al, "Insect flight control by neural stimulation of pupae-implanted flexible multisite electrodes," Electrical Engineering and Computer Science, Massachusetts Institute of Technology, Department of Biology, University of Washington, and Arizona Research Laboratories, University of Arizona, 2008.

Understanding Carrier IQ Technology, a 19-page report by the company providing the technology which describes what the software is and isn't capable of doing.

United States Attorneys' *Criminal Resource Manual* 1077 "Electronic Surveillance" provides interpretation of U.S. legal statutes.

United States Navy Radio Frequency Identification (RFID) Implementation Plan, Navy AIT Program Office, Naval Supply Systems Command (NAVSUP, Mechanicsburg, PA, December 8, 2005.

U.S. Customs and Border Protection, *Performance and Accountability Report, Fiscal Year 2011,* under the direction of Alan D. Bersin, Commissioner, March 2012.

U.S. Department of Home Security, "Iris and face technology demonstration and evaluation (IFTDE)," *Privacy Impact Assessment,* August 12, 2010.

Williams, D.; Ahmed, J., "The relationship between antisocial stereotypes and public CCTV systems: exploring fear of crime in the modern surveillance society," *Psychology, Crime & Law,* 15(8), Oct. 2009, pp 743–758.

URLs of interest.

European Group on Ethics in Science and New Technologies, an independent, multidisciplinary, pluralist body that advises the European Commission on ethics in science and new technologies.

> *www.ec.europa.eu/bepa/european-group-ethics/*

GRASP (General Robotics, Automation, Sensing & Perception) Laboratory at Penn Engineering.

> *https://www.grasp.upenn.edu/*

IEEE Robotics & Automation Society Technical Committee on Aerial Robotics and Unmanned Aerial Vehicles.

> *http://www.flyingrobots.org/*

Forensic Science Communications, a publication of the Federal Bureau of Investigation. Information.

> *www.fbi.gov/about-us/lab/forensic-science-communications/*

Library of Congress Thomas Register, searchable Congressional bills.

> http://thomas.loc.gov/

National Security Agency controversy over wiretapping issues.

> *https://en.wikipedia.org/wiki/NSA_warrantless_surveillance_controversy*

Public Assistance Reporting Information System.

> *http://www.acf.hhs.gov/programs/paris/*

Robotics and Computer Vision Laboratory, located in the Electrical Engineering Building, KAIST, Daejeon, Korea, was established in 1993 to develop high-performance computer vision systems modeled after human eye/brain cognition.

> *http://rcv.kaist.ac.kr/v2/*

UNITAR/UNOSAT surveillance imagery and maps of humanitarian projects.

> *http://www.unitar.org/*

United States Courts *Wiretap Report for 2010.*

> *http://www.uscourts.gov/Statistics/WiretapReports/*

U.S. Federal Trade Commission

> *http://www.ftc.gov/reports/privacy2000/privacy2000.pdf*

U.S. Geological Services satellite imagery and other imaging products available for download.

> *http://www.usgs.gov/*

U.S. Navy SPAWAR activities which include a variety of surveillance technologies and systems deployment.

> *http://www.public.navy.mil/SPAWAR/*

MEDIA

There are many good videos online describing and illustrating surveillance, far too many to list, but a sampling is included here. Customs and Border Protection posts many examples of interceptions that can be found by searching CBP on Youtube.com.

DNA Profiling

Court TV Canada discussion by Joseph Neuberger regarding a proposal for mandatory DNA sampling by Toronto, Ontario, police. Duration 1:44.

> *http://www.youtube.com/watch?v=wBstIzITiVs*

Robotics

Video: "Vijay Kumar, Robots that fly and team up," *March 2012 TED conference* video by CNN.

> *http://www.cnn.com/2012/03/04/opinion/ted-kumar-flying-robots/index.html*

Smart Cameras

Quick introduction to "smart cameras."

> *http://www.youtube.com/watch?v=tcSqe57Pd5E*

Introduction to facial-recognition software. Duration 1:40.

http://www.youtube.com/watch?v=EVSkhYHk6TQ

Video: Professor Jim Davis and his research team describe a system for using principles of human recognition to determine if trouble can be recognized by surveillance cameras.

http://www.youtube.com/watch?v=RYSSdra2XJE

VOA Special English Technology Report on a study of facial recognition systems (appropriate for hearing-impaired listeners). Duration 4:00.

http://www.youtube.com/watch?v=bqy7XnJ8IP0

Facial recognition software to research anthropological artifacts to determine how they may have looked. This kind of technology could also have applications for forensic reconstruction.

http://www.youtube.com/watch?v=LoCr9AEYpCo&feature=related

CBS 2010 news report on SuperBowl surveillance.

http://www.youtube.com/watch?v=c56BnV-YrKw

GLOSSARY

AAAR	Autonomous Aerial Robots
ADSB	Automatic Dependent Surveillance-Broadcast, a system in which the location of new aerial vehicles will broadcast location-monitoring information once per second
AFIS	automated fingerprint identification system
AIDC	automatic identification and data capture
AIT	auto-identification technology (see also RFID)
ALPR	automatic license-plate recognition
ANPR	automatic number-plate recognition
APIS	Advance Passenger Information System (CBP)
ASP	aerial surveillance platform
BIC	Border Intelligence Center (CBP)
CAS	collision avoidance system
CBR	chemical, radiological, biological
ATF	Bureau of Alcohol, Tobacco, Firearms and Explosives
CIA	Central Intelligence Agency
CJIS	Criminal Justice Information Services
COTS	commercial off-the-shelf
DHS	Department of Homeland Security
DIA	Defense Intelligence Agency
DoJ	Department of Justice
DVR	digital video recorder
ECD	electronic control device
EPC card	electronic product code card, a low-resource card suitable for including information on items such as retail products or other objects that don't require large amounts of information or long-term storage
FAVIAU	Forensic Audio, Video, and Image Analysis Unit of the FBI's Operation Technology Division
FBI	Federal Bureau of Investigation
FED	field emission display
FFL	Federal Firearms Licensees
FPP	Fraud Prevention Program
function creep	the propensity for surveillance systems installed for one purpose to gradually be used for other purposes outside the original mandate

HAM	HAM radio operators, amateur radio operators—named after three *H*arvard *AM*ateurs named *H*yman, *A*lmy, and *M*urray. For the history, see the congressional debate on the proposed *Wireless Regulation Bill*, 1911.
IC3	Internet Crime Complaint Center *www.ic3.gov*
ICB	Information Coordination Branch
IDENT	Automated Biometric Identification System
IMS	ion mobile spectrometry, a method used in chemical detectors
InSAR	interferometric synthetic aperture radar
IOFS	Intelligence Operations Framework System
IRTPA	Intelligence Reform and Terrorism Prevention Act of 2004
ISP	image signal processor
ISR	intelligence, surveillance, and reconnaissance
IVHS	Intelligent Vehicle/Highway System
JSP	Joint Security Program
LEARN	Law Enforcement Archive Reporting Network
LEOKA	Law Enforcement Officers Killed and Assaulted
LPR	license-plate reader
MEDS	Multifunction Electronic Display System
MEMS	micro-electromechanical systems
MSR	magnetic-stripe reader
MSS	Mobile Surveillance System (CBP)
MVSS	Mobile Video Surveillance System
NFC	near-field communication (short-range communication)
NECMEC	National Center for Missing and Exploited Children
NICS	National Instant Criminal Background Check System
NSA	National Security Agency
NTC-C	National Targeting Center-Cargo (CBP)
NTC-P	National Targeting Center-Passenger (CBP)
NTIA	National Telecommunications and Information Administration
NVLS	National Vehicle Location Service
OBP	Office of Border Patrol (CBP)
OIOC	Office of Intelligence Operation Coordination
OIT	Office of Information and Technology
OIUS	optical infrared ultraviolet surveillance

ONVIF	Open Network Video Interface Forum
OPSG	Operation Stonegarden
OTA	Office of Technology Assessment
pen register	a device to log what numbers or destinations were 'dialed' over telecommunications devices. Contemporary pen registers may log other information, such as time and duration.
POC	point of contact, as in access to a national database
PTZ	pan, tilt, zoom
RFID	radio-frequency identification
RIID	Radiation Isotope Identification Devices
RITA	Research and Innovative Technology Administration, an arm of the U.S. Department of Transportation
RPAS	remotely piloted aerial systems
RPM	Radiation Portal Monitor
RVSS	Remote Video Surveillance Systems
SBI	Secure Border Initiative
SBUV	solar-blind ultraviolet—a form of UV detection that works in daylight
SENTRI	Secure Electronic Network for Traveler Rapid Inspection
SOC	system on chip
SUAV	small UAV, small uninhabited aerial vehicle
subpoena	a court-ordered request to supply information, such as documents, tangible evidence (which could include biometrics samples), or a personal report of matters related to a legal investigation
TACCOM	Tactical Communication Modernization Program
TECS	The Enforcement Communication System
TI	tactical infrastructure
TIDE	Terrorist Identities Datamart Environment
TOC	technology readiness level
tracing	in the context of audio surveillance, determining the destination point of a call and sometimes also the route through which the connection was linked
tracking	monitoring the movements of an animal, object, communication, or cosmic phenomenon through two or more locations within a proximate or specified time period
trapping	in the context of audio surveillance, seizing a call so that the caller cannot terminate the connection. This is sometimes done to facilitate a trace.

TSDB	Terrorist Screening Data Base
TWL	Terrorist Watch List
UAV	uninhabited/unmanned aerial vehicle
UCR	uniform crime reports
UVG	uninhabited/unmanned ground vehicle
VERTIS	Vehicle, Road, and Traffic Intelligence Society (Japan)
VICS	Vehicle Information and Communication System
VPU	video processing unit

Index